Acceleration of Biodiesel Production

Acceleration of Biodiesel Production

Editors

Indra Neel Pulidindi
Aharon Gedanken

 Basel • Beijing • Wuhan • Barcelona • Belgrade • Novi Sad • Cluj • Manchester

Editors
Indra Neel Pulidindi
Department of Chemical Sciences
G S F C Univeristy
Vadodara
India

Aharon Gedanken
Department of Chemistry
Bar Ilan University
Ramat Gan
Israel

Editorial Office
MDPI
St. Alban-Anlage 66
4052 Basel, Switzerland

This is a reprint of articles from the Special Issue published online in the open access journal *Bioengineering* (ISSN 2306-5354) (available at: www.mdpi.com/journal/bioengineering/special_issues/acceleration_biodiesel_production).

For citation purposes, cite each article independently as indicated on the article page online and as indicated below:

Lastname, A.A.; Lastname, B.B. Article Title. *Journal Name* **Year**, *Volume Number*, Page Range.

ISBN 978-3-0365-9393-7 (Hbk)
ISBN 978-3-0365-9392-0 (PDF)
doi.org/10.3390/books978-3-0365-9392-0

© 2023 by the authors. Articles in this book are Open Access and distributed under the Creative Commons Attribution (CC BY) license. The book as a whole is distributed by MDPI under the terms and conditions of the Creative Commons Attribution-NonCommercial-NoDerivs (CC BY-NC-ND) license.

Contents

About the Editors vii

Preface ix

Fahed Javed, Muhammad Waqas Saif-ul-Allah, Faisal Ahmed, Naim Rashid, Arif Hussain and William B. Zimmerman et al.
Kinetics of Biodiesel Production from Microalgae Using Microbubble Interfacial Technology
Reprinted from: *Bioengineering* 2022, 9, 739, doi:10.3390/bioengineering9120739 1

Fahed Javed, Muhammad Rizwan, Maryam Asif, Shahzad Ali, Rabya Aslam and Muhammad Sarfraz Akram et al.
Intensification of Biodiesel Processing from Waste Cooking Oil, Exploiting Cooperative Microbubble and Bifunctional Metallic Heterogeneous Catalysis
Reprinted from: *Bioengineering* 2022, 9, 533, doi:10.3390/bioengineering9100533 18

Gang Li, Yuhang Hao, Tenglun Yang, Wenbo Xiao, Minmin Pan and Shuhao Huo et al.
Enhancing Bioenergy Production from the Raw and Defatted Microalgal Biomass Using Wastewater as the Cultivation Medium
Reprinted from: *Bioengineering* 2022, 9, 637, doi:10.3390/bioengineering9110637 34

Sandro L. Barbosa, David Lee Nelson, Lucas Paconio, Moises Pedro, Wallans Torres Pio dos Santos and Alexandre P. Wentz et al.
Environmentally Friendly New Catalyst Using Waste Alkaline Solution from Aluminum Production for the Synthesis of Biodiesel in Aqueous Medium
Reprinted from: *Bioengineering* 2023, 10, 692, doi:10.3390/bioengineering10060692 49

Dhurba Neupane
Biofuels from Renewable Sources, a Potential Option for Biodiesel Production
Reprinted from: *Bioengineering* 2022, 10, 29, doi:10.3390/bioengineering10010029 61

Sara Anelović, Marko Božinović, Željka Ćurić, Anita Šalić, Ana Jurinjak Tušek and Kristina Zagajski Kučan et al.
Deep Eutectic Solvents for Biodiesel Purification in a Microextractor: Solvent Preparation, Selection and Process Optimization
Reprinted from: *Bioengineering* 2022, 9, 665, doi:10.3390/bioengineering9110665 90

Biagio Anderlini, Alberto Ughetti, Emma Cristoni, Luca Forti, Luca Rigamonti and Fabrizio Roncaglia
Upgrading of Biobased Glycerol to Glycerol Carbonate as a Tool to Reduce the CO_2 Emissions of the Biodiesel Fuel Life Cycle
Reprinted from: *Bioengineering* 2022, 9, 778, doi:10.3390/bioengineering9120778 111

Misael B. Sales, Pedro T. Borges, Manoel Nazareno Ribeiro Filho, Lizandra Régia Miranda da Silva, Alyne P. Castro and Ada Amelia Sanders Lopes et al.
Sustainable Feedstocks and Challenges in Biodiesel Production: An Advanced Bibliometric Analysis
Reprinted from: *Bioengineering* 2022, 9, 539, doi:10.3390/bioengineering9100539 124

Dheeraj Rathore, Surajbhan Sevda, Shiv Prasad, Veluswamy Venkatramanan, Anuj Kumar Chandel and Rupam Kataki et al.
Bioengineering to Accelerate Biodiesel Production for a Sustainable Biorefinery
Reprinted from: *Bioengineering* 2022, 9, 618, doi:10.3390/bioengineering9110618 151

Qiuyun Zhang, Jialu Wang, Shuya Zhang, Juan Ma, Jingsong Cheng and Yutao Zhang
Zr-Based Metal-Organic Frameworks for Green Biodiesel Synthesis: A Minireview
Reprinted from: *Bioengineering* **2022**, *9*, 700, doi:10.3390/bioengineering9110700 . . . . **174**

Omojola Awogbemi and Daramy Vandi Von Kallon
Application of Tubular Reactor Technologies for the Acceleration of Biodiesel Production
Reprinted from: *Bioengineering* **2022**, *9*, 347, doi:10.3390/bioengineering9080347 . . . . **186**

About the Editors

Indra Neel Pulidindi

Dr. Indra Neel Pulidindi received his Ph.D. degree from the Indian Institute of Technology Madras in 2010 under the supervision of Professor T. K. Varadarajan and Professor (Em.) B. Viswanathan. He worked in the laboratory of Professor Aharon Gedanken from 2010 to 2016 in Israel and worked on the biomass conversion of biofuels and biochemicals. From 2016 to 2017, he worked in the laboratory of Professor Tae Hyun Kim at Hanyang University and made systematic studies on biomass composition analysis and the conversion of biomass to biochemicals. Subsequently, he worked in the laboratory of Prof. Xinling Wang at Shanghai Jiao Tong University on carbon-fiber-reinforced materials. Dr. Neel has 49 research papers, one patent, four patent applications, 3 books, 1 ebook, 7 book chapters, 5 ebook chapters to his credit. Dr. Neel guided several Ph.D., masters, undergraduate, and freshmen students in their academic research curriculum and helped them earn their degrees. Dr. Neel has worked as an assistant professor at GSFC University, Vadodara, India, from 19 September 2022 to 18 October 2023. He is currently working as a scientific consultant at JSCIAR, India. His researcher interests include CO_2 conversion, biomass conversion and energy conversion and storage.

Aharon Gedanken

Prof. Aharon Gedanken is an outstanding and legendary scientist with over 900 publications in peer-reviewed journals of international repute and high impact factor. He has 37 patent applications, a book on biofuels, and over 10 book chapters to his credit. Prof. Gedanken made remarkable contributions to the fields of sonochemistry and microwave technology and their application to nanomaterials, biomaterials, and biofuels. He served as a faculty member in the Department of Chemistry at Bar Ilan University for over 34 years (1975–2009) and has been an emeritus professor at the same institute for over 13 years (2009–to date). Prof. Gedanken's research metrics, namely, H-index and citations, are 117 and 51510, respectively. His research interests include solid-state chemistry, catalysis, energy, materials science, and biochemistry.

Preface

Among various renewable energy sources, biodiesel is a potential alternative to fossil-based diesel, useful for transportation as well as industrial applications. Among various biofuels, biodiesel is on the verge of commercialization because of the availability of technology for its large-scale demand-based production and distribution. Intense research and development activity has gone into the development of biodiesel as a substitute to conventional fossil-based diesel. Various facets of this new energy source include the following: sustainable feedstock (i.e., edible oil, non-edible oil, animal fats, and algal biomass); selective and effective extraction of lipids from algal biomass; unconventional fields of activation (i.e., microwave irradiation, sonochemical irradiation, fast stirring, microbubble technology, hydrodynamic cavitation, and special reactor design); use of enzymes, solid bases, solid acids, and nanomaterials as catalysts; use of unconventional solvents (i.e., deep eutectic solvents); methods of effective preheating of the raw materials; separation of the unreacted alcohol; and effective utilization of the by-product, glycerol. The advanced technology mentioned above is the use of microwave irradiation in the presence of solid-base catalysts, which has remarkably reduced the reaction times to a few seconds. Such a technology can indeed cater to the demands of the current century. Ever since the launch of the Special Issue titled "Acceleration of Biodiesel Production" on 10 January 2022, researchers across the globe have contributed enthusiastically, resulting in the publication of six research papers and five review articles. The editors place on record their grateful thanks to each of the research groups that have contributed to the advancement of knowledge in this field. Indebtedness is due to Mr. Adonis Tao, the managing editor of the journal *Bioengineering*, for his steadfast support. We do hope the collection of eleven state-of-the-art papers on the single subject, namely, "Acceleration of Biodiesel Production", and making the same freely accessible to the research fraternity will accelerate the use of biodiesel in the day-to-day life in the service of mankind for the next few decades.

Dedicated to "My LORD and my God, Jesus Christ." John 20:28. "My grace is sufficient for thee. 2 Corinthians 12:9".

Indra Neel Pulidindi and Aharon Gedanken
Editors

Article

Kinetics of Biodiesel Production from Microalgae Using Microbubble Interfacial Technology

Fahed Javed [1], Muhammad Waqas Saif-ul-Allah [2], Faisal Ahmed [2], Naim Rashid [3], Arif Hussain [1], William B. Zimmerman [4] and Fahad Rehman [1,*]

1. Microfluidics Research Group, Department of Chemical Engineering, COMSATS University Islamabad, Lahore Campus, Lahore 54000, Pakistan
2. Process and Energy Systems Engineering Center-PRESTIGE, Department of Chemical Engineering, COMSATS University Islamabad, Lahore Campus, Lahore 54000, Pakistan
3. Division of Sustainable Development, College of Science and Engineering, Hamad Bin Khalifa University, Qatar Foundation, Doha 34110, Qatar
4. Department of Chemical and Biological Engineering, The University of Sheffield, Sheffield S10 2TN, UK
* Correspondence: frehman@cuilahore.edu.pk; Tel.: +92-42-111-001-007; Fax: +92-42-9203100

Abstract: As an alternative to fossil fuels, biodiesel can be a source of clean and environmentally friendly energy source. However, its commercial application is limited by expensive feedstock and the slow nature of the pretreatment step-acid catalysis. The conventional approach to carry out this reaction uses stirred tank reactors. Recently, the lab-scale experiments using microbubble mediated mass transfer technology have demonstrated its potential use at commercial scale. However, all the studies conducted so far have been at a lab scale~100 mL of feedstock. To analyze the feasibility of microbubble technology, a larger pilot scale study is required. In this context, a kinetic study of microbubble technology at an intermediate scale is conducted (3 L of oil). Owing to the target for industrial application of the process, a commercial feedstock (*Spirulina*), microalgae oil (MO) and a commercial catalyst para-toluene sulfonic acid (PTSA) are used. Experiments to characterize the kinetics space (response surface, RSM) required for up-scaling are designed to develop a robust model. The model is compared with that developed by the gated recurrent unit (GRU) method. The maximum biodiesel conversion of 99.45 ± 1.3% is achieved by using these conditions: the molar ratio of MO to MeOH of 1:23.73 ratio, time of 60 min, and a catalyst loading of 3.3 wt% MO with an MO volume of 3 L. Furthermore, predicted models of RSM and GRU show proper fits to the experimental result. It was found that GRU produced a more accurate and robust model with correlation coefficient R^2 = 0.9999 and root-mean-squared error (RSME) = 0.0515 in comparison with RSM model with R^2 = 0.9844 and RMSE = 3.0832, respectively. Although RSM and GRU are fully empirical representations, they can be used for reactor up-scaling horizontally with microbubbles if the liquid layer height is held constant while the microbubble injection replicates along the floor of the reactor vessel—maintaining the tessellation pattern of the smaller vessel. This scaling approach maintains the local mixing profile, which is the major uncontrolled variable in conventional stirred tank reactor up-scaling.

Keywords: biodiesel; kinetics; esterification; RSM; microbubble technology

Citation: Javed, F.; Saif-ul-Allah, M.W.; Ahmed, F.; Rashid, N.; Hussain, A.; Zimmerman, W.B.; Rehman, F. Kinetics of Biodiesel Production from Microalgae Using Microbubble Interfacial Technology. *Bioengineering* 2022, 9, 739. https://doi.org/10.3390/bioengineering9120739

Academic Editors: Indra Neel Pulidindi, Aharon Gedanken and Reeta Rani Singhania

Received: 20 October 2022
Accepted: 25 November 2022
Published: 29 November 2022

Publisher's Note: MDPI stays neutral with regard to jurisdictional claims in published maps and institutional affiliations.

Copyright: © 2022 by the authors. Licensee MDPI, Basel, Switzerland. This article is an open access article distributed under the terms and conditions of the Creative Commons Attribution (CC BY) license (https://creativecommons.org/licenses/by/4.0/).

1. Introduction

The renewable energy generation is an important factor in reducing harmful effects on the environment caused by the excess use of conventional fuels. Biodiesel derived from inexpensive feedstocks, such as microalgae, can be a suitable alternative for replacing fossil fuel. Biodiesel is generally produced from various resources, such as waste cooking oil, animal fats, energy corps, yeast lipids, and microalgae [1,2]. Biodiesel consists of long chains of carboxylic acids of alkyl ester produced from esterification and transesterification of lipids or oils [3,4].

For sustainable production of biodiesel, feedstock cost is the key parameter affecting overall economics of the process. Since biodiesel production through refined oils feedstocks (soybean oil or sunflower oil) is expensive, low-cost unrefined oil (animal or microalgae oil) can be utilized as a cheaper substitute for biodiesel production [5–7]. However, the unrefined feedstock's generally consists of a large quantity of FFA. The presence of these FFA in feedstock are not suitable for transesterification using base catalysts due to the soap formation [8,9]. Therefore, acid catalyzed esterification is employed to reduce FFA content by converting them into FAMEs [10,11]. The acid-catalysis is significantly slower than transesterification due to the low-miscibility of MeOH with oil reducing overall mass transfer; as a result reaction rate decreases. The slower reaction rate directly affects the overall economics of the process [12–14]. However, the development of an economical method for commercial biodiesel production would be a revolutionary milestone for the fuel industry. Large scale biodiesel production could address many global challenges, such as waste management, energy supply, and environmental pollution. The major challenges that hinder commercialization of biodiesel include (1) cost-intensive methods of acid catalyzed esterification of biodiesel feedstock, (2) inability of the present technologies to scale-up, (3) expensive two stage production process, i.e., acid catalysis followed by base catalysis, and (4) expensiveness of various feedstocks [15–17].

Recently, microbubble mediated mass transfer technology has proven to increase the reaction rate by injecting one of the reactants in the vapor phase [18–20]. Microbubbles provide a large interfacial area, low buoyancy force, and high contact time on the bubble surface, which facilitate the rate of reaction of the system [21,22]. Furthermore, a smaller radius causes the increase of pressure inside the microbubble, as stated by Young–Laplace's law. Hence, the temperature inside the bubble could be predicted to be higher as compared with the boiling point of the alcohol. This also increases the surface energy of the bubble. All of these factors have yielded an unprecedented higher rate and conversion of the esterification reaction [18,20,23]. For example, Fahed et al. (2019) investigated the effect of microbubble on the esterification reaction by producing ethyl acetate and achieved 79.95% in 35 min compared with the conventional method, which achieved 64% conversion of esterification reaction in 350 min for ethyl acetate production [23]. In another study, Naveed et al. (2019) reported a 97% conversion of oleic acid into biodiesel in 30 min using microbubble technology which is a higher conversion than the conventional method, which achieved 80% conversion in 312 min using H_2SO_4 as a catalyst [20]. The major focus of these studies was to develop microbubble technology for esterification reaction using single component feedstock. In this context, Fahed et al. (2021) validated the effectiveness of microbubble technology in an unrefined feedstock (chicken fat oil) and showed an overall process conversion of 89.90% in 30 min [18]. To further investigate the effect of microbubble technology, Fahed et al. (2022) investigated the effect of microbubble technology by integrating microbubble technology with heterogeneous catalyst using waste cooking oil to further increase the rate of reaction and achieved an 85% conversion in 20 min [24]. The higher conversion was achieved for both chicken fat oil and waste cooking oil in a shorter period of time, indicating that microbubble technology is an economical method for biodiesel production.

However, all previous studies on microbubble technology mainly focused on lab-scale experiments~100 mL of oil and without significant control over vapor pressure of MeOH. The major focus of this study is to scale-up the microbubble mediated mass transfer technology from lab-scale to semi-pilot scale with up to 3 L volume of oil. The experiments were design using response surface methodology (RSM) and compared with gated recurrent unit (GRU). There are many studies in the literature that use RSM for optimization for biodiesel production [25,26]. However, the current study is the first study that implemented both RSM and GRU using microbubble technology. Furthermore, several technologies have been developed to manipulate the reaction equilibrium to achieve a higher conversion. These reactions are usually slow and limited by reaction kinetics and mass transfer, which are key constraints, such as esterification reactions. However, an entirely different technique has been developed using microbubble mediated mass transfer technology: an inherent liquid–liquid reaction is converted into liquid–vapor reaction, which entirely changes the reaction kinetics. The kinetics of a reaction are entirely based on two hypotheses, i.e., (1) by increasing interfacial area, mass transfer of the process increases to which rate of reaction is also enhanced. (2) Simultaneous removal of a reactant could also increase the reaction kinetics in the forward direction. Moreover, this study also investigated reaction kinetics on the semi-pilot scale the first time to understand the feasibility and compatibility of microbubble technology for further scale-up.

Keeping the semi-pilot scale nature of the study, a commercial feedstock (biomass of *Spirulina*) and a commercial catalyst p-Toluenesulfonic acid (p-TSA) is chosen to investigate microbubble mediated mass transfer. Oil from *Spirulina* was derived using solvent extraction method. Experiments were designed using response surface methodology (RSM), a robust model derived from RSM compared with another model developed using gated recurrent unit (GRU). This is the first study that demonstrated the potential scale-up of microbubble technology and compared both RSM and GRU models to increase the commercial feasibility of the process.

2. Material and Methods

2.1. Materials

Spirulina-biomass was purchased from Sentron Asia Company in Lahore, Pakistan. p-Toluenesulfonic acid (p-TSA) was purchased from Sigma Aldrich, St. Louis, MO, USA. MeOH of 99% analytical grade and 99% analytical grade n-Hexane were purchased from DAEJUNG chemicals, Siheung-si, South Korea.

2.2. Lipids Extraction from Spirulina Biomass

Microalgae oil/lipids (MO) were extracted from dry biomass using hexane and MeOH in a ratio of 7:3 by vol %. The biomass of *Spirulina* and solvent were stirred at 1000 rpm for 6 h at room temperature. Afterward, biomass was separated through filtration (whatman filter paper 42), and oil was recovered by evaporating solvent using vacuum evaporation (Buchi R-210, BUCHI Corporation, New Castle, DE, USA) at 60 °C [27]. The gravimetric method determined the oil yield [27]. Physiochemical properties and lipid composition of derived MO are presented in Table 1.

Table 1. Physicochemical properties of microalgae oil.

Parameters	Units	Value
FFA content	%	32.5 ± 2
Density (25 °C)	$Kg\ m^3$	920 ± 5
Kinematic viscosity (40 °C)	$mm^2\ s^{-1}$	30.06 ± 3
MO composition		
Mystic acid (C14:0)	%	1.90 ± 0.5
Palmitic acid (C16:0)	%	35.67 ± 3
Palmitoleic acid (C16:1)	%	6.11 ± 2
Linoleic acid (C18:2)	%	48.55 ± 2
Linolenic acid (C18:3)	%	2.17 ± 0.5
Stearic acid (C18:0)	%	5.60 ± 2

2.3. Pilot-Scale Experimental Setup for Biodiesel Production

The esterification reaction between MO and MeOH was performed using p-TSA as a catalyst. The schematic diagram of the pilot scale process was shown in Figure 1. In the current pilot-scale process, MeOH vapors were formed using a local fabricated digitally controlled vaporizer purchased from EES Technologies, Lahore, Pakistan. The vaporizer provided MeOH vapor at control/desired vapor flowrate, pressure, and temperature. A customized bubble reactor was fabricated using grade 3 sintered borosilicate glass diffuser.

The total volume of the bubble reactor was 3.5 L (radius = 43.18 mm and height = 609 mm), and the working volume was up to 3 L. The experiments were designed using RSM (Design Expert 11). The details of the RSM model are given in the next section. The bubble reactor was filled with different volumes of microalgae MO according to the response surface methodology RSM model. The MeOH vapors were produced in a vaporizer then passed through a borosilicate glass diffuser to form microbubbles. The temperature of the reactor was measured through a thermocouple (Digital thermometer, Jiangsu, China). The sample was collected continuously at a regular interval of 10 min. Once, the reaction had been run for the given time, the samples were filtered (whatman filter paper 42) and washed with deionized water. The samples were dried using a vacuum evaporator (Buchi R-210, BUCHI Corporation, New Castle, DE, USA) and stored for further analysis. All the experiments were performed in triplicate, and their average values with standard deviation was reported.

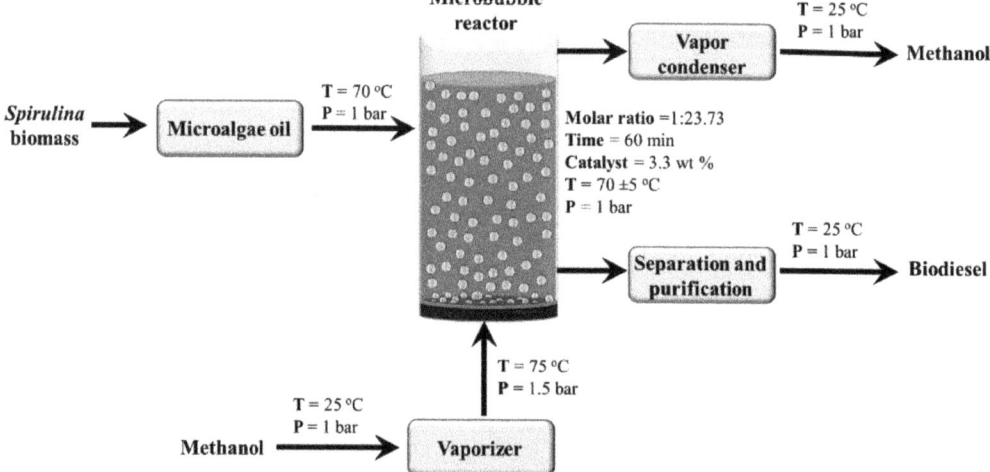

Figure 1. Process flow diagram of scale-up microbubble reactor.

2.4. Modeling and Experimental Design through Response Surface Methodology

RSM is a statistical and mathematical tool that uses multiple variables to design experiments for optimization [28]. In the current study, BBD was used to design experiments. BBD was used to design a process with more than two factors; this method provides fewer experiments than factorial design. Furthermore, BBD follows a cubical design edge using a midpoint with three levels each (−1, 0, +1) [29].

BBD was used with three factors and five center points in the current study. Three factors used in this study are: A (molar ratio of oil to MeOH = 1:5 to 1:25), B (catalyst dosage = 0 to 5 g wt% of MO), and C (TIme-10 to 90 min). According to this design, a total of 17 runs were conducted to evaluate the current process feasibility. The designed experiments and their response with predicted values of RSM and GRU are shown in Table S-I (Supplementary Information) and the RSM final conversion equation with coded value was given in Equation (1) [30,31]. The goodness-of-fit summary provided by RSM shows that quadric model is best suited for current experimental design Table S-II. The suggested model is best suited for current experimental responses, and the current suggested model is also assessed through analysis of variance (ANOVA) as shown in Table S-III.

$$Conversion = 88.68 + 3.27A + 26.13B + 7.97C + 3.24AB + 3.49AC + 3.98BC - 1.74A^2 - 29.76B^2 - 5.81C^2 \quad (1)$$

2.5. Modeling through Gated Recurrent Unit (GRU)

Commonly used artificial neural networks (ANN) involve three main layers, such as (1) input layer, (2) hidden layer (3) output layer and are indicated as x, h, and y. Recurrent neural network, a variant of ANN, was proposed for modeling time series and sequential data [32]. However, the vanishing gradient issue with large sequential data limits the application of RNNs. To solve the vanishing gradient problem, long short-term memory (LSTM), a variant of RNN, was introduced by Hochreiter and Schmidhuber [33]. This new variant of RNN, LSTM, was incorporated with four gates to replace the original hidden state in the memory cell. In 2014, GRU was introduced by adding a gated mechanism to the recurrent neural network [34]. Unlike LSTM, GRU merged the input gate and the forget gate into the update gate, as shown in Figure 2.

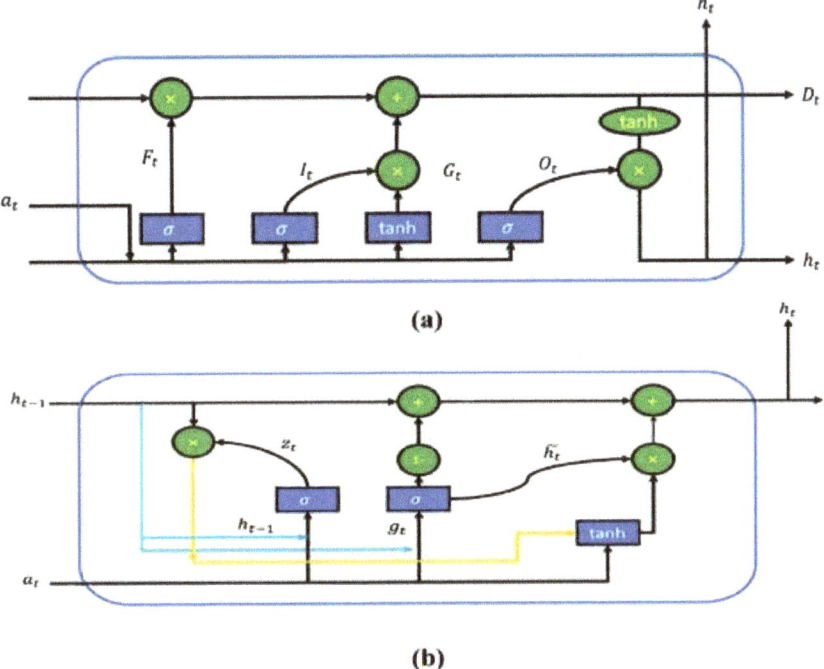

Figure 2. Comparison parameter between (**a**) LSTM and (**b**) GRU framework [35].

Hidden unit activation (r_t) is processed at a time step using Equation (2):

$$r_t = \sigma(W_r h_{t-1} + U_r x_t) \tag{2}$$

Here, σ depicts logistic sigmoid function, W_r and U_r represents weight matrices. After that, by using tanh type layer $\tilde{h}_t$ is calculated using r_t:

$$\tilde{h}_t = \tanh(W(r_t \times h_{t-1}) + U x_t) \tag{3}$$

Equation (4) is the equation that distinguishes GRU from the LSTM. Here, z_t combined the remember gate along with forget gate in LSTM. z_t is calculated as follows:

$$z_t = \sigma(W_z h_{t-1} + U_z x_t) \tag{4}$$

Lastly, the hidden state (h_t) in GRU is calculated using Equation (5):

$$h_t = (1 - z_t)(h_{t-1}) + (z_t)(\tilde{h}_t) \tag{5}$$

This study incorporated artificial intelligence modeling for the prediction of response surface, conversion in this case. Gated recurrent unit (GRU) was utilized as a sequence learning deep learning technique that requires suitable value of its hyper-parameters. Suitable architecture of the GRU model was obtained by varying hyper-parameters values (Table 2). To study the comparison between GRU and RSM, different parameters were considered, such as R^2, RMSE, MAPE, and mean absolute error MAE.

Table 2. Hyperparameters for gated recurrent unit model.

Hyperparameters	Bounds	Set Values
Number of hidden units	Positive integers	50
Gradient threshold	0–1	0.1
Initial learning rate	0–1	0.01
Learn rate drop factor	0–1	0.2
Learn rate drop period	Positive integers	100
Training Epochs	Positive integers	150

2.6. Biodiesel Analysis

Biodiesel analysis was performed using gas chromatography (GC) and ASTM method of FFA analysis. Briefly, in GC (Shimadzu GC-2014, Shimadzu Europa, Duisburg, Germany) the system was equipped with a flame ionization detector with column EN14103 (30 m × 0.32 mm id. × 0.25 µm film thickness). Nitrogen was introduced as a carrier with an initial temperature of 523 K and a split ratio of 50:1 [20]. For FFA analysis, the AOCS standard titration method was used [36,37]. To determine the FFA of the solution, the following Equations (6) and (7) were used [27]:

$$\text{Acid value} \left(\frac{\text{mgKOH}}{\text{g biodiesel}} \right) = \frac{(F_A - F_B) \times N \times 56.11}{W} \quad (6)$$

$$\text{Free fatty acid (FFA) (\%)} = \frac{1}{2} \times AV \quad (7)$$

3. Results and Discussion

3.1. Effect of Different Parameters on Free Fatty Acid Conversion

3.1.1. Effect of Catalyst Loading and Molar Ratio on Free Fatty Acid Conversion

The simultaneous effect of the molar ratio of oil: MeOH and catalyst loading are shown in Figure 3. The graph indicates that increasing catalyst loading from 0 to 5 wt% of MO significantly increases the conversion of the process. Increasing the catalyst loading enhances the protonation of FFA in MO. As the degree of protonation increases, the conversion of FFA and the reaction rate also increases. A further increase in catalyst loading after a certain point conversion of FFA was not increased due to insufficient active sites of MO. Furthermore, Perturbation plot of RSM indicate that most dominate factor in the current study is catalyst loading as show in Figure S-II.

On the other hand, by increasing the molar ratio of oil and MeOH, the FFA conversion also increases due to the contact time of FFA with MeOH vapors increasing. At a lower molar ratio, less volume of MeOH passes through the MO and leaves the system and vice versa. However, increasing the molar ratio after a certain limit, a small amount of MeOH can start to accumulate in the reactor, slightly reducing the reaction rate, as indicated by the results. RSM's optimized condition was molar ratio: 1:23.73 Oil: (MeOH) and catalyst loading 3.3 wt% of MO.

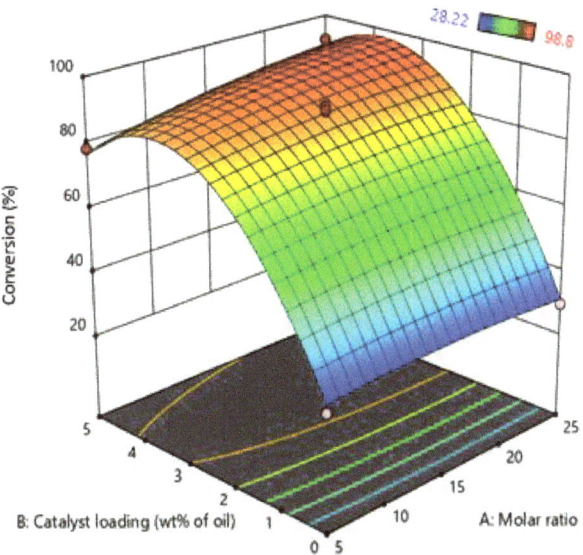

Figure 3. Effect of molar ratio (oil: MeOH) and catalyst loading on biodiesel production.

3.1.2. Effect of Reaction Time and Molar Ratio on Free Fatty Acid Conversion

The effect of both time and molar ratio was investigated and are shown in Figure 4. Time is another important parameter to study the rate of reaction. The results indicate that increasing time conversion and the rate of reaction increase due to an increase in reaction time increases the contact time of oil molecules with MeOH. As a result, conversion of FFA increases. However, a higher molar ratio and less reaction time decrease the reaction rate due to the bubble having less contact time with MeOH, as the flowrate of MeOH vapors is too high. As a result, the result MeOH vapors leave the system unreacted [18,20,23].

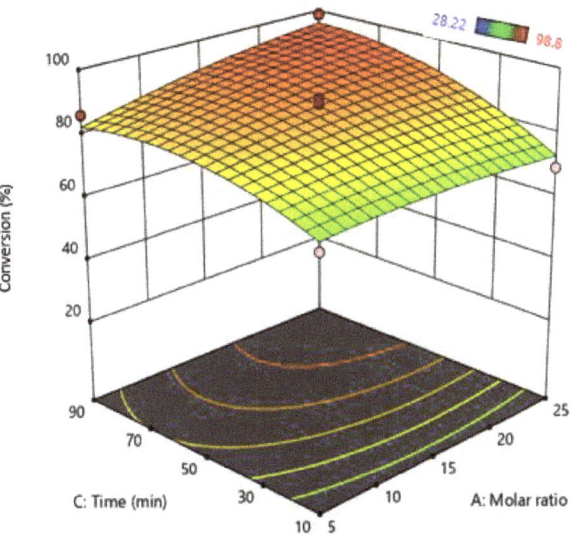

Figure 4. Interactive effect of molar ratio and reaction time.

Furthermore, an increase in MeOH flowrate also increases the formation of macrobubbles at higher flow rates and tends to produce larger bubbles [38]. Macrobubbles have a large buoyancy-force and less residence time compared to microbubbles. As they rise significantly faster, MeOH does not come into contact with FFA and, as a result, MeOH leaves the system unreacted. Both optimize reaction time and molar ratio can also provide an optimized flow rate to increase FFA conversion. The optimized time and molar ratio according to RSM were molar ratios: 1:23.73 Oil: (MeOH) and Time = 59.79~60 min.

3.1.3. Effect of Reaction Time and Catalyst Loading on Free Fatty Acid Conversion

The simultaneous effect of catalyst loading and reaction time were important parameters for scaling up of biodiesel process. These parameters greatly affect the cost and energy of the biodiesel process. The interactive effect of catalyst loading and reaction time on response surface are shown in Figure 5. FFA conversion tends to increase with the reaction time. However, the effect of reaction time is masked by the effect of catalyst loading. The 3D plot also shows that catalyst loading directly relates to FFA conversion due to an increase in catalyst loading increasing the number of available reaction sides; as a result, higher conversion of FFA was achieved.

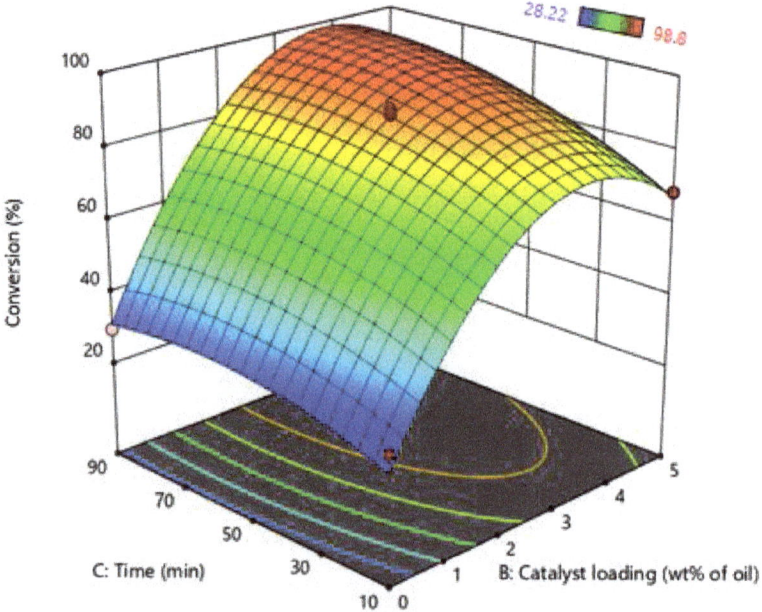

Figure 5. Interactive effect of reaction time and catalyst loading.

3.2. Scale-Up of Microbubble Reactor

The main goal of this work is to investigate the scale-up capacity of microbubble technology at optimized conditions provided by RSM and GRU and verify that provided conditions are suitable for converting FFA into biodiesel. For scale-up of microbubble reactor, different experiments were performed at different volumes of reactor varying from 1 to 3 L (Conversion: 1 L = 99.12%, 2 L = 99.55%, 3 L = 99.45%), as shown in Figure 6. It was observed that an increase in the volume in the reactor has negligible effect on the FFA conversion. An increase in the volume of oil also increased the pressure head of the microbubble reactor. By increasing the volume of oil, microbubbles stay in contact with the oil for a longer period of time. However, it is observed that an equilibrium is achieved at 200 mm and does not change if it is increased any further. This could be explained on the basis of mass transfer occurring across the bubble interface and kinetics of the reaction. At 200 mm, a boundary layer would have formed at the bubble interface stopping the mass transfer of MeOH from inside the bubble to the oil working as a "solid sphere". Several studies in hot microbubble distillation/stripping show that increasing the liquid layer heat can decrease separation efficiency because the non-equilibrium driving force diminishes [39]. Thus, further upscaling of the reactor should be horizontal by maintaining the microbubble tessellation pattern in the reactor. However, this scaling approach maintains the local mixing profile, which is the major uncontrolled variable in conventional stirred tank reactor upscaling.

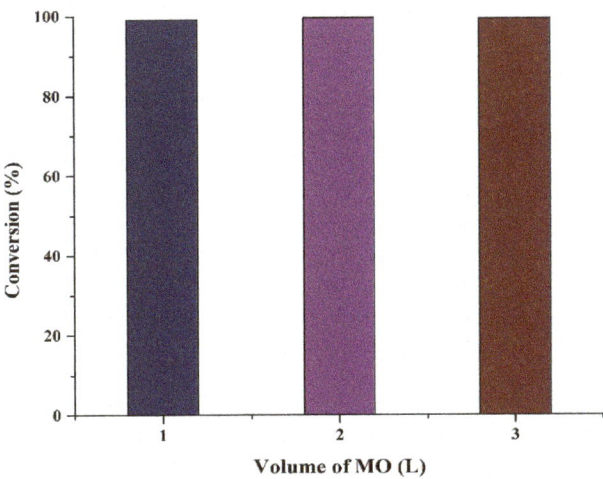

Figure 6. Effect on reaction conversion by increasing microalgae oil volume.

Conventionally, biodiesel is produced by mixing both reactants in liquid phase in a batch reactor. The mass transfer is limited by low miscibility of reactants, apart from castor oil/MeOH, which has an OH at C-12, reducing the overall conversion. However, the current process MeOH was injected in the form of vapors (bubbles) and as a result a vapor–liquid system is formed. In vapor–liquid system, conversion of MO is also increased as diffusion of vapor–liquid is higher than the liquid–liquid reaction. Furthermore, microbubbles exhibit less buoyancy force, increasing the residence time of MeOH bubble. FFA is premixed with the catalyst and is already protonated. As the microbubbles of MeOH rise, the reaction between MeOH and protonated MO starts instantaneously. The amount of alcohol available in the interface is in excess as compared to the available MO pushing the reaction in a forward direction. As the bubble rises, the alcohol is transferred into the MO. As the temperature of the reactor is maintained higher than the boiling point of MeOH, it does not condenses in the reactor and leaves as vapor [24]. The vapors of MeOH can be collected, condensed and recycled to improve the process economics. Figure 7 clearly exhibits three different slopes. The highest rate is achieved in the first 10 min of the process. Afterword, the rate slightly slows down and an overall conversion of 97% is achieved in the next 30 min (overall 40 min). Since almost all of the MO has already reacted, only 2% conversion is achieved in the last 20 min. To enhance feasibility and hence the economics, the reaction could be stopped at 10 min and product could be separated using a suitable separation technique, such as distillation or else 99% pure product could be obtained in 60 min. Comparison of current MO study with other conventional acid catalyst based biodiesel production is shown in Table 3.

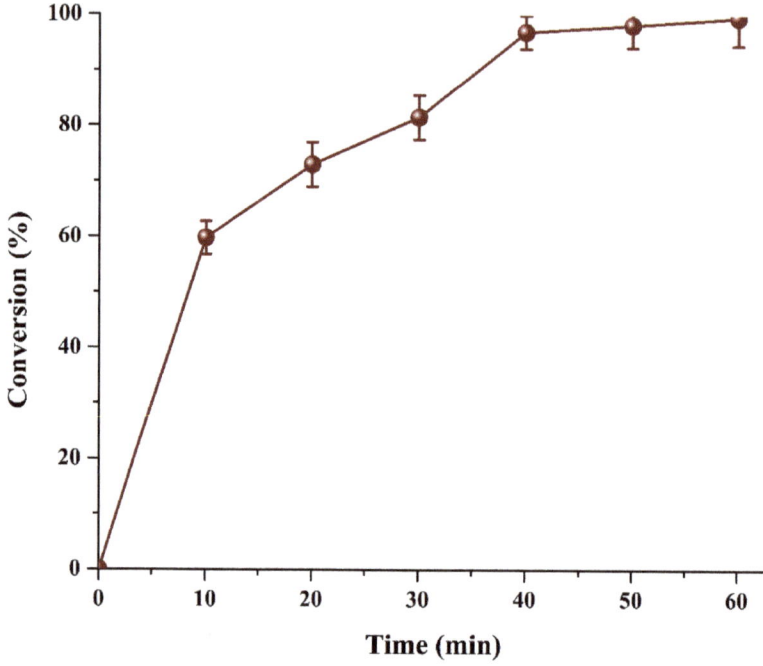

Figure 7. Effect microalgae oil conversion with respect to reaction time.

Table 3. The comparison of current up-scaling study with other conventional catalyst based biodiesel production.

Feedstock	Catalyst	Reaction Time (min)	Conversion (%)	Reference
Conventional method	H_2SO_4	120	78	[40]
Microbubble Technology	H_2SO_4	30	98	[20]
Microbubble Technology	p-TSA	30	97	[20]
Microbubble Technology	p-TSA	30	89.90	[18]
Microbubble Technology	Sr/ZrO_2	20	85	[24]
Microbubble Technology (semi pilot-scale)	p-TSA	60	99.45 ± 1.3	This study

3.3. Reaction Kinetics of Biodiesel Conversion

Scale-up results already show that by increasing the liquid layer height, the conversion does not change and has been identified; the reaction kinetics for a microbubble mediated esterification system can be upscaled via scaling out the reactor horizontally, maintaining the same aerator pattern along the reactor bottom surface. This is a logical conclusion from the seminal paper of Al-Mashhadani et al. (2015) [41]. In their paper, the geometry of the placement of the internal baffle in an airlift loop reactor is systematically varied, yet the hydrodynamics of the phase distribution is invariant to the baffles position, basically demonstrating that it is insignificant. Only the height above the aerator injection point is shown to matter. This follows logically from the fact that microbubbles that are injected in laminar flow and maintain laminar flow only rise vertically. By tessellating the aerators to provide downcomer regions in between, this configuration replicates the micromixing and bulk mixing profiles. This horizontal tessellation/upscaling approach has been found in all pilot scale studies for microbubble distillation/hot microbubble stripping [39]. Rees-Zimmerman and Chaffin (2021) [42] found that in modeling the hydrodynamics of tall bioreactors with variable bubble size, there is no mass transfer limitation with microbubbles, but the reactor height above a fixed level is immaterial. Hence, studying the kinetics for this critical layer height is sufficient for upscaling horizontally.

The kinetics of this study were investigated under optimized condition of RSM, i.e., the molar ratio of MO to MeOH of 1:23.73 ratio, time of 60 min, and a catalyst loading of 3.3 wt% MO. The results show that the difference between predicted value of RSM (99.10% as shown in Figure S-I) and actual value (99.45%) was less than 0.5% which indicate the validation of the current RSM model. To evaluate the reaction behavior of vapor–liquid, Ha was calculated using Equation (8) to assess the whether a reaction occurs on the bubble's bulk or surface [43,44].

$$Ha = \frac{\sqrt{(M_{g/l})_T\, k\, C_b}}{k_{bl}} \tag{8}$$

$$(M_{g/l})_{25°C} = 6.02 \times 10^{-5} \left(\frac{V_l^{0.36}}{\mu_l^{0.61} V_g^{0.64}} \right) \tag{9}$$

$$(M_{g/l})_T = 4.996 \times 10^3 (M_{g/l})_{25°C} \exp\left(\frac{-2539}{T} \right) \tag{10}$$

$M_{g/l}$ was calculated at 25 and 70 °C by Equations (9) and (10) [45]. For a bubble size less than 2 mm, Equation (11) was used to calculate k_{bl} [46].

$$k_{bl} = 0.31 \left(\frac{(D_{g/l})^2 \rho_l g}{\mu_l} \right)^{\frac{1}{3}} \tag{11}$$

The calculated value of Ha is greater than 1, which indicates that the reaction occurs on the bubble surface, due to which bubble size is a crucial parameter of controlling reaction kinetics. The order of reaction was calculated using E by Equation (12) [43].

$$E = Ha\left(1 - \frac{Ha - 1}{2E_i} \right) \tag{12}$$

E_i was determined by using Equation (13) [43]

$$E_i = 1 + (M_{g/l})_T \left(\frac{C_b H}{b\, P_g} \right) \tag{13}$$

The value of Ha and E show that the current reaction is pseudo first order due to the value of both of them I almost equal. The rate of reaction was calculated by using Equation (14).

$$-r_A = \frac{1}{\frac{1}{k_g \sigma} + \frac{H}{a\sqrt{(D_{g/l})_T k C_b}}} P_g \tag{14}$$

The values used to calculate for k_g, σ, k_{bl}, and H are 5.32×10^{-3} kmol s^{-1} m^{-3}, 1.24×10^{-4} ms^{-1} and 43.05 kmol s^{-1}.m^{-3} Pa, respectively. The final reaction rate was determine using Equation (15);

$$-r_A = \left(1.32 \times 10^{-5} \right)(P_A \times 101{,}325)\left(\sqrt{C_b} \right) \tag{15}$$

To determine the current activation energy (E_A) in the scale-up reactor, the Arrhenius equation was used. The experiments were conducted by varying the reaction temperature (70–90 °C) [43]. The Arrhenius equation was used to develop the relations with rate constants to develop the equation to determine E_A using Equation (16) [47]. An Arrhenius plot between $ln\,(k)$ and $1/T$ is shown in Figure 8.

$$ln\, k = -\frac{E_A}{RT} + ln\, A° \tag{16}$$

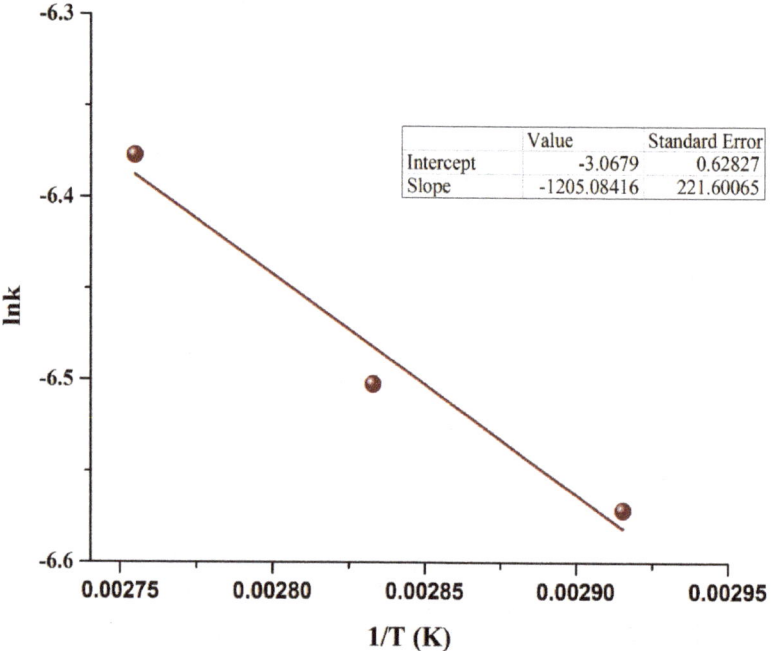

Figure 8. Arrhenius plot for microalgae oil based biodiesel reaction.

The E_A of the esterification reaction in the current scale-up microbubble reactor is calculated to be 10.01 ± 0.3 kJ mol^{-1}. The current E_A was significantly less than conventional processes, as shown in Table 4. This low E_A indicates that reactions occur on the surface of bubbles [20]. It also shows that less energy is needed for MO to pass the barrier to form a product. In addition, the latent heat of MeOH was also freely available as free energy, making the reaction nature more exergonic, enhancing the rate of reaction and reducing the E_A. The current reaction kinetics and reduced E_A successfully show the implementation of current scale-up reactor to industrial level.

Table 4. Comparison of activation energy with different biodiesel feedstocks.

Feedstock	Method	Scale of Experiments	Catalyst	E_A (kJ mol^{-1})	Reference
Jatropha	Conventional method	Lab-scale	1% H_2SO_4 and 1% NaOH	87.808	[48]
Microalgae	Supercritical method	Lab-scale	No catalyst	105	[49]
Chlorella	Conventional method	Lab-scale	HCl	38.892	[50]
Spirulina platensis	Single stage extraction–transesterification process	Lab-scale	H_2SO_4	14.518	[51]
Oleic acid	Microbubble technology	Lab-scale	7% H_2SO_4	26.37	[20]
Chicken fat oil	Microbubble technology	Lab-scale	7% PTSA	24.9	[18]
Spirulina	Microbubble technology	Semi pilot scale	3.3% PTSA	10.01 ± 0.3	This study

3.4. Gated Recurrent Unit and Response Surface Methodology Comparison

RSM model has reported 0.9844 correlation coefficient R^2 that shows good fitting efficiency, as shown in Figure 9. However, GRU has shown superior efficiency (Figure 10) and reported 0.9999 R^2. The values of mean absolute percentage error (MAPE), root mean square error (RMSE) and mean absolute error (MAE) for RSM model and GRU model have been reported in Table 5.

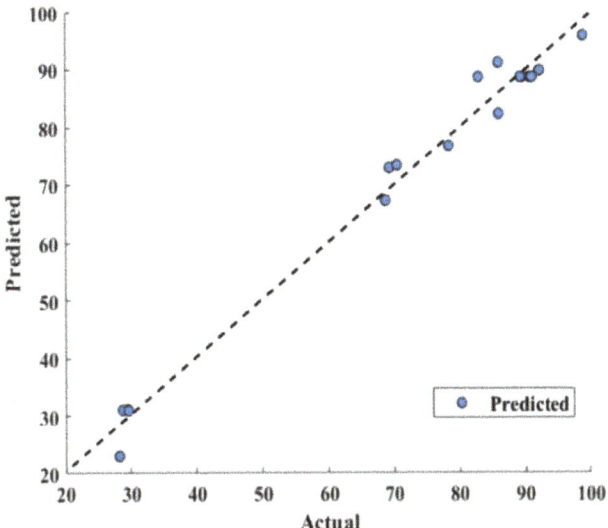

Figure 9. Response surface methodology model performance prediction (actual versus predicted).

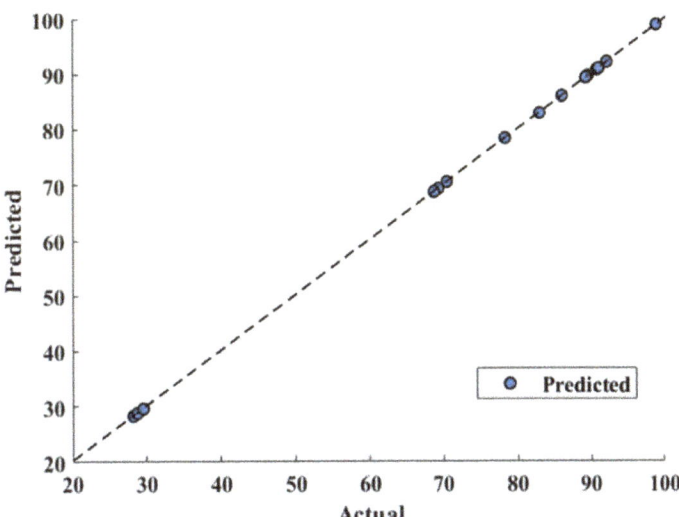

Figure 10. Gated recurrent unit model prediction.

Table 5. Performance criteria comparison (response surface methodology vs. gated recurrent unit).

Criteria	Conversion % Prediction Performance	
	RSM Model	GRU Model
R^2	0.9844	0.9999
MAPE	0.0465	0.00083
RMSE	3.0832	0.0515
MAE	2.6847	0.045

Table 5 confirms the superiority of the GRU model over RSM as its predicted values are in good agreement with actual values, as shown in Figure 10. It can also be confirmed that GRU reported performance criterions used, such as MAPE, RMSE, and MAE, lesser than that of the RSM model. From Figure 9, deviating RSM model prediction values from actual values are clearly visible, causing lower R^2 (0.9844) compared to that of GRU (0.9999), as well as a reported larger MAPE, RMSE and MAE compared to that of the GRU model, as show in Figure 10. The MAE for the RSM model is 2.6847, which is approximately 60 times higher compared to that of the GRU model (0.045). Furthermore, MAPE for the GRU model (0.00083) confirms its superiority over the RSM model (0.0465). Furthermore, in terms of RMSE, GRU reported an RMSE of 0.0515 and RSM reported 3.0832.

Furthermore, a 45-degree perfect line was added to the plot to better understand the prediction accuracy of the models. The prediction performance increases as the scatter points approach toward a 45-degree perfect line and hence decreases the error. The plot depicted that the GRU model prediction of conversion% were following the 45-degree perfect line more strictly than that of the RSM model and hence reported lesser RMSE, MAPE, and MAE. Furthermore, a higher R^2 value for the GRU model (0.9999) depicts outperforming performance compared to the R^2 value for the RSM model (0.9844). This scatter plot comparison also pointed out that the GRU model has performed much better than the RSM model to predict conversion.

4. Conclusions

The current study successfully developed a robust model which shows a high feasibility of microalgae-based biodiesel on a semi-pilot scale. Both models were in line with experimental observation. In addition, the comparison of both RSM and GRU showed that GRU was more accurate than RSM to predict the conversion. Furthermore, predicted models of RSM and GRU show a proper fit to the experimental result. It was found that GRU produced a more accurate and robust model with correlation coefficient R^2 = 0.9999 and root-mean-squared error (RSME) = 0.0515 in comparison with the RSM model with R^2 = 0.9844 and RMSE = 3.0832, respectively. Furthermore, the kinetics of the pilot scale microbubble reactor revealed that more than 99.45 ± 1.3% conversion of FFA into biodiesel was achieved in 60 min, and the current reaction follows pseudo first order kinetics with respect to MO. Additionally, a lower E_A of 10.01 ± 0.3 kJ mol^{-1} indicates that less energy was required for reactant to jump the barrier to form product. Although RSM and GRU are fully empirical representations, they can be used for reactor upscaling horizontally with microbubbles. Horizontal out-scaling maintains the liquid layer height, while the microbubble injection replicates along the floor of the reactor vessel—keeping the tessellation pattern of the smaller vessel with the same bubble flux per unit area. This scaling approach maintains the local mixing profile, which is the major uncontrolled variable in conventional stirred tank reactor upscaling. This study should prove to be a milestone in future studies for further scale-up of microbubble mass transfer technology. Further research needs to be carried out in terms of reactor development and finding new material for reactor formation, keeping the process cost to a minimum. Additionally, life cycle analysis should be carried out to calculate the environmental impact of the microbubble technology. Life cycle analysis will provide overall insight toward sustainability of the developed approach to analyze manufacturing, ecological effect, and energy expenditures of current technology.

Supplementary Materials: The following supporting information can be downloaded at: https://www.mdpi.com/article/10.3390/bioengineering9120739/s1, Table S-I: The designed experiments and their response with predicted values of RSM and GRU; Table S-II: Goodness-of-Fit summary generated through RSM; Table S-III: Statistical ANOVA analysis of current RSM model; Figure S-I: Predicted values from RSM; Figure S-II: Perturbation plot of experimental parameters.

Author Contributions: Conceptualization, F.R. and N.R.; methodology, F.J.; validation, F.R., A.H. and W.B.Z.; formal analysis, F.A.; data curation, M.W.S.-u.-A.; writing—original draft preparation, F.J.; writing—review and editing, F.J.; supervision, F.R.; funding acquisition, F.R. and W.B.Z. All authors have read and agreed to the published version of the manuscript.

Funding: This research was funded by Higher Education Commission Pakistan (HEC), grant number 20-7924/17, 20-7924/17 and Engineering and Physical Sciences Research Council (EPSRC) grant number EP/S031421/1 and EP/N011511/1.The APC was funded by Engineering and Physical Sciences Research Council (EPSRC) grant number EP/S031421/1 and EP/N011511/1.

Institutional Review Board Statement: Not applicable.

Informed Consent Statement: Not applicable.

Data Availability Statement: The data presented in this study are available on request from the corresponding author. The data are not publicly available due to confidentiality of work.

Acknowledgments: FJ and FR want to acknowledge National Research Program for Universities (NRPU) provided by higher education commission project number 20-7924/17 and project number 20-3982/14. In addition, WZ would like to acknowledge EPSRC grant numbers EP/S031421/1 and EP/N011511/1.

Conflicts of Interest: The authors declare no conflict of interest.

Abbreviations

Words	Abbreviations
Activation energy	E_A
Absolute percentage error	MAPE
Acid value	AV
Box–Behnken design	BBD
Concentration of MO	C_b
Enhancement factor	E
Free Fatty Acid	FFA
Fatty Acid Methyl Esters	FAMEs
Gas constant	R
Gated recurrent unit	GRU
Gas diffusion coefficient	$M_{g/l}$
Henry constant	H
Hatta number	Ha
Interfacial area	σ
Infinite enhancement factor	E_i
Liquid film coefficient	k_{bl}
Mass of biodiesel (g)	W
Microalgae oil	MO
Methanol	MeOH
Molar volume of MO	v_l
Molar volume of MeOH	v_g
Mean absolute error	MAE
Normality of KOH	N
Partial pressure of MeOH	p_g
Pressure of MeOH (bar)	P_A
Pre-exponential factor	$A°$
Rate constant	K
Rate of reaction	r_a
Root mean square error	RMSE
Response surface methodology	RSM
Volume of KOH used for titration (mL)	F_A
Volume of KOH used for blank titration (mL)	F_B

References

1. Malekghasemi, S.; Kariminia, H.-R.; Plechkova, N.K.; Ward, V.C. Direct transesterification of wet microalgae to biodiesel using phosphonium carboxylate ionic liquid catalysts. *Biomass Bioenergy* **2021**, *150*, 106126. [CrossRef]
2. Javed, F.; Rehman, F.; Khan, A.U.; Fazal, T.; Hafeez, A.; Rashid, N. Real textile industrial wastewater treatment and biodiesel production using microalgae. *Biomass Bioenergy* **2022**, *165*, 106559. [CrossRef]
3. Sitepu, I.R.; Garay, L.A.; Sestric, R.; Levin, D.; Block, D.E.; German, J.B.; Boundy-Mills, K.L. Oleaginous yeasts for biodiesel: Current and future trends in biology and production. *Biotechnol. Adv.* **2014**, *32*, 1336–1360. [CrossRef]
4. Soccol, C.R.; Neto, C.J.D.; Soccol, V.T.; Sydney, E.B.; da Costa, E.S.F.; Medeiros, A.B.P.; de Souza Vandenberghe, L.P. Pilot scale biodiesel production from microbial oil of Rhodosporidium toruloides DEBB 5533 using sugarcane juice: Performance in diesel engine and preliminary economic study. *Bioresour. Technol.* **2017**, *223*, 259–268. [CrossRef] [PubMed]
5. Banković-Ilić, I.B.; Stojković, I.J.; Stamenković, O.S.; Veljkovic, V.B.; Hung, Y.-T. Waste animal fats as feedstocks for biodiesel production. *Renew. Sustain. Energy Rev.* **2014**, *32*, 238–254. [CrossRef]
6. Nath, B.; Das, B.; Kalita, P.; Basumatary, S. Waste to value addition: Utilization of waste Brassica nigra plant derived novel green heterogeneous base catalyst for effective synthesis of biodiesel. *J. Clean. Prod.* **2019**, *239*, 118112. [CrossRef]
7. Faisal, A.; Javed, F.; Hassan, M.; Gorji, M.; Akram, S.; Rashid, N.; Rehman, F. Experimental and mathematical nonlinear rheological characterization of chicken fat oil-a sustainable feedstock for biodiesel. *Biomass Convers. Biorefinery* **2021**, 1–8. [CrossRef]
8. Tang, Z.-E.; Lim, S.; Pang, Y.-L.; Shuit, S.-H.; Ong, H.-C. Utilisation of biomass wastes based activated carbon supported heterogeneous acid catalyst for biodiesel production. *Renew. Energy* **2020**, *158*, 91–102. [CrossRef]
9. Singh, N.K.; Singh, Y.; Sharma, A. Optimization of biodiesel synthesis from Jojoba oil via supercritical methanol: A response surface methodology approach coupled with genetic algorithm. *Biomass Bioenergy* **2022**, *156*, 106332. [CrossRef]
10. Meher, L.C.; Sagar, D.V.; Naik, S. Technical aspects of biodiesel production by transesterification—A review. *Renew. Sustain. Energy Rev.* **2006**, *10*, 248–268. [CrossRef]
11. Zimmerman, W.B.; Kokoo, R. Esterification for biodiesel production with a phantom catalyst: Bubble mediated reactive distillation. *Appl. Energy* **2018**, *221*, 28–40. [CrossRef]
12. Tran, D.-T.; Chang, J.-S.; Lee, D.-J. Recent insights into continuous-flow biodiesel production via catalytic and non-catalytic transesterification processes. *Appl. Energy* **2017**, *185*, 376–409. [CrossRef]
13. Talebian-Kiakalaieh, A.; Amin, N.A.S.; Mazaheri, H. A review on novel processes of biodiesel production from waste cooking oil. *Appl. Energy* **2013**, *104*, 683–710. [CrossRef]
14. Li, X.; Wang, Q.; Wu, Y.; Chen, J.; Li, S.; Ye, Y.; Wang, D.; Zheng, Z. Optimization of key parameters using RSM for improving the production of the green biodiesel from FAME by hydrotreatment over Pt/SAPO-11. *Biomass Bioenergy* **2022**, *158*, 106379. [CrossRef]
15. Wang, Y.-T.; Fang, Z.; Yang, X.-X. Biodiesel production from high acid value oils with a highly active and stable bifunctional magnetic acid. *Appl. Energy* **2017**, *204*, 702–714. [CrossRef]
16. Vardon, D.R.; Moser, B.R.; Zheng, W.; Witkin, K.; Evangelista, R.L.; Strathmann, T.J.; Rajagopalan, K.; Sharma, B.K. Complete utilization of spent coffee grounds to produce biodiesel, bio-oil, and biochar. *ACS Sustain. Chem. Eng.* **2013**, *1*, 1286–1294. [CrossRef]
17. Narkhede, N.; Singh, S.; Patel, A. Recent progress on supported polyoxometalates for biodiesel synthesis via esterification and transesterification. *Green Chem.* **2015**, *17*, 89–107. [CrossRef]
18. Javed, F.; Shamair, Z.; Hafeez, A.; Fazal, T.; Aslam, R.; Akram, S.; Rashid, N.; Zimmerman, W.B.; Rehman, F. Conversion of Poultry-Fat Waste to a Sustainable Feedstock for Biodiesel Production via Microbubble Injection of Reagent Vapor. *J. Clean. Prod.* **2021**, *311*, 127525. [CrossRef]
19. Khan, Z.; Javed, F.; Shamair, Z.; Hafeez, A.; Fazal, T.; Aslam, A.; Zimmerman, W.B.; Rehman, F. Current developments in esterification reaction: A review on process and parameters. *J. Ind. Eng. Chem.* **2021**, *103*, 80–101. [CrossRef]
20. Ahmad, N.; Javed, F.; Awan, J.A.; Ali, S.; Fazal, T.; Hafeez, A.; Aslam, R.; Rashid, N.; Rehman, M.S.U.; Zimmerman, W.B. Biodiesel production intensification through microbubble mediated esterification. *Fuel* **2019**, *253*, 25–31. [CrossRef]
21. Zimmerman, W.B.; Tesar, V.; Butler, S.; Bandulasena, H.C. Microbubble generation. *Recent Pat. Eng.* **2008**, *2*, 1–8. [CrossRef]
22. Zimmerman, W.B.; Al-Mashhadani, M.K.; Bandulasena, H.H. Evaporation dynamics of microbubbles. *Chem. Eng. Sci.* **2013**, *101*, 865–877. [CrossRef]
23. Javed, F.; Shamair, Z.; Ali, S.; Ahmad, N.; Hafeez, A.; Fazal, T.; Rehman, M.S.U.; Zimmerman, W.B.; Rehman, F. "Pushing and pulling" the equilibrium through bubble mediated reactive separation for ethyl acetate production. *React. Chem. Eng.* **2019**, *4*, 705–714. [CrossRef]
24. Javed, F.; Rizwan, M.; Asif, M.; Ali, S.; Aslam, R.; Akram, M.S.; Zimmerman, W.B.; Rehman, F. Intensification of Biodiesel Processing from Waste Cooking Oil, Exploiting Cooperative Microbubble and Bifunctional Metallic Heterogeneous Catalysis. *Bioengineering* **2022**, *9*, 533. [CrossRef] [PubMed]
25. Hariram, V.; Bose, A.; Seralathan, S. Dataset on optimized biodiesel production from seeds of Vitis vinifera using ANN, RSM and ANFIS. *Data Brief* **2019**, *25*, 104298. [CrossRef] [PubMed]
26. Manojkumar, N.; Muthukumaran, C.; Sharmila, G. A comprehensive review on the application of response surface methodology for optimization of biodiesel production using different oil sources. *J. King Saud Univ.-Eng. Sci.* **2020**, *34*, 198–208. [CrossRef]

27. Park, J.-Y.; Nam, B.; Choi, S.-A.; Oh, Y.-K.; Lee, J.-S. Effects of anionic surfactant on extraction of free fatty acid from Chlorella vulgaris. *Bioresour. Technol.* **2014**, *166*, 620–624. [CrossRef] [PubMed]
28. Behera, S.K.; Meena, H.; Chakraborty, S.; Meikap, B. Application of response surface methodology (RSM) for optimization of leaching parameters for ash reduction from low-grade coal. *Int. J. Min. Sci. Technol.* **2018**, *28*, 621–629. [CrossRef]
29. Yolmeh, M.; Jafari, S.M. Applications of response surface methodology in the food industry processes. *Food Bioprocess Technol.* **2017**, *10*, 413–433. [CrossRef]
30. Jiao, W.; Yu, L.; Feng, Z.; Guo, L.; Wang, Y.; Liu, Y. Optimization of nitrobenzene wastewater treatment with O_3/H_2O_2 in a rotating packed bed using response surface methodology. *Desalin. Water Treat.* **2016**, *57*, 19996–20004. [CrossRef]
31. Ruhul, A.; Kalam, M.; Masjuki, H.; Fattah, I.R.; Reham, S.; Rashed, M. State of the art of biodiesel production processes: A review of the heterogeneous catalyst. *RSC Adv.* **2015**, *5*, 101023–101044. [CrossRef]
32. Cai, X.; Zhang, N.; Venayagamoorthy, G.K.; Wunsch, D.C. Time series prediction with recurrent neural networks using a hybrid PSO-EA algorithm. In Proceedings of the 2004 IEEE International Joint Conference on Neural Networks (IEEE Cat. No.04CH37541), Budapest, Hungary, 25–29 July 2004; Volume 1642, pp. 1647–1652.
33. Hochreiter, S.; Schmidhuber, J. Long Short-Term Memory. *Neural Comput.* **1997**, *9*, 1735–1780. [CrossRef]
34. Cho, K.; Merrienboer, B.v.; Gülçehre, Ç.; Bahdanau, D.; Bougares, F.; Schwenk, H.; Bengio, Y. Learning Phrase Representations Using RNN Encoder Decoder for Statistical Machine Translation. In Proceedings of the EMNLP, Doha, Qatar, 25–29 October 2014.
35. Saif-ul-Allah, M.W.; Qyyum, M.A.; Ul-Haq, N.; Salman, C.A.; Ahmed, F. Gated Recurrent Unit Coupled with Projection to Model Plane Imputation for the PM2.5 Prediction for Guangzhou City, China. *Front. Environ. Sci.* **2022**, *9*, 753. [CrossRef]
36. Marchetti, J.; Errazu, A. Comparison of different heterogeneous catalysts and different alcohols for the esterification reaction of oleic acid. *Fuel* **2008**, *87*, 3477–3480. [CrossRef]
37. Yin, J.; Chen, L.; Wang, Z.; Qu, R.; Liu, X.; Xu, Q.; Ren, S. Biodiesel production from esterification of oleic acid over aminophosphonic acid resin D418. *Fuel* **2012**, *102*, 499–505. [CrossRef]
38. Rehman, F.; Medley, G.J.; Bandulasena, H.; Zimmerman, W.B. Fluidic oscillator-mediated microbubble generation to provide cost effective mass transfer and mixing efficiency to the wastewater treatment plants. *Environ. Res.* **2015**, *137*, 32–39. [CrossRef] [PubMed]
39. Desai, P.D.; Turley, M.; Robinson, R.; Zimmerman, W.B. Hot microbubble injection in thin liquid film layers for ammonia separation from ammonia rich-wastewater. *Chem. Eng. Process.-Process Intensif.* **2021**, *180*, 108693. [CrossRef]
40. Tashtoush, G.M.; Al-Widyan, M.I.; Al-Jarrah, M.M. Experimental study on evaluation and optimization of conversion of waste animal fat into biodiesel. *Energy Convers. Manag.* **2004**, *45*, 2697–2711. [CrossRef]
41. Al-mashhadani, M.K.; Wilkinson, S.J.; Zimmerman, W.B. Laboratory preparation of simulated sludge for anaerobic digestion experimentation. *J. Eng.* **2015**, *21*, 131–145.
42. Rees-Zimmerman, C.R.; Chaffin, S.T. Modelling the effect of bioreactor height on stripping fermentation products from the engineered bacterium Geobacillus thermoglucosidasius. *Biochem. Eng. J.* **2021**, *176*, 108195. [CrossRef]
43. Levenspiel, O. *Chemical Reaction Engineering*, 3rd ed.; John Wiley & Sons: Hoboken, NJ, USA, 1998.
44. Kierzkowska-Pawlak, H. Determination of kinetics in gas-liquid reaction systems. An overview. *Ecol. Chem. Eng. S* **2012**, *19*, 175–196. [CrossRef]
45. Díaz, M.; Vega, A.; Coca, J. Correlation for the estimation of gas-liquid diffusivity. *Chem. Eng. Commun.* **1987**, *52*, 271–281. [CrossRef]
46. Kimweri, H.T.H. Enhancement of Gas-Liquid Mass Transfer in Hydrometallurgical Leaching Systems. Ph.D. Thesis, University of British Columbia, Vancouver, BC, Canada, 2001.
47. Roman, F.F.; Ribeiro, A.E.; Queiroz, A.; Lenzi, G.G.; Chaves, E.S.; Brito, P. Optimization and kinetic study of biodiesel production through esterification of oleic acid applying ionic liquids as catalysts. *Fuel* **2019**, *239*, 1231–1239. [CrossRef]
48. Jain, S.; Sharma, M. Kinetics of acid base catalyzed transesterification of Jatropha curcas oil. *Bioresour. Technol.* **2010**, *101*, 7701–7706. [CrossRef]
49. Song, E.-S.; Lim, J.-W.; Lee, H.-S.; Lee, Y.-W. Transesterification of RBD palm oil using supercritical methanol. *J. Supercrit. Fluids* **2008**, *44*, 356–363. [CrossRef]
50. Ahmad, A.; Yasin, N.M.; Derek, C.; Lim, J. Kinetic studies and thermodynamics of oil extraction and transesterification of Chlorella sp. for biodiesel production. *Environ. Technol.* **2014**, *35*, 891–897. [CrossRef]
51. Nautiyal, P.; Subramanian, K.; Dastidar, M. Kinetic and thermodynamic studies on biodiesel production from Spirulina platensis algae biomass using single stage extraction–transesterification process. *Fuel* **2014**, *135*, 228–234. [CrossRef]

Article

Intensification of Biodiesel Processing from Waste Cooking Oil, Exploiting Cooperative Microbubble and Bifunctional Metallic Heterogeneous Catalysis

Fahed Javed [1], Muhammad Rizwan [2], Maryam Asif [1], Shahzad Ali [1], Rabya Aslam [3], Muhammad Sarfraz Akram [2], William B Zimmerman [4] and Fahad Rehman [1,*]

1. Microfluidics Research Group, Department of Chemical Engineering, COMSATS University Islamabad, Lahore 54000, Pakistan
2. Institute of Energy and Environmental Engineering, University of Punjab, Lahore 54000, Pakistan
3. Institute of Chemical Engineering and Technology, University of Punjab, Lahore 54000, Pakistan
4. Department of Chemical & Biological Engineering, The University of Sheffield, Sheffield S1 3JD, UK
* Correspondence: frehman@cuilahore.edu.pk; Tel.: +92-42-111-001-007; Fax: +92-42-9203100

Citation: Javed, F.; Rizwan, M.; Asif, M.; Ali, S.; Aslam, R.; Akram, M.S.; Zimmerman, W.B.; Rehman, F. Intensification of Biodiesel Processing from Waste Cooking Oil, Exploiting Cooperative Microbubble and Bifunctional Metallic Heterogeneous Catalysis. *Bioengineering* 2022, 9, 533. https://doi.org/10.3390/bioengineering9100533

Academic Editors: Indra Neel Pulidindi and Aharon Gedanken

Received: 2 September 2022
Accepted: 4 October 2022
Published: 8 October 2022

Publisher's Note: MDPI stays neutral with regard to jurisdictional claims in published maps and institutional affiliations.

Copyright: © 2022 by the authors. Licensee MDPI, Basel, Switzerland. This article is an open access article distributed under the terms and conditions of the Creative Commons Attribution (CC BY) license (https://creativecommons.org/licenses/by/4.0/).

Abstract: Waste resources are an attractive option for economical the production of biodiesel; however, oil derived from waste resource contains free fatty acids (FFA). The concentration of FFAs must be reduced to below 1 wt.% before it can be converted to biodiesel using transesterification. FFAs are converted to fatty acid methyl esters (FAMEs) using acid catalysis, which is the rate-limiting reaction (~4000 times slower than transesterification), with a low conversion as well, in the over biodiesel production process. The study is focused on synthesizing and using a bifunctional catalyst (7% Sr/ZrO_2) to carry out esterification and transesterification simultaneously to convert waste cooking oil (WCO) into biodiesel using microbubble-mediated mass transfer technology. The results reveal that a higher conversion of 85% is achieved in 20 min using 7% Sr/ZrO_2 for biodiesel production. A comprehensive kinetic model is developed for the conversion of WCO in the presence of a 7% Sr/ZrO_2 catalyst. The model indicates that the current reaction is pseudo-first-order, controlled by the vapor–liquid interface, which also indicates the complex role of microbubble interfaces due to the presence of the bifunctional catalyst. The catalyst could be recycled seven times, indicating its high stability during biodiesel production. The heterogeneous bifunctional catalyst is integrated with microbubble-mediated mass transfer technology for the first time. The results are unprecedented; furthermore, this study might be the first to use microbubble interfaces to "host" bifunctional metallic catalysts. The resulting one-step process of esterification and transesterification makes the process less energy-intensive and more cost-efficient, while also reducing process complexity.

Keywords: biodiesel; waste cooking oil; transesterification; bifunctional catalyst; microbubble technology

1. Introduction

The current standard of living is substantially dependent on energy. Its generation is a measure of progress for a developing country. Energy is consumed in domestic and industrial sectors generated primarily by fossil fuels and other sources. Biodiesel production is considered a potential alternative to fossil fuels due to its numerous advantages, such as nontoxicity and biodegradability, with low harmful emissions during combustion. Generally, biodiesel is produced via two routes: (1) esterification reaction of free fatty acids (FFA), and (2) transesterification of triglycerides, both with alcohol [1,2]. Homogeneous catalysts such as NaOH, KOH, H_3PO_4, H_2SO_4, and HCL are usually preferred due to their higher degree of interaction because of superior miscibility [3,4]. However, homogenous catalysts dissolved in the reaction mixture cause numerous challenges during the downstream separation and purification stages [5,6]. Biodiesel production through heterogeneous catalysts is an advantageous alternative to homogeneous catalysts as it provides numerous advantages, including reduced unit operation requirement, non-corrosiveness, low waste generated, reusability, recyclability, and ease of product separation [7,8]. Basic catalysts are generally used for converting triglycerides into biodiesel, but they also saponify FFAs in the feedstock, resulting in emulsions that add steps for purification, while reducing yield [9,10]. Acidic catalysts are generally used to convert FFA through esterification and transform triglycerides via hydrolysis into diglycerides, which can further convert them into FFA and cause catalyst deactivation through leaching [11,12]. To overcome the drawbacks related to the individual acidic and basic heterogeneous catalysts, bifunctional catalysts were introduced. Bifunctional catalysts can combine the characteristics of both acidic catalysts that can transform FFAs and basic catalysts that tackle triglycerides in the feedstock [13,14]. Furthermore, saponification generated by FFAs and water can be completely avoided using a bifunctional catalyst. An effective bifunctional catalyst with amphoteric base material on which acidic or basic promotors can further be modified is required. Zirconium oxide (ZrO_2) is amphoteric in nature. It has been reported to be an efficient base for heterogeneous catalysts due to its mechanical strength, corrosion resistance, chemical stability, and high water retention [6,15]. Recently, ZrO_2 has been modified to yield acidic forms, i.e., tungstate zirconia alumina (Al_2O_3/ZrO_2, TiO_2/ZrO_2) and basic forms such as CaO, and La_2O_3 [16,17]. Thus, ZrO_2 can be modified and improved to design specific catalysts with desired properties.

The use of solid heterogeneous catalysts enhances the ability of biodiesel production from WCO without additional treatment [18,19]. However, a long reaction time is still a major challenge in biodiesel production. Jitputti et al. investigated the comparison of zirconia and sulfated zirconia and achieved conversion of 49.3% and 86.3% of crude palm kernel oil, respectively, in almost 4 h [20]. Jamil et al. investigated the Mn@MgOZrO$_2$ bifunctional catalyst for waste *Phoenix dactylifera* L. oil and achieved 96.4% biodiesel conversion in 4 h [21]. Current results collectively indicate that bifunctional catalysts have enhanced performance, but an efficient process method is needed that reduces reaction time and increases the reaction rate [22,23].

Microbubble technology was recently introduced for biodiesel production, yielding improved results by enhancing the mass transfer and rate of reaction via alcohol injection within the microbubble phase [24,25]. Microbubbles have less buoyancy force, high surface energy, high temperature, and high residence time due to their smaller size than macrobubbles. When the bubble rises in the laminar regime due to its smaller size and provides internal mixing at the bubble and oil interface, homogeneity is achieved within a millisecond. Microbubble formation at a low flow rate is more favorable for smaller bubble formation, reduced coalescence, and higher surface energy as a result mass transfer increases, which directly increase the conversion of the process [2,26]. Ahmed et al. reported an increase in the reaction rate and reduced reaction time of oleic acid and methanol (MeOH) using microbubbles and achieved approximately 96% conversion of biodiesel in just 0.5 h [27]. Javed et al. studied acid esterification using chicken fat oil and MeOH and achieved 89% conversion in 0.5 h [28]. These results illustrate that microbubble technology promises an increasing rate of reaction by converting the liquid–liquid bulk reaction into a gas–liquid interfacial reaction. All these studies provide sufficient evidence that the reaction occurs at the MeOH/oil interface. However, the current work emphasizes another amphoteric characteristic (bifunctional catalyst) present on the bubble interface. The competition of microbubbles and particles at interfaces is a well-known industrial process due to dissolved air flotation, which is also intensified by fluidic oscillated microbubbles [29,30]. The current study is based on synthesizing a heterogeneous bifunctional catalyst and integrating it with microbubble-mediated mass transfer technology to enhance the reaction rate and overall process conversion. To the best of the authors' knowledge, this is the first work that uses microbubble interfaces to "host" bifunctional metallic catalysts. The closest work used bimetallic catalysts to enhance the hydroxyl radical production from ozone microbubbles, but illustrated the mechanism as bubble–pellet transient collisions, where the pellets are much larger than the microbubbles [31]. There is only one similar study for single-metal catalysis, which implemented ozone microbubble dissociation into hydroxyl radicals for oxidation reactions [32]. Here, the traditional two-step bio-diesel production process was converted into a single-step process by carrying out esterification and transesterification simultaneously, making the process more energy- and cost-efficient, while also reducing process complexity. It should be noted that esterification reactions are, in general, equilibrium reactions that achieve 60–80% conversion without reactive separation. The high conversions demonstrated in this paper give credence to the suggestion that the microbubble interface, populated by bimetallic catalyst particles, serves as the heterogeneous catalyst interface, while the vapor-phase product (water) is simultaneously extracted.

In the current study, efforts are made to present a sustainable approach of generating energy in the form of biodiesel from waste. Strontium zirconium oxide (7% Sr/ZrO_2) was synthesized and integrated with rapidly developing microbubble technology from WCO. The synthesized heterogeneous catalyst was characterized using various analytical techniques. To increase its applicability on a commercial scale and to optimize the process, additional parameters were studied. Furthermore, a detailed analysis of the reaction kinetics, catalytic mechanism, and process economics was also conducted. Hence, this study can contribute additional insight into the integration of heterogeneous catalysts with acidic and basic active sites using microbubble technology to produce high-quality biodiesel.

2. Materials and Methods

2.1. Materials

Analytical-grade FAMEs, hexane, zirconium(IV) butoxide, strontium nitrate, and poly(ethylene glycol)-block-poly(propylene glycol)-block-poly(ethylene glycol) ((EO)20-(PO)70(EO)20) triblock copolymer (TBC) were purchased from Sigma Aldrich. Analytical-grade MeOH (99%) and butanol were procured from DAEJUNG chemicals, Siheung-si, South Korea. WCO was collected from the university cafeteria in Lahore, Pakistan. All other chemicals used for catalyst preparation were purchased from Sigma Aldrich, St. Louis, MO, USA.

2.2. Oil Purification

WCO was filtered and then washed five times with distilled water at 60 °C until neutral pH was achieved. Afterward, WCO was dried over an anhydrous sodium sulfate bed to remove traces of water. The dried WCO was then stored in a sealed bottle, and its composition was characterized using GC–MS analysis. The properties of WCO and its oil composition are presented in Table 1.

Table 1. WCO properties and its oil composition.

Parameters	Units	Values
Viscosity	mPa·s	31.16
Density	kg·m^{-3}	919
FFA	%	9
Oil composition		
Linoleic acid	wt.%	9
Linolenic acid	wt.%	62
Palmitic acid	wt.%	12
Lignoceric acid	wt.%	17

2.3. Catalyst Preparation

Zirconium(IV) butoxide and strontium nitrate ($Sr(NO_3)_2$) were used as the precursors to synthesize the mesoporous specimen. Approximately 5.0 g of TBC was dissolved in 50.0 mL of ethanol and left to stir for 4 h at room temperature. Then, 80 mmol of zirconium(IV) n-butoxide (80 wt.% solution in 1-butanol) was dissolved in 20 mL of 68–70 wt.% nitric acid and 50.0 mL of ethanol. Once dissolved, a calculated amount of strontium metal solution (1.0 M) was added to a flask and stirred for 2 h. The pH was carefully maintained at 12 using 2 M NaOH solution. The solution was heated under continuous slow stirring at room temperature for 4 h. Subsequently, the two solutions were combined, and 30.0 mL of distilled water for complete transfer of the solutions. The combined solution was stirred for 5 h at room temperature. The solvent was removed at 100 °C for 24 h in the oven. Lastly, the catalyst was calcined under air in the furnace at 550 °C for 5 h. The maximum conversion of biodiesel was achieved using 7% Sr loading during the preliminary experiments, due to which 7% Sr was used throughout the study to optimize the amount to impregnate ZrO_2.

2.4. Characterization of Catalyst

Fourier-transform infrared spectroscopy (FTIR) analysis with a Thermo-Nicolet 6700P Spectrometer, Thermo Fisher Scientific, Waltham, MA, USA, was used to detect the functional groups of 7% Sr/ZrO_2. The wavelength of FTIR was set in the range of 800 to 4000 cm^{-1}. To study the surface morphology of the catalyst, scanning electron microscopy (SEM) (FEI Nova 450 NanoSEM, Thermo Fisher Scientific, Waltham MA, USA,) was used. To identify the effect of strontium on the crystallinity of ZrO_2, X-ray diffraction (XRD) was used. For XRD (Equinox 2000, Thermo Fisher Scientific, USA), the range of $2\theta = 2°–116°$ was selected using Cu-Kα radiation ($\lambda = 0.145$ nm). For surface area and porosity analysis, a Micromeritics TriStar II-3020 analyzer (Micromeritics, Norcross, GA, USA) was used to obtain N_2 adsorption/desorption isotherms of the catalyst at 77.3 K. The Brunauer–Emmett–Teller (BET) adsorption approach was used to infer the surface area. The pore surface area and volume were determined using the t-plot method.

2.5. Experimental Procedure

Biodiesel was produced through both esterification and transesterification, during which WCO reacted with MeOH in the presence of 7% Sr/ZrO_2. Initially, MeOH was heated using a heating mantle around its boiling point in a round-bottom flask. Meanwhile, WCO and 7% Sr/ZrO_2 were premixed in a separate beaker for a specific time interval. The pre-mixed solution was then transferred to the microbubble reactor. The microbubble reactor consisted of sintered-borosilicate diffuser (40 to 16 µm pore size) with a total reactor volume of 500 mL. MeOH was injected from the bottom of the reactor in the form of vapor. The temperature of the reactor was maintained at 70 °C using a brisk heater. The reaction was terminated when the desired quantity of MeOH passed through the reactor.

Afterward, biodiesel samples were centrifuged at 3500 rpm for 10 min to separate the catalyst from biodiesel, and the samples were washed with deionized water to remove impurities. For complete water removal, samples were dried using a rotary evaporator (Buchi R-210, BUCHI Corporation, New Castle, DE, USA). For parametric study, the molar ratio of WCO and MeOH was varied from 1:5 to 1:25. Catalyst loading ranged from 0 to 3 wt.% WCO. The temperature was varied from 70 to 90 °C to study the effect of biodiesel production and activation energy. All experiments were carried out three times to calculate the standard error. A diagram of the process is given in Figure 1.

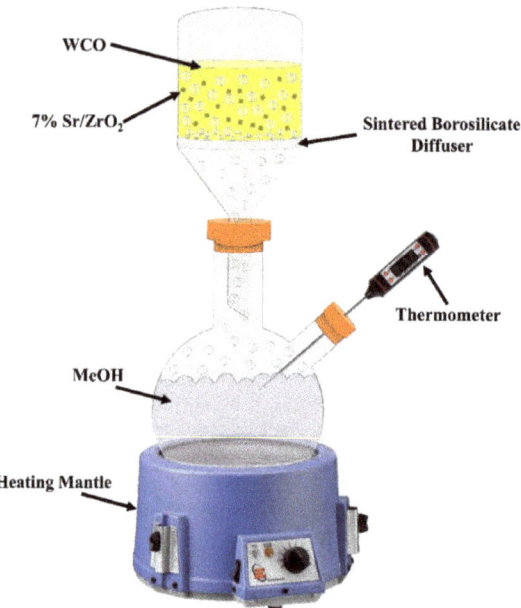

Figure 1. Systematic diagram of microbubble technology for biodiesel production.

2.6. Biodiesel Analysis

The analysis process of biodiesel using GC was taken from the literature [27,28]. Briefly, a Shimadzu GC-2014, Shimadzu Europa, Duisburg, Germany, was used to analyze the biodiesel with a GC-FID and column Agilent J&W EN14103 Column, Agilent Technologies, Santa Clara, CA, USA, with 30 m length, 0.32 mm id, and 0.25 μm film thickness. Nitrogen gas with a flowrate of 1.5 mL/min was used as the carrier gas. The conditions were kept constant for every sample to achieve comparable readings. The sample was injected at 523 K and a split ratio of 50:1 [27,28].

3. Results and Discussion

3.1. Catalyst Analysis

The characterization of 7% Sr/ZrO$_2$ is shown in Figure 2a–c,e. The FTIR spectrum of zirconia shows a broad peak band in the region of 3700–3400 cm^{-1}, which was attributed to asymmetric stretching of –OH groups. The band at 900 cm^{-1} was associated with ZrO. The peak at 1623 cm^{-1} corresponded to the C=O group due to SrO$_2$ in the catalyst. The weak absorption bands at 1603 cm^{-1} and 1318 cm^{-1} were attributed to the bending vibration of C–H bands. Pore size and surface area analyses showed that the catalyst formation was mesoporous, ranging from 3 to 11 nm. The XRD diffractograms of synthesized Sr/ZrO$_2$ revealed the crystal structure of pure ZrO$_2$ to be monoclinic. The structure of ZrO$_2$ was sustained after calcination at 550 °C for 5 h, with no peak broadening observed. The pattern of Sr/ZrO$_2$ predominantly showed peaks of ZrO$_2$ as the parent peak, with an additional peak at 31.750° for Sr/ZrO$_2$. The pore size distribution of the catalyst is shown in Figure 2b, and the pore size and surface area data estimated via BET analysis are shown in Table 2. The catalyst pore size demonstrated mesopores ranging between 3 and 11 nm.

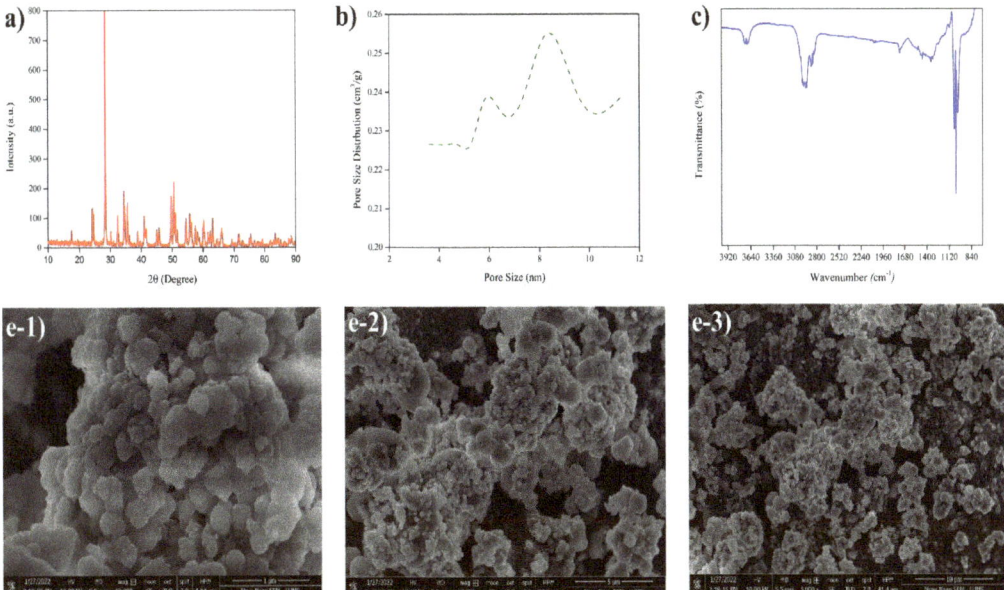

Figure 2. Characterization of Sr/ZnO₂: (**a**) XRD analysis, (**b**) BET analysis, (**c**) FTIR, and (**e-1–3**) SEM analysis.

Table 2. Structural properties of catalyst.

Catalyst	Surface Area (m^2/g)	BJH Area (m^2/g)	BJH Volume (cm^3/g)	BET Pore Diameter (nm)	BJH Pore Diameter (nm)
7% Sr/ZrO$_2$	119.80	117.34	0.2154	8.97	8.21

Figure 2e-1–3 show the 7% Sr/ZrO$_2$ surface morphology at different magnitudes using SEM. The modified ZrO$_2$ showed well-shaped crystalline particles after impregnation of Sr particles. The analysis revealed that the addition of Sr metal to ZrO$_2$ increased the amphoteric behavior of ZrO$_2$, which increased both the basic and the acidic active sites of the catalyst. The presence of both acidic and basic sites in 7% Sr/ZrO$_2$ facilitated both esterification and transesterification.

3.2. Parameter Optimization for Biodiesel Production

3.2.1. Effect of Molar Ratio on Biodiesel Production

To study the effect of molar ratio on WCO and MeOH for biodiesel production, different experiments were performed by changing the molar ratio (WCO/MeOH = 1:5 to 1:25), as shown in Figure 3. Initially, the results indicated that, by increasing the molar ratio from 1:5 to 1:15, the reaction conversion increased from 59% to 85%, before decreasing to 68% at a higher molar ratio of 1:25. Increasing the molar ratio also increased the process conversion. By increasing the MeOH quantity, the reaction time of the system also increased, due to which the reaction moved further in the forward direction. Furthermore, the system operated at a temperature higher than the boiling point of MeOH, which resulted in unreacted MeOH leaving via the top of reactor. However, the MeOH volume could be increased by changing the molar ratio to an extent that a limited amount of MeOH stayed in the reactor, as reported by Javed et al. (2021) [28]. The presence of unreacted MeOH in the system diminished the active catalyst sites due to oil dilution, which reduced the catalyst activity [33,34]. Hence, the process efficiency decreased at higher molar ratios, as observed from the current result. The optimal molar ratio was found to be 1:15.

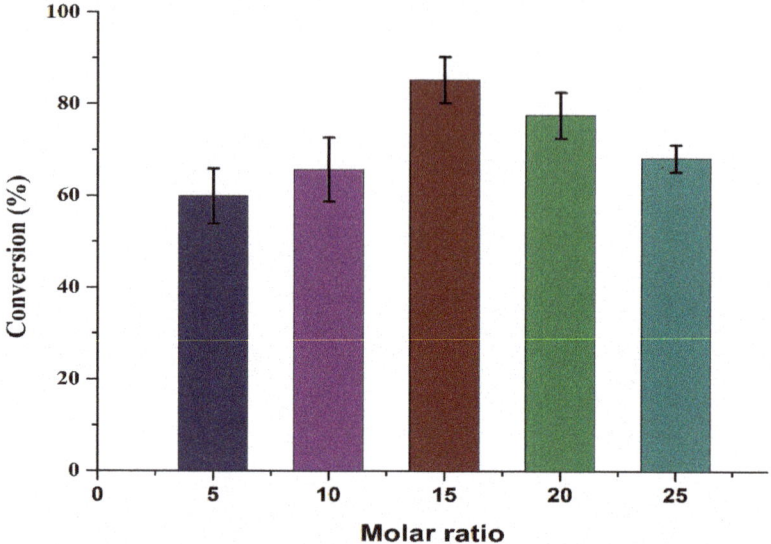

Figure 3. Effect of molar ratio (WCO/MeOH = 1:5 to 1:25) on biodiesel production.

3.2.2. Influence of Catalyst on the Conversion of WCO

The effect of catalytic activity using 7% Sr/ZrO$_2$ is shown in Figure 4. The microbubble process yielded a low conversion of 34% in the absence of catalyst during the transesterification reaction. The conversion of the transesterification reaction was related to the active sites of the 7% Sr/ZrO$_2$ catalyst. The conversion of the process increased from 64% to 90% upon increasing catalyst loading. Increasing the catalyst loading seemingly enhanced the entanglement of WCO and 7% Sr/ZrO$_2$ catalyst. The degree of entanglement of WCO and 7% Sr/ZrO$_2$ affected the conversion of WCO, whereby a higher degree of entanglement led to a higher conversion of WCO and a higher rate of reaction, and vice versa. However, the results illustrate that, beyond 1% catalyst loading, only a 5% increase in conversion was obtained upon doubling catalyst loading. One possible reason is that increasing the catalyst loading increased the number of active sites, whereas the reactive side of WCO remained constant during entanglement, such that excess loading of 7% Sr/ZrO$_2$ did not drastically enhance the conversion of the transesterification reaction. Furthermore, the results demonstrate that 1% catalyst loading was optimal for the current study.

3.2.3. Effect of Temperature on Biodiesel Production

The effect of temperature is a vital parameter controlling the reaction rate of the process. However, raising the temperature also increases the processing costs and renders the process unviable for commercial scale. The temperature was varied from 70 to 90 °C to study the impact of temperature on the process (Figure 5). The results indicate that the conversion of WCO was not significantly affected by the change in temperature, as the initial temperature of the reactor was above the boiling point of MeOH. When bubbles rose, they reacted with WCO, while unreacted MeOH left the system, thereby not affecting the overall conversion of the process. However, at the start of the reaction, the high-temperature system showed a higher conversion of 58% in 5 min at 90 °C than that achieved at 70 °C (42%). A possible explanation for this behavior is that, at high temperature, the viscosity of oil decreased, due to which the miscibility of 7% Sr/ZrO$_2$ and MeOH increased with WCO. Furthermore, by increasing the temperature, the collision frequency of bubbles with

WCO also increased, which also enhanced the rate of reaction. Hence, the optimal and most economical temperature for this study was 70 °C.

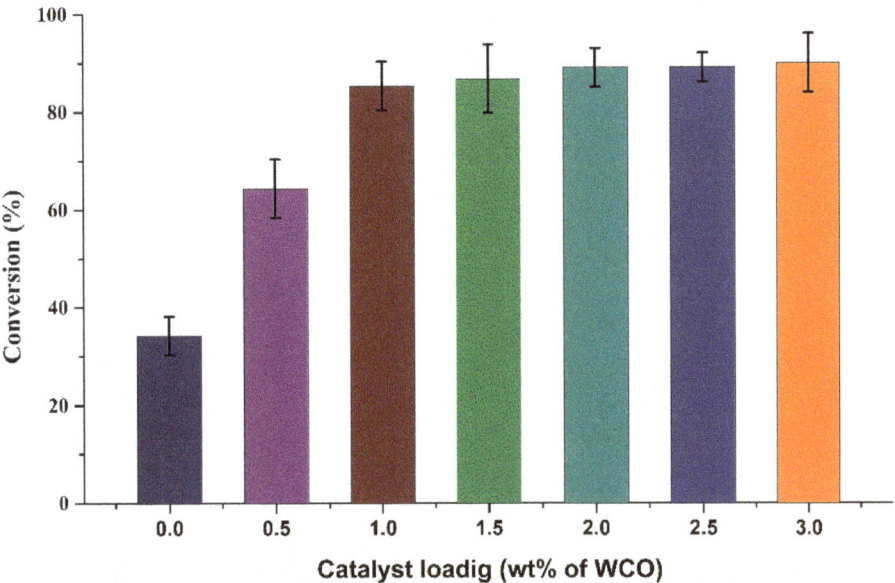

Figure 4. Influence of catalyst loading on the conversion of WCO.

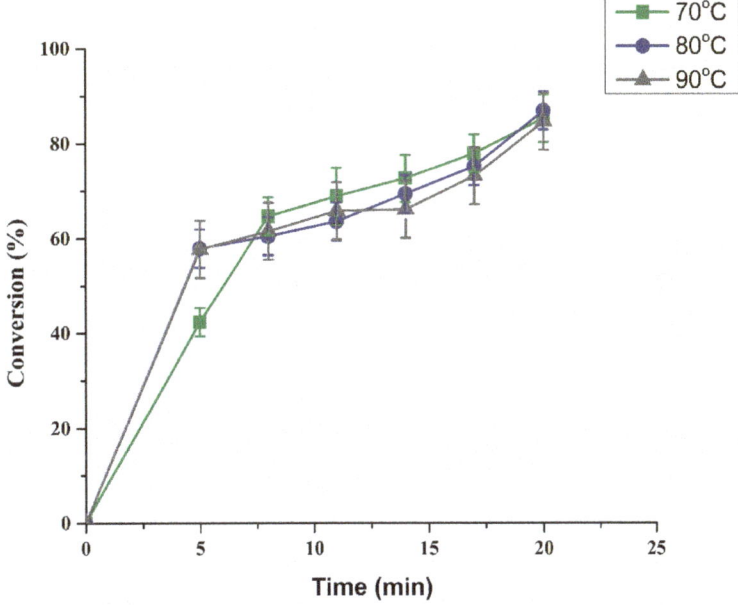

Figure 5. Effect of different temperatures on WCO conversion into biodiesel.

3.2.4. Effect of Reaction Time on Biodiesel Production

The biodiesel production was monitored in a reaction using WCO and MeOH as the model system. Before measuring biodiesel production using heterogeneous catalysts in the microbubble system, it was necessary to perform a control experiment using 7% Sr/ZrO_2 under the same conditions by mimicking the conventional process with a beaker and magnetic stirring. For both processes, 100 g of WCO, 72 mL of MeOH, and 1 wt.% of 7% Sr/ZrO_2 were used. Only 8% conversion was achieved in 20 min, eventually reaching 93% in 240 min. However, in the current microbubble system, MeOH was introduced in the microbubble phase, enhancing the diffusion rate of both reactants, in addition to increasing the system mass transfer efficiency [27,28]. As indicated from the microbubble results (Figure 6), a higher conversion of 85% was achieved in 20 min.

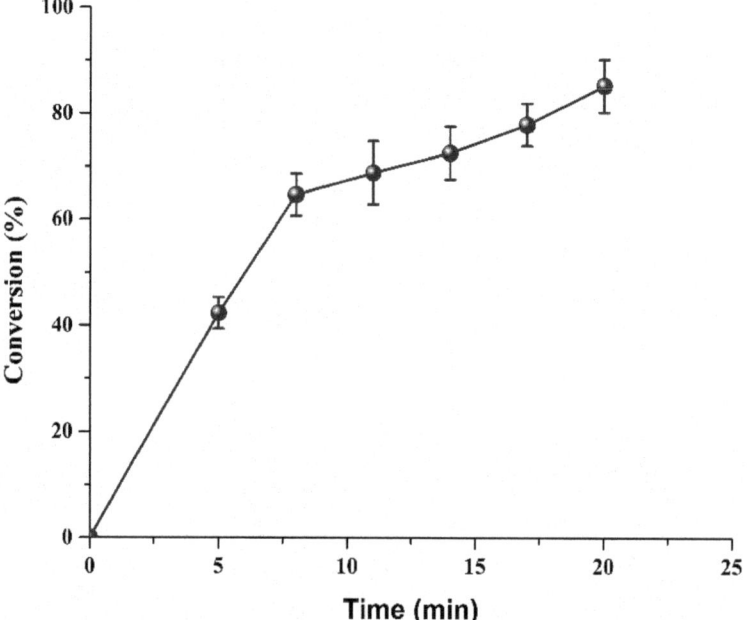

Figure 6. Biodiesel production over time using microbubble technology.

The first 8 min of the reaction occurred spontaneously, as illustrated from the steep curve, achieving 65% conversion, before gradually increasing to 85% in 20 min. The possible reason for this is that, at the start of the reaction, the WCO and catalyst had readily available active sites for reaction as time increased, and the concentration of WCO was reduced by more than 65% via conversion into biodiesel; accordingly, bubbles had a lower chance of interaction with the freshly available biodiesel. Consequently, the rate of reaction slowed down during the last 12 min. A comparison of the study with other heterogeneous catalysts is shown in Table 3. The current study clearly achieved a higher reaction rate in a shorter period than previously reported studies.

Table 3. Comparison of the current study with other heterogeneous catalysts used for biodiesel production.

Catalyst	Temperature (°C)	Time (min)	Catalyst Loading (wt.%)	Conversion (%)	Reference
Li-Al HTA	65	60	3	83	[35]
KI/SiO	70	480	5	91	[36]
KF/Al_2O_3	60	480	3	90	[37]
3% La_2O_3–ZrO_2	65	300	6	56	[38]
ZrO_2/SiO_2	120	120	10	48.6	[6]
Li/ZrO_2	65	180	3	98.2	[39]
21% La_2O_3/ZrO_2	200	480	5	84.9	[40]
7% Sr/ZrO_2	70	20	1	85	This study

3.3. Reaction Kinetics and Mechanism of WCO-Based Biodiesel

3.3.1. Proposal of a Reaction Mechanism for Biodiesel Production Using 7% SR/ZRO$_2$

The current study proposes that 7% Sr/ZrO_2 performed both esterification and transesterification simultaneously to produce biodiesel. ZrO_2 is amphoteric and possesses both basic and acidic active sites. The active side was further strengthened by modifying it with Sr, which further improved the catalyst activity. The proposed reaction mechanism for this study is shown in Figure 7, which further elaborates how esterification and transesterification occurred on the acidic and basic active sites of the catalyst. The chemical reaction proposed through the heterogeneous catalyst is based on three basic steps: adsorption of the reactant on active sites of the catalyst, reaction between active sites of the catalyst, and desorption of product from active sites of the catalyst.

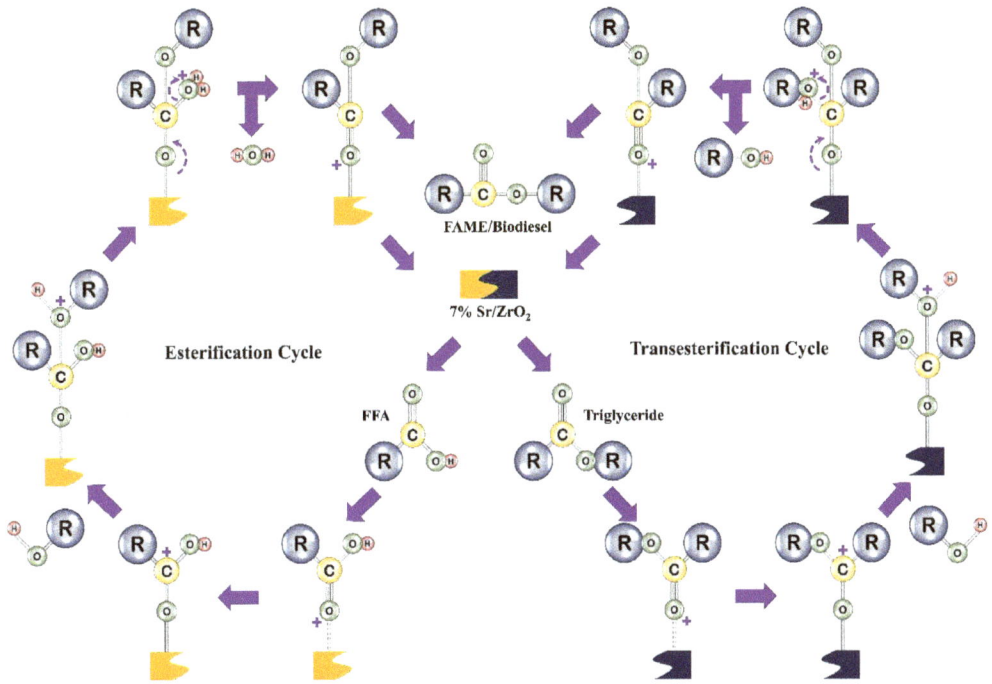

Figure 7. Proposed reaction of 7% Sr/ZrO_2 for converting WCO into biodiesel.

In biodiesel production, both FFA and MeOH molecules get absorbed onto the acidic and basic catalyst sites. The FFA molecules change to carbocations, with oxygen anions from MeOH. In the second step, a nucleophilic attack of the carbocation and oxygen anion is directed toward the molecules of triglycerides and FFA, thereby supporting both esterification and transesterification. Furthermore, tetrahedral intermediates are also formed due to nucleophilic attacks. In the last step, –OH and –C–O– bonds break. As a result, hydroxyl group and alkyl triglycerides are desorbed. Desorption of these molecules provides us with a final product known as biodiesel (mono-alkyl ester). After desorption of biodiesel, both acidic and basic sites of the catalyst are again available for another cycle. This process continues until the reaction is completed; moreover, water and glycerol are produced as byproducts of this process.

3.3.2. Kinetics Analysis and Activation Energy of WCO-Based Biodiesel

To investigate the reaction kinetics of the vapor–liquid system using heterogeneous catalysts, different parameters were optimized in the above experiments, and the kinetics were determined in optimal conditions. To validate the current hypothesis of the catalyst facilitating the reaction on the bubble surface while moving the bubble upward, the Hatta number (Ha) was calculated using Equation (1) [41,42]. Ha signifies whether the reaction occurred on the surface of the bubble or in bulk. If the value of Ha is greater than 1, then the reaction occurred on the surface of the bubble, and the controlling factor was the reaction kinetics. If the reaction occurred in bulk, then mass transfer was the controlling factor, with intense mixing becoming the dominant factor.

$$Ha = \frac{\sqrt{(D_o)_T \, k \, C_b}}{k_{bl}}, \tag{1}$$

where k is the rate constant, k_{bl} is a liquid film coefficient, and D_o is a coefficient of diffusion of MeOH in WCO at different temperatures (see Equations (2) and (3) [43], where v_l and v_g are molar volumes of WCO and MeOH, and μ_l is the viscosity of WCO).

$$(D_o)_{25°C} = 6.02 \times 10^{-5} \left(\frac{V_l^{0.36}}{\mu_l^{0.61} V_g^{0.64}} \right). \tag{2}$$

$$(D_o)_T = 4.996 \times 10^3 (D_o)_{25°C} \exp\left(\frac{-2539}{T} \right). \tag{3}$$

The value of k_{bl} was calculated using Equation (4) for bubble sizes less than 2 mm [44].

$$k_{bl} = 0.31 \left(\frac{(D_{g/l})^2 \rho_l g}{\mu_l} \right)^{\frac{1}{3}}. \tag{4}$$

In the current study, a calculated value of $Ha > 1$ implied that the reaction occurred on the surface of the bubble and that the catalyst active site was induced by the MeOH bubble. To calculate the order of reaction of the vapor–liquid system, the enhancement factor (E) was determined using Equation (5) [41], where E_i is the infinite enhancement factor, calculated using Equation (6).

$$E = Ha \left(1 - \frac{Ha - 1}{2E_i} \right). \tag{5}$$

$$E_i = 1 + (D_o)_T \left(\frac{C_l H}{b \, P_g} \right), \tag{6}$$

where C_l is the WCO concentration (kmol·m^{-3}), P_g is the partial pressure of vapors, and H is Henry's constant. The values of E and Ha indicate that the reaction is pseudo-first-order due to the similarity of both values. The rate was determined using Equation (7).

$$-r_A = \frac{1}{\frac{1}{k_a \sigma} + \frac{H}{a\sqrt{(D_o)_T k C_b}}} P_a, \tag{7}$$

where k_a is the gas film coefficient, and P_a (bar) is the pressure of the system. The calculated values of k_a with the interfacial area ($k_a\sigma$), k_{bl}, and H were 0.011 kmol·s^{-1}·m^{-3}, 1.04 × 10^{-4} ms^{-1}, and 3.72 kmol·m^{-3}·Pa^{-1}, respectively. The overall rate of reaction is shown in Equation (8).

$$-r_A = \left(8.730 \times 10^{-5}\right)(P_a \times 101325)\left(\sqrt{C_b}\right). \tag{8}$$

The current kinetics show that the order of the reaction was pseudo-first-order, and that the reaction occurred on the microbubble surface, with the concentration gradient of WCO and MeOH remaining high throughout the reaction. The high concentration gradient and high surface area of catalyst facilitated biodiesel production in a shorter time.

To further investigate the kinetics of the vapor–liquid system, the activation energy (E_A) was determined using the Arrhenius equation as a function of the effect of temperature on the rate of reaction [28,41]. Numerous studies determined the E_A of processes using the Arrhenius equation [27,45]. The relationship of k as a function of pre-exponential factor is used to calculate E_A using Equation (9) [46], where $A°$ is the pre-exponential factor and R is the gas constant (8.314 J·mol^{-1}·K^{-1}) (Figure 8).

$$\ln k = -\frac{E_A}{RT} + \ln A°. \tag{9}$$

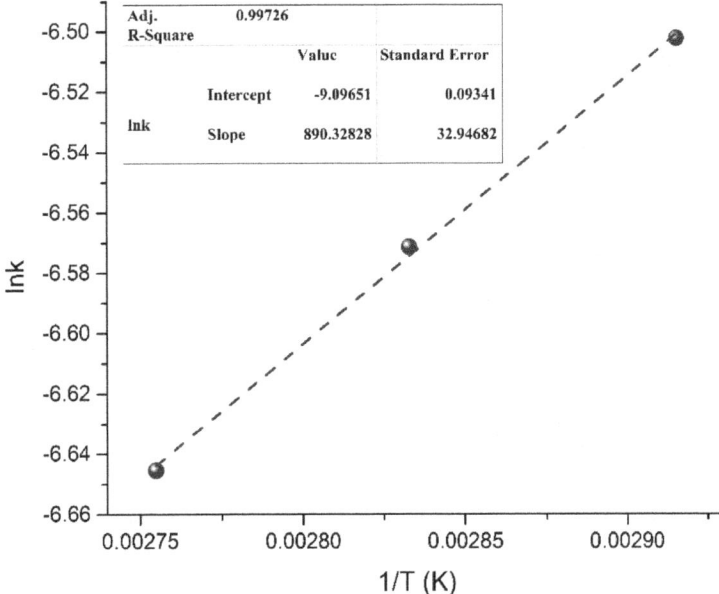

Figure 8. Arrhenius plot for determining E_A required to convert WCO into biodiesel.

The E_A of the 7% Sr/ZrO$_2$-based biodiesel production process using microbubble technology was estimated as 7.4 kJ·mol^{-1}. The achieved E_A is lower than that of other biodiesel processes with different catalysts (Table 4). The low E_A also implies that 7% Sr/ZrO$_2$ in the system facilitated the vapor–liquid system and enhanced the overall reaction rate. Furthermore, the low E_A indicates that less energy was needed for the reactant to pass the activation barrier.

Furthermore, 7% Sr/ZrO$_2$ has a high surface area and microbubbles have high surface energy, which collectively facilitated the rate of reaction and reduced the E_A of the system. Moreover, MeOH was injected in the form of vapor, indicating that the latent heat of MeOH was also available in the form of free energy, making the nature of the reaction more exergonic. The high reaction rate and reduced E_A demonstrate the potential for implementation of both the heterogeneous catalyst and the vapor–liquid system on a commercial level.

Table 4. The comparison of E_A different biodiesel processes.

Feedstock	Type of Transesterification	Catalyst	Activation Energy (kJ.mol^{-1})	Reference
Waste cooking oil	Ultra-Sonication	Calcium diglyceroxide	119.23	[34]
Waste cooking oil	Supercritical method	No catalyst	50.5	[47]
Waste cooking oil	Microwave technology	Calcium diglyceroxide	26.56	[48]
Waste cooking oil	Conventional method	CaO/SiO$_2$	66.27	[49]
Waste cooking oil	Conventional method	Cs$_{2.5}$H$_{0.5}$PW$_{12}$O$_{40}$	36	[50]
Stearic acid	Conventional method	ZrO$_2$/SiO$_2$	47	[51]
Rapeseed oil	Solvent-free method	Sulfated zirconia	22.5	[52]
Levulinic acid	Conventional method	SO$_4^{2-}$/ZrO$_2$	14.61	[53]
Oleic acid	Microbubble process	H$_2$SO$_4$	26.37	[27]
Chicken fat oil	Microbubble process	PTSA	24.9	[28]
Waste cooking oil	Microbubble process	7% Sr/ZrO$_2$	7.4	This study

3.4. Reusability and Reactivation of the Sr/ZrO$_2$

The reusability of the catalyst was evaluated under the optimized conditions obtained in the current study. The 7% Sr/ZrO$_2$ catalyst was evaluated for five cycles, with the conversion dropping after each cycle, as shown in Figure 9. After each cycle, the catalyst was centrifuged and washed with MeOH and acetone before reintroducing it into a new cycle. The results show that, after the third cycle, the conversion decreased by less than 10%, whereas, after the seventh cycle, the conversion reached 48%. A possible reason for this behavior is the leaching of Sr ions into the reaction medium, reducing the catalyst activity [54,55]. However, the catalyst can be regenerated after four cycles by loading a certain amount of Sr ions, followed by calcination of the catalyst. Hence, the catalyst can be used for up to four cycles in the current system, after which catalyst reactivation is required.

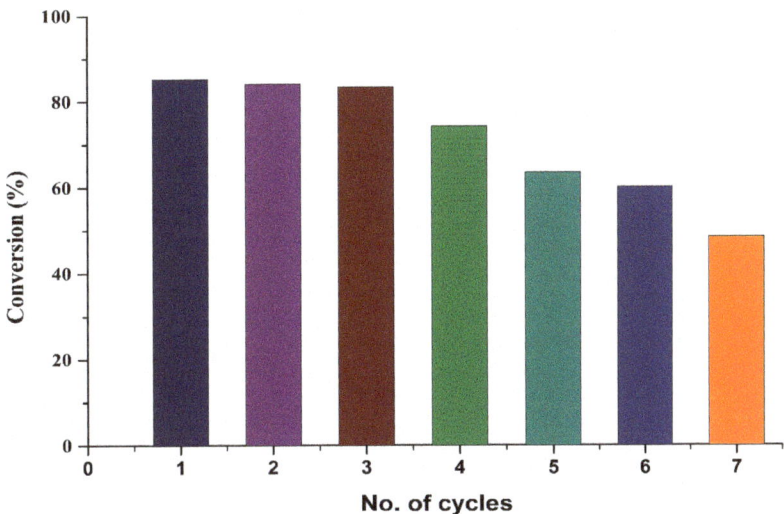

Figure 9. Reusability of the catalyst in the current microbubble technology.

4. Conclusions

A ZrO$_2$-based bifunctional heterogeneous catalyst was successfully prepared using strontium nitrate. The physicochemical properties of the catalyst enabled the interaction between ZrO$_2$ and strontium nitrate. Moreover, the bifunctional heterogeneous catalyst improved catalytic activity when combined with microbubble technology. The results achieved 85% conversion in 20 min, which is higher than previously reported bifunctional catalysts. The activation energy of the current process was 7.4 kJ·mol^{-1}, highlighting the effect of the catalyst on increasing the process efficiency. The catalyst also showed substantial chemical and thermal stability, as it could be reused at least four times without losing biodiesel production activity. The current study provides sufficient evidence for the presence of bifunctional metallic catalysts on the interface of microbubbles in the form of biodiesel processing with a high reaction rate and low activation energy. This study again supports that the use of microbubble technology is a viable alternative for the production of low-cost biodiesel for sustainable energy production.

Author Contributions: Conceptualization, F.R. and R.A.; methodology, F.J.; validation, F.R., M.S.A. and W.B.Z.; formal analysis, M.A.; investigation, M.R.; data curation, S.A.; writing—original draft preparation, F.J.; writing—review and editing, F.J. and S.A.; supervision, F.R.; funding acquisition, F.R and W.B.Z. All authors have read and agreed to the published version of the manuscript.

Funding: This research was funded by Higher Education Commission Pakistan (HEC), grant number 20-7924/17 and Engineering and Physical Sciences Research Council (EPSRC) grant number EP/S031421/1 and EP/N011511/1.The APC was funded by Engineering and Physical Sciences Research Council (EPSRC) grant number EP/S031421/1 and EP/N011511/1.

Institutional Review Board Statement: Not applicable.

Informed Consent Statement: Not applicable.

Data Availability Statement: The data presented in this study are available on request from the corresponding author. The data are not publicly available due to confidentiality of work.

Acknowledgments: The authors would like to acknowledge Higher Education Commission Pakistan (HEC) for the Project Number 20-7924/17. WZ would also like to acknowledge support from EPSRC grant number EP/S031421/1 and EP/N011511/1.

Conflicts of Interest: The authors declare no conflict of interest.

References

1. Javed, F.; Aslam, M.; Rashid, N.; Shamair, Z.; Khan, A.L.; Yasin, M.; Fazal, T.; Hafeez, A.; Rehman, F.; Rehman, M.S.U. Microalgae-based biofuels, resource recovery and wastewater treatment: A pathway towards sustainable biorefinery. *Fuel* **2019**, *255*, 115826. [CrossRef]
2. Khan, Z.; Javed, F.; Shamair, Z.; Hafeez, A.; Fazal, T.; Aslam, A.; Zimmerman, W.B.; Rehman, F. Current developments in esterification reaction: A review on process and parameters. *J. Ind. Eng. Chem.* **2021**, *103*, 80–101. [CrossRef]
3. Tan, S.X.; Lim, S.; Ong, H.C.; Pang, Y.L. State of the art review on development of ultrasound-assisted catalytic transesterification process for biodiesel production. *Fuel* **2019**, *235*, 886–907. [CrossRef]
4. Bashir, M.A.; Wu, S.; Zhu, J.; Krosuri, A.; Khan, M.U.; Aka, R.J.N. Recent development of advanced processing technologies for biodiesel production: A critical review. *Fuel Process. Technol.* **2022**, *227*, 107120. [CrossRef]
5. Rodríguez-Ramírez, R.; Romero-Ibarra, I.; Vazquez-Arenas, J. Synthesis of Sodium Zincsilicate (Na_2ZnSiO_4) and heterogeneous catalysis towards biodiesel production via Box-Behnken Design. *Fuel* **2020**, *280*, 118668. [CrossRef]
6. Ibrahim, M.M.; Mahmoud, H.R.; El-Molla, S.A. Influence of support on physicochemical properties of ZrO_2 based solid acid heterogeneous catalysts for biodiesel production. *Catal. Commun.* **2019**, *122*, 10–15. [CrossRef]
7. Aransiola, E.F.; Ojumu, T.V.; Oyekola, O.; Madzimbamuto, T.; Ikhu-Omoregbe, D. A review of current technology for biodiesel production: State of the art. *Biomass Bioenergy* **2014**, *61*, 276–297. [CrossRef]
8. Al-Sakkari, E.; El-Sheltawy, S.; Attia, N.; Mostafa, S. Kinetic study of soybean oil methanolysis using cement kiln dust as a heterogeneous catalyst for biodiesel production. *Appl. Catal. B Environ.* **2017**, *206*, 146–157. [CrossRef]
9. Anuar, M.R.; Abdullah, A.Z. Challenges in biodiesel industry with regards to feedstock, environmental, social and sustainability issues: A critical review. *Renew. Sustain. Energy Rev.* **2016**, *58*, 208–223. [CrossRef]
10. de Lima, A.L.; Ronconi, C.M.; Mota, C.J. Heterogeneous basic catalysts for biodiesel production. *Catal. Sci. Technol.* **2016**, *6*, 2877–2891. [CrossRef]
11. Su, F.; Guo, Y. Advancements in solid acid catalysts for biodiesel production. *Green Chem.* **2014**, *16*, 2934–2957. [CrossRef]
12. Chai, M.; Tu, Q.; Lu, M.; Yang, Y.J. Esterification pretreatment of free fatty acid in biodiesel production, from laboratory to industry. *Fuel Process. Technol.* **2014**, *125*, 106–113. [CrossRef]
13. Farooq, M.; Ramli, A.; Subbarao, D. Biodiesel production from waste cooking oil using bifunctional heterogeneous solid catalysts. *J. Clean. Prod.* **2013**, *59*, 131–140. [CrossRef]
14. Alhassan, F.H.; Rashid, U.; Taufiq-Yap, Y. Synthesis of waste cooking oil-based biodiesel via effectual recyclable bi-functional $Fe_2O_3MnOSO_2{}^{2-}/ZrO_2$ nanoparticle solid catalyst. *Fuel* **2015**, *142*, 38–45. [CrossRef]
15. Ranjbar, M.; Yousefi, M.; Lahooti, M.; Malekzadeh, A. Preparation and characterization of tetragonal zirconium oxide nanocrystals from isophthalic acid-zirconium (IV) nanocomposite as a new precursor. *Int. J. Nanosci. Nanotechnol.* **2012**, *8*, 191–196.
16. Li, H.; Fang, Z.; Smith, R.L., Jr.; Yang, S. Efficient valorization of biomass to biofuels with bifunctional solid catalytic materials. *Prog. Energy Combust. Sci.* **2016**, *55*, 98–194. [CrossRef]
17. Raia, R.Z.; da Silva, L.S.; Marcucci, S.M.P.; Arroyo, P.A. Biodiesel production from *Jatropha curcas* L. oil by simultaneous esterification and transesterification using sulphated zirconia. *Catal. Today* **2017**, *289*, 105–114. [CrossRef]
18. Zhang, Y.; Duan, L.; Esmaeili, H. A review on biodiesel production using various heterogeneous nanocatalysts: Operation mechanisms and performances. *Biomass Bioenergy* **2022**, *158*, 106356. [CrossRef]
19. Mukhtar, A.; Saqib, S.; Lin, H.; Shah, M.U.H.; Ullah, S.; Younas, M.; Rezakazemi, M.; Ibrahim, M.; Mahmood, A.; Asif, S. Current status and challenges in the heterogeneous catalysis for biodiesel production. *Renew. Sustain. Energy Rev.* **2022**, *157*, 112012. [CrossRef]
20. Jitputti, J.; Kitiyanan, B.; Rangsunvigit, P.; Bunyakiat, K.; Attanatho, L.; Jenvanitpanjakul, P. Transesterification of crude palm kernel oil and crude coconut oil by different solid catalysts. *Chem. Eng. J.* **2006**, *116*, 61–66. [CrossRef]
21. Jamil, F.; Myint, M.T.Z.; Al-Hinai, M.; Al-Haj, L.; Baawain, M.; Al-Abri, M.; Kumar, G.; Atabani, A. Biodiesel production by valorizing waste *Phoenix dactylifera* L. Kernel oil in the presence of synthesized heterogeneous metallic oxide catalyst (Mn@MgO-ZrO_2). *Energy Convers. Manag.* **2018**, *155*, 128–137. [CrossRef]
22. Mathew, G.M.; Raina, D.; Narisetty, V.; Kumar, V.; Saran, S.; Pugazhendi, A.; Sindhu, R.; Pandey, A.; Binod, P. Recent advances in biodiesel production: Challenges and solutions. *Sci. Total Environ.* **2021**, *794*, 148751. [CrossRef] [PubMed]
23. Maheswari, P.; Haider, M.B.; Yusuf, M.; Klemeš, J.J.; Bokhari, A.; Beg, M.; Al-Othman, A.; Kumar, R.; Jaiswal, A.K. A review on latest trends in cleaner biodiesel production: Role of feedstock, production methods, and catalysts. *J. Clean. Prod.* **2022**, *355*, 131588. [CrossRef]
24. Zimmerman, W.B.; Tesar, V.; Butler, S.; Bandulasena, H.C. Microbubble generation. *Recent Pat. Eng.* **2008**, *2*, 1–8. [CrossRef]
25. Zimmerman, W.B.; Al-Mashhadani, M.K.; Bandulasena, H.H. Evaporation dynamics of microbubbles. *Chem. Eng. Sci.* **2013**, *101*, 865–877. [CrossRef]
26. Zimmerman, W.B.J.; Tesar, V. Bubble generation for aeration and other purposes. U.S. Patent No EP2081666B1, 19 October 2011.
27. Ahmad, N.; Javed, F.; Awan, J.A.; Ali, S.; Fazal, T.; Hafeez, A.; Aslam, R.; Rashid, N.; Rehman, M.S.U.; Zimmerman, W.B. Biodiesel production intensification through microbubble mediated esterification. *Fuel* **2019**, *253*, 25–31. [CrossRef]
28. Javed, F.; Shamair, Z.; Hafeez, A.; Fazal, T.; Aslam, R.; Akram, S.; Rashid, N.; Zimmerman, W.B.; Rehman, F. Conversion of poultry-fat waste to a sustainable feedstock for biodiesel production via microbubble injection of reagent vapor. *J. Clean. Prod.* **2021**, *311*, 127525. [CrossRef]

29. Hanotu, J.O.; Bandulasena, H.; Zimmerman, W.B. Aerator design for microbubble generation. *Chem. Eng. Res. Des.* **2017**, *123*, 367–376. [CrossRef]
30. Hanotu, J.; Bandulasena, H.H.; Chiu, T.Y.; Zimmerman, W.B. Oil emulsion separation with fluidic oscillator generated microbubbles. *Int. J. Multiph. Flow* **2013**, *56*, 119–125. [CrossRef]
31. Jothinathan, L.; Cai, Q.; Ong, S.; Hu, J. Fe-Mn doped powdered activated carbon pellet as ozone catalyst for cost-effective phenolic wastewater treatment: Mechanism studies and phenol by-products elimination. *J. Hazard. Mater.* **2022**, *424*, 127483. [CrossRef]
32. Khuntia, S.; Majumder, S.K.; Ghosh, P. Catalytic ozonation of dye in a microbubble system: Hydroxyl radical contribution and effect of salt. *J. Environ. Chem. Eng.* **2016**, *4*, 2250–2258. [CrossRef]
33. Li, X.-F.; Zuo, Y.; Zhang, Y.; Fu, Y.; Guo, Q.-X. In situ preparation of K_2CO_3 supported Kraft lignin activated carbon as solid base catalyst for biodiesel production. *Fuel* **2013**, *113*, 435–442. [CrossRef]
34. Gupta, A.R.; Yadav, S.V.; Rathod, V.K. Enhancement in biodiesel production using waste cooking oil and calcium diglyceroxide as a heterogeneous catalyst in presence of ultrasound. *Fuel* **2015**, *158*, 800–806. [CrossRef]
35. Shumaker, J.L.; Crofcheck, C.; Tackett, S.A.; Santillan-Jimenez, E.; Crocker, M. Biodiesel production from soybean oil using calcined Li–Al layered double hydroxide catalysts. *Catal. Lett.* **2007**, *115*, 56–61. [CrossRef]
36. Samart, C.; Sreetongkittikul, P.; Sookman, C. Heterogeneous catalysis of transesterification of soybean oil using KI/mesoporous silica. *Fuel Process. Technol.* **2009**, *90*, 922–925. [CrossRef]
37. Boz, N.; Kara, M.; Sunal, O.; Alptekin, E.; Değirmenbaşi, N. Investigation of the fuel properties of biodiesel produced over an alumina-based solid catalyst. *Turk. J. Chem.* **2009**, *33*, 433–442. [CrossRef]
38. Salinas, D.; Sepúlveda, C.; Escalona, N.; Gfierro, J.; Pecchi, G. Sol-gel La_2O_3–ZrO_2 mixed oxide catalysts for biodiesel production. *J. Energy Chem.* **2018**, *27*, 565–572. [CrossRef]
39. Ding, Y.; Sun, H.; Duan, J.; Chen, P.; Lou, H.; Zheng, X. Mesoporous Li/ZrO_2 as a solid base catalyst for biodiesel production from transesterification of soybean oil with methanol. *Catal. Commun.* **2011**, *12*, 606–610. [CrossRef]
40. Georgogianni, K.; Katsoulidis, A.; Pomonis, P.; Manos, G.; Kontominas, M. Transesterification of rapeseed oil for the production of biodiesel using homogeneous and heterogeneous catalysis. *Fuel Process. Technol.* **2009**, *90*, 1016–1022. [CrossRef]
41. Levenspiel, O. *Chemical Reaction Engineering*, 3rd ed.; John Wiley & Sons: Hoboken, NJ, USA, 1998.
42. Kierzkowska-Pawlak, H. Determination of kinetics in gas-liquid reaction systems. An overview. *Ecol. Chem. Eng. S* **2012**, *19*, 175–196. [CrossRef]
43. Díaz, M.; Vega, A.; Coca, J. Correlation for the estimation of gas-liquid diffusivity. *Chem. Eng. Commun.* **1987**, *52*, 271–281. [CrossRef]
44. Kimweri, H.T.H. *Enhancement of Gas-Liquid Mass Transfer in Hydrometallurgical Leaching Systems*; University of British Columbia: Vancouver, BC, Canada, 2001.
45. Neumann, K.; Werth, K.; Martín, A.; Górak, A. Biodiesel production from waste cooking oils through esterification: Catalyst screening, chemical equilibrium and reaction kinetics. *Chem. Eng. Res. Des.* **2016**, *107*, 52–62. [CrossRef]
46. Roman, F.F.; Ribeiro, A.E.; Queiroz, A.; Lenzi, G.G.; Chaves, E.S.; Brito, P. Optimization and kinetic study of biodiesel production through esterification of oleic acid applying ionic liquids as catalysts. *Fuel* **2019**, *239*, 1231–1239. [CrossRef]
47. Aboelazayem, O.; Gadalla, M.; Saha, B. Biodiesel production from waste cooking oil via supercritical methanol: Optimisation and reactor simulation. *Renew. Energy* **2018**, *124*, 144–154. [CrossRef]
48. Gupta, A.R.; Rathod, V.K. Calcium diglyceroxide catalyzed biodiesel production from waste cooking oil in the presence of microwave: Optimization and kinetic studies. *Renew. Energy* **2018**, *121*, 757–767. [CrossRef]
49. Putra, M.D.; Irawan, C.; Ristianingsih, Y.; Nata, I.F. A cleaner process for biodiesel production from waste cooking oil using waste materials as a heterogeneous catalyst and its kinetic study. *J. Clean. Prod.* **2018**, *195*, 1249–1258. [CrossRef]
50. Li, L.; Zou, C.; Zhou, L.; Lin, L. Cucurbituril-protected $Cs_{2.5}H_{0.5}PW_{12}O_{40}$ for optimized biodiesel production from waste cooking oil. *Renew. Energy* **2017**, *107*, 14–22. [CrossRef]
51. Mahmoud, H.R.; El-Molla, S.A.; Ibrahim, M.M. Biodiesel production via stearic acid esterification over mesoporous ZrO_2/SiO_2 catalysts synthesized by surfactant-assisted sol-gel auto-combustion route. *Renew. Energy* **2020**, *160*, 42–51. [CrossRef]
52. Rattanaphra, D.; Harvey, A.P.; Thanapimmetha, A.; Srinophakun, P. Kinetic of myristic acid esterification with methanol in the presence of triglycerides over sulfated zirconia. *Renew. Energy* **2011**, *36*, 2679–2686. [CrossRef]
53. Unlu, D.; Ilgen, O.; Hilmioglu, N.D. Biodiesel additive ethyl levulinate synthesis by catalytic membrane: SO_4^{-2}/ZrO_2 loaded hydroxyethyl cellulose. *Chem. Eng. J.* **2016**, *302*, 260–268. [CrossRef]
54. Jabeen, M.; Munir, M.; Abbas, M.M.; Ahmad, M.; Waseem, A.; Saeed, M.; Kalam, M.A.; Zafar, M.; Sultana, S.; Mohamed, A. Sustainable production of biodiesel from novel and non-edible ailanthus altissima (Mill.) seed oil from green and recyclable potassium hydroxide activated ailanthus cake and cadmium sulfide catalyst. *Sustainability* **2022**, *14*, 10962. [CrossRef]
55. Munir, M.; Ahmad, M.; Mubashir, M.; Asif, S.; Waseem, A.; Mukhtar, A.; Saqib, S.; Munawaroh, H.S.H.; Lam, M.K.; Khoo, K.S. A practical approach for synthesis of biodiesel via non-edible seeds oils using trimetallic based montmorillonite nano-catalyst. *Bioresour. Technol.* **2021**, *328*, 124859. [CrossRef] [PubMed]

Article

Enhancing Bioenergy Production from the Raw and Defatted Microalgal Biomass Using Wastewater as the Cultivation Medium

Gang Li [1], Yuhang Hao [1], Tenglun Yang [1], Wenbo Xiao [1], Minmin Pan [2], Shuhao Huo [3] and Tao Lyu [4,*]

1. School of Artificial Intelligence, Beijing Technology and Business University, Beijing 100048, China
2. Department for Solar Materials, Helmholtz Centre for Environmental Research GmbH-UFZ, Permoserstraße 15, 04318 Leipzig, Germany
3. School of Food and Biological Engineering, Jiangsu University, Zhenjiang 212013, China
4. School of Water, Energy and Environment, Cranfield University, College Road, Cranfield MK43 0AL, UK
* Correspondence: t.lyu@cranfield.ac.uk

Citation: Li, G.; Hao, Y.; Yang, T.; Xiao, W.; Pan, M.; Huo, S.; Lyu, T. Enhancing Bioenergy Production from the Raw and Defatted Microalgal Biomass Using Wastewater as the Cultivation Medium. *Bioengineering* 2022, 9, 637. https://doi.org/10.3390/bioengineering9110637

Academic Editors: Indra Neel Pulidindi and Aharon Gedanken

Received: 29 September 2022
Accepted: 31 October 2022
Published: 2 November 2022

Publisher's Note: MDPI stays neutral with regard to jurisdictional claims in published maps and institutional affiliations.

Copyright: © 2022 by the authors. Licensee MDPI, Basel, Switzerland. This article is an open access article distributed under the terms and conditions of the Creative Commons Attribution (CC BY) license (https://creativecommons.org/licenses/by/4.0/).

Abstract: Improving the efficiency of using energy and decreasing impacts on the environment will be an inevitable choice for future development. Based on this direction, three kinds of medium (modified anaerobic digestion wastewater, anaerobic digestion wastewater and a standard growth medium BG11) were used to culture microalgae towards achieving high-quality biodiesel products. The results showed that microalgae culturing with anaerobic digestate wastewater could increase lipid content (21.8%); however, the modified anaerobic digestion wastewater can boost the microalgal biomass production to 0.78 ± 0.01 g/L when compared with (0.35–0.54 g/L) the other two groups. Besides the first step lipid extraction, the elemental composition, thermogravimetric and pyrolysis products of the defatted microalgal residues were also analysed to delve into the utilisation potential of microalgae biomass. Defatted microalgae from modified wastewater by pyrolysis at 650 °C resulted in an increase in the total content of valuable products (39.47%) with no significant difference in the content of toxic compounds compared to other groups. Moreover, the results of the life cycle assessment showed that the environmental impact (388.9 mPET$_{2000}$) was lower than that of raw wastewater (418.1 mPET$_{2000}$) and standard medium (497.3 mPET$_{2000}$)-cultivated groups. Consequently, the method of culturing microalgae in modified wastewater and pyrolyzing algal residues has a potential to increase renewable energy production and reduce environmental impact.

Keywords: biodiesel; waste recovery; renewable energy; microalgal technology; life cycle assessment

1. Introduction

In total, 84% of the world's energy demand is still met by non-renewable energy sources, and the ever-growing energy consumption is depleting the total fossil energy storage [1]. In the meantime, the consequences of the large utilization of fossil fuels have led to serious environmental issues, such as toxic compound discharge, global warming, the extinction of species, desertification etc. [2]. Renewable energy could contribute significantly to adjusting the global energy structure, thus, has been regarded as the inevitable choice for sustainable development. Renewable energy includes biomass energy, wind energy, solar energy, water energy, geothermal energy, etc., and has the characteristics of wide resource distribution, great utilization potential and small environmental impacts. Among these, biomass energy is a significant source of renewable carbon that can be transformed into endless conventional solid, liquid, and gaseous fuels [3].

Traditional biomass energy sources, e.g., crops, are always criticized due to the concern of land use competition for food production and slower rates of plant growth [4]. Microalgae, as a major class of biomass energy, are outstanding attributed to their various advantages, e.g., rapid growth, high lipid content, carbon sequestration ability, and low demands of environmental conditions (barren land, saline- and waste-water) [5]. The esters and glycerol rich in microalgae are good raw materials for the preparation of liquid fuels. The calorific value of bio-oil prepared by microalgae pyrolysis is relatively high. Moreover, the nutrients from wastewater can be effectively used by microalgae for its biomass growth. Consequently, the concept of utilizing wastewater for the cultivation of microalgae has drawn much attention in terms of the simultaneous treatment of wastewater and producing renewable energy.

Effluent from anaerobic digestion reactors has been seen as an ideal medium for microalgal growth regarding the reduced level of organics and variety of residual nutrients (e.g., N, P) in the liquid [6]. Although the challenge of lacking key elements in the wastewater can bring a negative impact on the growth of microalgae [7], the approach of artificial wastewater adjustment, e.g., addition of a certain amount of missing elements, has been deployed and proven to be an efficient way to address this problem to cultivate a good amount of algal biomass [8]. When considering the entire life cycle, barriers such as algal residues after lipid extraction have hitherto hindered the application of microalgal biomass energy. Algal residues account for about 70% of the dry weight of microalgae biomass and still contain large amounts of carbohydrates and proteins [9]. In most studies, the subsequent utilization of algae residues was not mentioned after culturing microalgae and extracting lipids. The energy held by photosynthesis in the biomass may be recovered quickly and cleanly using the pyrolysis method, which outperforms biological methods in terms of effectiveness, cost, and energy balance [10]. Therefore, quantifying the potential of biofuel generation from defatted algal biomass would be important for renewable energy production.

Besides the biofuel generation from microalgal biomass, the environmental impact of the pyrolysis process needs to be assessed to identify any trade-offs [11]. Some drawbacks have been reported; for example, the relative contents of nitrogen compounds in the residues of defatted microalgae could increase during the pyrolysis [12]. Furthermore, the addition of a catalyst would increase the contents of CO in pyrolysis products of *Haematococcus pluvialis* residues [13]. Moreover, the pyrolysis of *Isochrysis* after oil extraction required 2591 kJ/kg of energy [14]. The whole process of bioenergy production from microalgae residues can be fully understood by using life cycle assessment (LCA) [15,16]. In different culture conditions, such as using different wastewaters or modified wastewater, the compositions of microalgae and subsequent microalgae residues would be impacted [17]. Consequently, the properties of the compositions of microalgae have a direct effect on pyrolysis products [18]. The assessment of the pyrolysis of microalgae residues gained by different culture methods has not hitherto been sufficiently investigated.

In this study, the algae strain *Desmodesmus* sp. was chosen as the model microalgal species and cultured in modified anaerobic digestion wastewater (MAW), original anaerobic digestion wastewater (AW), and BG11 media, respectively. The lipid was firstly extracted from the microalgae cultivated in three culture medias, then, the algae residues were pyrolyzed at a temperature range of 350 °C to 750 °C. The lipid contents and compositions of microalgae were compared, and the differences in thermal decomposition behaviour of three kinds of microalgae residues were evaluated. Through the evaluation of the pyrolysis products, the optimum pyrolysis temperature was determined, and the life cycle assessment was carried out. This study provides a new insight into the thorough utilization of microalgae to produce high-quality biodiesel through the improvement of the microalgae culture medium and the reuse of the defatted microalgal biomass. A follow-up LCA analysis provides an environmentally sustainable view of the proposed strategy. The results have great significance to the improvement of the microalgae culture medium, the optimization of the technological process of biofuel production, and the strategy of environmental control in the pyrolysis process.

2. Materials and Methods

2.1. Algal Strain and Culture Conditions

Desmodesmus sp. EJ 8-10 (hereinafter referred to as EJ 8-10) is an algal strain that was identified in a river in Beijing, China. With an initial inoculum ratio of 10% (v/v), EJ 8-10 was pre-cultivated in flasks (250 mL). The following steps were the precise cultivation conditions: BG11 medium (autoclaved) (Appendix A, Table A1); lighting intensity: 120 ± 2 mmol/m^2/s; temperature: 27 ± 1 °C; lighting period: 14 h:10 h (light:dark); pH: 7.5 [19]. Anaerobic digestion wastewater (AW) was gathered from a pig farm in the Shunyi District of Beijing. The supernatant was collected for microalgae cultivation after centrifugation (10,000 rpm, 15 min). The composition is shown in Table A2 (Appendix B). High concentrations of NH_4^+–N could reduce microalgal vitality; the collected supernatant was diluted to 10% with deionized water [20]. To address the nutritional deficit in wastewater, the modified anaerobic digestion wastewater (MAW) medium was created by additionally adding ammonium ferric citrate ($C_6H_8FeNO_7$), dipotassium hydrogenphosphate (K_2HPO_4), and magnesium sulfate heptahydrate ($MgSO_4 \cdot 7H_2O$) [21]. The standard medium of BG11 was used as a control group, and 0.1 OD_{680} of initial microalgal biomass was inoculated into three mediums (AW, MAW, and BG11) and cultured for 14 days under identical circumstances as previously indicated. The experiments were run three times for each group.

2.2. Lipid Extraction and Fatty Acid Analysis

After culture, microalgae were collected by centrifuging at 10,000 rpm for 10 min. The total lipid content was then determined by using an improved approach based on Abou-Shanab et al. [22]. The mixed solvent (volume ratio of chloroform, methanol, and water was 1:2:0.8) was added to ground algae powder (mass ratio of algae powder to quartz sand was 1:3) and then oscillated for 5 min [23]. After standing for 15 min, the mixture was centrifuged (6000 rpm, 2 min) and the upper extract was collected. After the above operation was repeated for precipitation two times, all extracts were merged. Extracts were mixed thoroughly with chloroform, methanol, and water until the final volume ratio was 1:1:0.9. Chloroform solution was gathered following delamination of the combination. Solvent chloroform was evaporated in a rotary evaporator (vacuum, 60 °C), and the obtained lipid was weighed. We collected algae residues (hereinafter referred to as AR) and used them for subsequent pyrolysis to prepare bio-oil. The experiments were carried out three times for each group. The lipid contents were calculated by the following formula:

$$C = W1/Wb \times 100\% \quad (1)$$

where the lipid mass (mg) is $W1$, the algae mass (mg) is Wb, and the lipid content (%) is C.

Preparative FAMEs and gas chromatography–mass spectrometry (GC–MS) analyses were used to assess the fatty acid composition. The preparation process of FAMEs refers to a method of Wang et al. [24]. We added 10 mL of the mixture (methanol, concentrated sulfuric acid, and chloroform had a volume ratio of 4.25:0.75:5) to a screw-top glass bottle (25 mL) containing 0.1 g of the sample [25]. In a 90 °C water bath (Cole-Parmer, Vernon Hills, IL, USA), transesterification was performed for 90 min. A meticulous collection of the FAME-containing chloroform layer was made for GC–MS analysis. A flame ionization detector and an RTX-Wax capillary column (30 m 0.32 mm 0.25 mm; Restek Corp., Bellefonte, PA, USA) were placed in the GC (QP2010; Shimadzu Corp., Kyoto, Japan). The oven's temperature was initially set at 100 °C (held for three minutes), then was increased to 200 °C at 4 °C/min, and increased to 250 °C (held for five minutes) at 3 °C/min. The carrier gas (helium) flow rate was regulated at 30 mL/min, and the injector temperature was fixed at 230 °C. The NIST Mass Spectral Database was used to identify the FAME compounds, and the peak regions of the compounds were compared to those of the external standard (C18:2) (Sigma Aldrich, Saint Louis, MN, USA) to determine their amounts [19]. The experiments were conducted three times for each group.

2.3. Elemental Analysis, Thermogravimetric Analysis (TGA), and Pyrolysis of Algal Residues

Each lipid-extracted AR sample's primary elements composition (C, H, N, and S) was determined using an elemental analyzer (EA; Flash EA-1112, Thermo Scientific, Waltham, MA, USA). The experiments were performed three times for each group. Using Equations (2)–(4), the higher heating values (HHV) of AR samples were determined [26,27]:

$$HHV(OLS) = 1.87C^2 - 144C - 2082H + 63.8C \times H + 129N + 20147 \quad (2)$$

$$HHV(PLS) = 5.22C^2 - 319C - 1674H + 38.6C \times H + 133N + 21028 \quad (3)$$

$$\begin{aligned} HHV &= [HHV(OLS) + HHV(PLS)]/2 \\ &= (3.55C^2 - 232C - 2230H + 51.2C \times H + 131N + 20600) \times 10^{-3} \end{aligned} \quad (4)$$

where C, H, and N stand for the sample's respective carbon, hydrogen, and nitrogen contents (%), respectively.

In the TGA procedure, nitrogen (99.999% purity, 100 mL/min) was used as a shielding gas while 2–4 mg samples were pyrolyzed from 50 °C to 800 °C at 20 °C/min. The samples were pyrolyzed and examined using pyrolysis–gas chromatography–mass spectrometry (Py–GC–MS), which consists of a rapid pyrolyzer (Frontier Labs 3030i, Koriyama, Fukushima, Japan) coupled with a GC–MS (Agilent 7890A/5975C, Santa Clara, CA, USA) in order to describe thermal decomposition behaviour and pyrolysis products of AR.

The pyrolysis products at various pyrolysis temperatures (350 °C, 450 °C, 550 °C, 650 °C, and 750 °C) were examined to find the best pyrolysis conditions for AR. The GC–MS was operating under the same circumstances as before. Pyrolysis products were located by scanning the NIST11 database (National Institute of Standards and Technology, Gaithersburg, MD, USA) for the resulting mass spectra [28].

2.4. Life Cycle Assessment (LCA)

To achieve a more comprehensive overview of energy consumption and environmental impacts caused by the pyrolysis process (under corresponding optimal temperature) of ARs obtained under different culture conditions, an LCA investigation was conducted [11]. The study's chosen objectives and field of inquiry were compliant with the international standards for life cycle assessments, i.e., ISO 14040 [29].

2.4.1. LCA Goals and System Boundaries

Utilizing LCA serves the objective of assessing the environmental impact of AR gathered in various media on the manufacture of the best pyrolysis products. Figure 1 depicts the boundary of the system, and the energy consumption during system operation is regarded as the input. The depreciation of pyrolysis equipment, the energy consumption of adding additional nutrients to the modified medium, and the impact of microalgae growth on the environment were excluded [30]. It is worth noting that the pollutants in the process were not treated as extras.

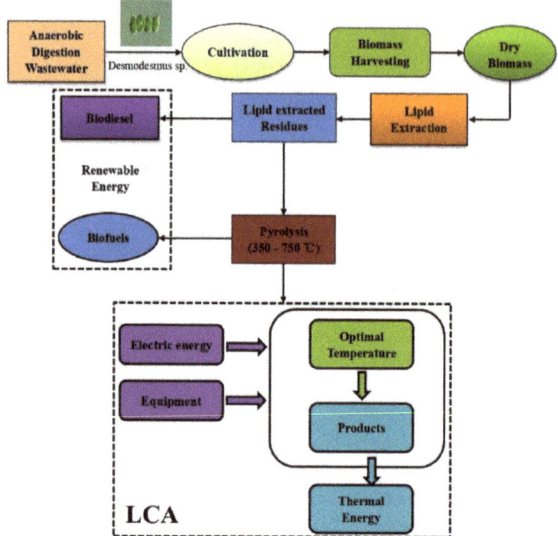

Figure 1. Process route and system boundaries of the LCA model for pyrolysis products of microalgal residues.

2.4.2. Selected Parameters to Describe the Environmental Impacts

The evaluation details four aspects of environmental impact, including photochemical ozone synthesis (kg VOC, CO, CH_4-eq), acidification (kg SO_2, NO_X-eq), eutrophication (kg PO_4, NO_X-eq), and global warming (kg CO_2, CH_4, NO_X, CO-eq) [31]. The energy consumption and environmental impact was determined using 1 kg biomass dry weight EJ 8–10 residue for assessment.

2.4.3. LCA Stages

In order to examine the pyrolysis stage of AR, this study's assessment ignored the energy transfer in the pyrolyzer and instead focused on the energy consumption of the pyrolysis furnace and online analysis [30]. The average energy consumption is 2.2 kWh for each real-time analysis. Based on the pyrolysis product results, the energy required to raise the temperature from ambient (25.6 °C) to each group's optimum conditions was estimated, and air pollutant emissions were determined using data from a prior study [11].

2.4.4. LCA Model

The formula below was used to determine environmental impact.:

$$EI = \sum [Q_i \times F_i] \quad (5)$$

where EI stands for the environmental impact, Q_i for the ith emission's quantity, and F_i for the influence of the ith emission on the environment as a whole [11].

Equation (6) was used to further standardize the EI for the comparative assessment of various sorts of impacts [30]:

$$S_{EI} = EI \times R^{-1} \quad (6)$$

where R is the accepted benchmark and S_{EI} stands for the standardized environmental effect.

The weighting factor was determined by the method of target distance as the following equation (Equation (7)):

$$W = E \times EN^{-1} \quad (7)$$

the unit (mPET2000) of standardized environmental potential impact is represented by the standard person equivalent, and W is the weighting factor for each particular parameter. E is the overall regional environmental impact potential in 1990, whereas EN is the regional environmental impact potential in 2000 [31].

2.5. Plotting and Statistical Analysis

Origin 9.8 (OriginLab, Northampton, Massachusetts, USA) was used for plotting, and data analysis was done by using SPSS 19.0 (IBM Corporation, Armonk, NY, USA). Prior to statistical analysis, the data were examined for normality and homogeneity of variance. Nonparametric test methods were employed for the analysis if the variables were not normally distributed. If variables were normally distributed, the F-test (ANOVA) was used to assess the significant difference in data. The significance level was 0.05. Data are presented as mean ± standard deviation.

3. Results and Discussion

3.1. Microalgal Growth, Lipid Accumulation and Fatty Acid Composition

Overall, culturing microalgae with AW can obtain the highest lipid accumulation (21.8%, Figure 2). Furthermore, the microalgae cultured in the modified anaerobic digestion wastewater (MAW) had the highest biomass production (0.78 ± 0.01 g/L) compared with those (0.35–0.54 g/L) in the other two groups. The outcomes demonstrated that anaerobic digestion of wastewater could increase lipid accumulation when used to cultivate microalgae. Tan et al. [32] explained that the anaerobic wastewater had balanced nutrients with many trace elements that were not present in the medium, but the low content of trace elements such as P, Fe and Mg in AW could not guarantee the rapid growth of microalgae. Those nutrients are all necessary for photosynthesis at the growth stages of microalgae, therefore, the biomass production of microalgae with the addition of P and Mg was the highest in MAW [32]. However, the lipid content (14.2%) of microalgae decreased after culturing in MAW. The main elements added to the modified medium were Mg and P, and P could promote lipid accumulation [33]. However, Mg was related to photosynthesis, and microalgae might first choose to accumulate carbohydrates instead of lipids in the case of sufficient carbon sources [34]. Therefore, it was possible that lipid content was lower in modified wastewater. In this experiment, the lipid productivity of microalgae in MAW was higher than that found in the study by Chinnasamy et al. [35], which may be due to the nutrient shortage in the medium during the microalgae growth.

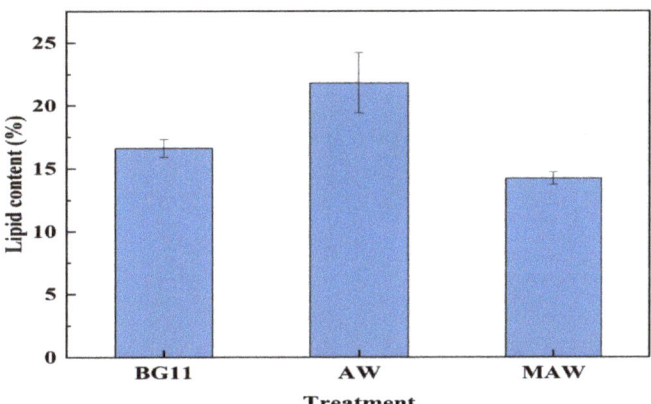

Figure 2. Lipid contents of three kinds of microalgae (obtained from BG11, AW and MAW medium, respectively).

The first step of lipid extraction was conducted to evaluate the content of different compositions. The AW group had the largest amounts of saturated fatty acids (SFAs), polyunsaturated fatty acids (PUFAs), and monounsaturated fatty acids (MUFAs) of the three groups (Appendix C, Table A3). These three fatty acids were frequently used in skin care products, and when cultured with AW, their production was boosted (Figure 3a). When compared with *Desmodesmus sp.* cultured with 10% original wastewater (48.37 mg/g) in the study of Li et al. [19], the total fatty acids content was lower than that of microalgae harvested from the AW medium (74.74 mg/g); higher P content in AW medium may contribute to the higher fatty acids content in the microalgae. Microalgae cultured in wastewater had advantages in their contents of pentadecanoic acid, heptadecanoic acid, heptadecenoic acid (cis-10), eicosenoic acid and linoleic acid (Figure 3a), among which pentadecanoic acid, margaric acid, paullinic acid and linoleic acid are widely used in medical, pharmaceutical, and nutritional fields [36]. Heptadecenoicacid (cis-10) could balance the low temperature resistance and combustion performance of biodiesel, and play an important role in the production of biodiesel. In addition, the contents of pentadecenoic acid, hexadecanoic acid and octadecadienoic acid (anti-9,12) were significantly increased ($p < 0.05$) after culturing microalgae with modified wastewater (Figure 3b). These fatty acids are important industrial raw materials for the preparation of medicines, emulsifiers, and detergents, respectively. Hexadecanoic acid and octadecadienoic acid (anti-9,12) were two of the most suitable biofuel sources extracted from microalgae, and ARs obtained from modified anaerobic digestion wastewater were advantageous for these molecules, which coincide with the research of Wang et al. [24]. The lipid contents of microalgae from MAW (129.20 mg/g) were higher than those cultured with original piggery effluent (48.37 mg/g); the MAW (4.91 mg/L) contained higher P contents compared with the original piggery effluent (3.10 mg/L), which could theoretically contribute to a higher accumulation of lipid content in the microalgae [19]. Similar to the research of Moradianetal et al. [3], microalgal biomass has received much attention and esteem because of their powerful capabilities in various aspects of life and industry.

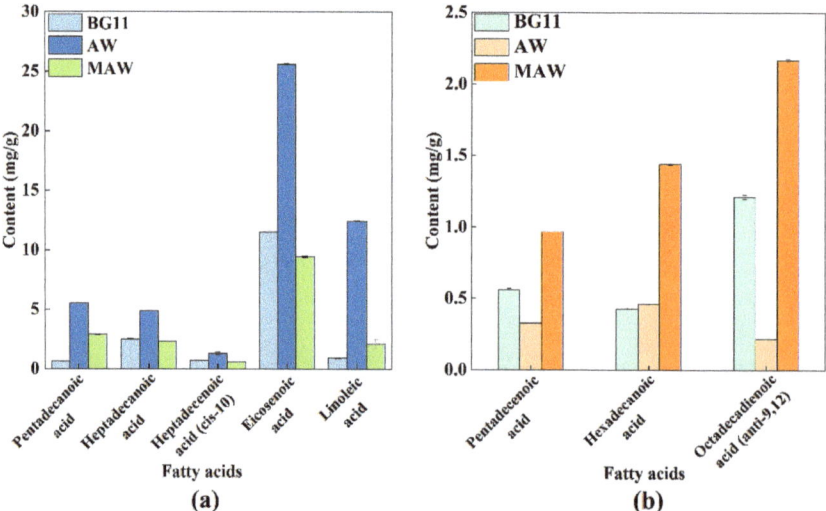

Figure 3. Contents of fatty acids in microalgae harvested by the three culture methods; (a) Pentadecanoic acid, Heptadecanoic acid, Heptadecenoic acid (cis-10), Eicosenoic acid and Linoleic acid, and (b) Pentadecenoic acid, Hexadecanoic acid and Octadecadienoic acid (anti-9,12).

3.2. Properties and Thermogravimetric Analysis of Microalgal Residues

Though with economic advantages, AW could significantly ($p < 0.05$) reduce the content of C, H, and N, leading to a decreased HHV when compared with the BG11 group. However, MAW could address this issue by increasing the C and H content to 47.24 ± 0.01% and 7.49 ± 0.28%, respectively. Notably, the contents of C and H were the key parameters determining the HHV of material, and the AR of MAW had a higher HHV, indicating a higher energy density, which would promote further pyrolysis products. The samples of Huang et al. [12] showed no significant difference in HHV before and after lipid extraction, indicating a similar energy potential of the material and residues of microalgae. It was clear from the results that further studies to convert microalgal residues into energy-related products are feasible.

TGA data, meanwhile (Figure 4a), added to the argument. In comparison to those harvested in BG11 and AW, microalgal residues harvested in MAW medium showed a slower rate of weight loss throughout the pyrolysis stage. Due to the continuing decompositions and carbonizations of AR, TG decreased slowly at the stage above 550 °C [37]. Microalgal residues obtained from the MAW had the highest contents of C (Table 1), which improved the thermal resistance of the AR and resulted in the highest amounts of thermal residues (46.75 wt%) [38]. Notably, the contents of N and S were also elevated in the group of MAW, indicating a potential higher production of harmful compounds, e.g., NO_x, SO_2, and HCN, during pyrolysis or subsequent combustion and upgrading processes. Therefore, a pyrolysis compound analysis for both valuable and toxic compounds was conducted.

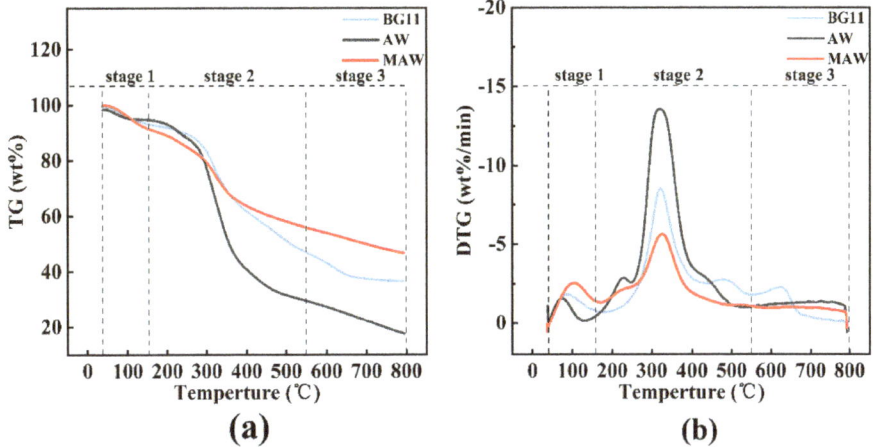

Figure 4. (a) TG and, (b) DTG analysis of pyrolysis process.

Table 1. Elemental analyses and the higher heating values (HHV) of microalgae biomass cultivated in BG11, AW and MAW media.

	C (%)	H (%)	N (%)	S (%)	HHV (MJ/kg)
BG11	35.01 ± 0.06	5.97 ± 0.01	8.04 ± 0.03	0.60 ± 0.01	15.27 ± 0.01
AW	21.87 ± 0.04	4.93 ± 0.05	5.81 ± 0.03	0.68 ± 0.11	12.51 ± 0.05
MAW	47.24 ± 0.01	7.49 ± 0.28	11.25 ± 0.16	0.95 ± 0.01	20.45 ± 0.07

3.3. Valuable and Toxic Pyrolysis Products of Algal Residues

Ingredients of the medium and pyrolysis temperatures could affect pyrolysis products, and forasmuch, the present study compared the pyrolysis products of microalgal residues from different microalgae media (BG11, AW and MAW) and different pyrolysis temperatures (over the range 350–750 °C). Consistent with other research results [39], aromatic hydrocarbons (benzenes, indenes, and their derivatives), aliphatic hydrocarbons (alkanes and olefins), phenols, fatty acids, nitrogen-containing compounds (amides, nitriles, and pyridines), polycyclic aromatic hydrocarbons (PAHs), and other trace components made up the majority of the pyrolysis products of ARs (ketones, alcohols, aldehydes, and furans).

3.3.1. Valuable Compounds

Among all pyrolysis products, aliphatic hydrocarbons are some of the products with economic value, which have great significance in the production of bio-oils. The percentage of valuable compounds in the pyrolysis products of microalgal residues from the MAW medium increased as the pyrolysis temperature rose from 350 to 750 °C and showed a single peak at 650 °C (39.47%, Figure 5a, Table A4 in Appendix D); compared with results of the BG11 (38.08%, 750 °C) and AW (36.05%, 750 °C) groups, MAW groups significantly reduced the pyrolysis temperature of optimal content and increased the relative content of valuable compounds in the pyrolysis products. The content of valuable compounds obtained by pyrolysis in microalgae residues from the MAW medium (39.47%) was higher than that from original anaerobically digested effluent medium (19.83%). The results agreed with the previous study that the content of valuable compounds from pyrolysis production in MAW medium increase by adding chemical components (ammonium ferric citrate, dipotassium hydrogenphosphate, and magnesium sulfate heptahydrate) [12]. Moreover, aromatic hydrocarbons in pyrolysis products are also high-value compounds, which can be used as transportation fuel additives in industry, and can also elevate octane numbers, thereby enhancing combustion efficiency [40]. The results showed that contents of aromatic hydrocarbons in the pyrolysis products of three groups of AR showed similar trends; aromatic hydrocarbons were the most abundant components in all pyrolysis products, and achieved the maximum value at 750 °C with the increase of pyrolysis temperature. The contents of aromatic hydrocarbons in the pyrolysis products of ARs in both BG11 (27.16%) and AW (24.54%) media were all lower than those in the MAW (32.07%) groups by comparison.

3.3.2. Toxic Compounds

Due to insufficient breakdown during pyrolysis, microalgae leftovers may potentially produce hazardous chemicals such as nitrogen-containing compounds and polycyclic aromatic hydrocarbons (PAHs), in addition to beneficial molecules, affecting the quality of bio-oil products and affecting environmental pollution with excessive emission of nitrogen oxides [41]. In the current investigation, there was a strong association (r = 0.93, 0.89, 0.87; $p < 0.01$) between pyrolysis temperatures and the relative concentrations of nitrogen-containing chemicals; the contents of nitrogen-containing compounds and PAHs produced by pyrolysis of microalgae residues in MAW groups (16.48%, 0%; 650 °C) were lower than those in the BG11 groups (26.60%, 2.15%; 750 °C) and AW groups (24.33%, 2.97%; 750 °C) at the temperature at which the maximum quantity of valuable pyrolysis products emerged. Except for nitrogen-containing compounds and PAHs, sulphides were also toxic products to consider. The ARs of MAW groups had the highest sulphur contents (0.95 ± 0.01%, Table 1), indicating that there might be a high potential for the production of sulphur dioxide, SO_2, during the pyrolysis process, and further improvement was needed [42].

In conclusion, when compared to BG11 and AW media, ARs from the MAW medium promoted higher levels of valuable compounds (such as aromatic hydrocarbons and aliphatic hydrocarbons), while the levels of toxic substrates (such as nitrogen-containing compounds and PAHs) were not significantly different ($p > 0.05$). It is important to note that choosing the ideal pyrolysis temperature involves more than just balancing the needs of the goal product with the trade-offs between harmful substances, valuable products, and energy consumption.

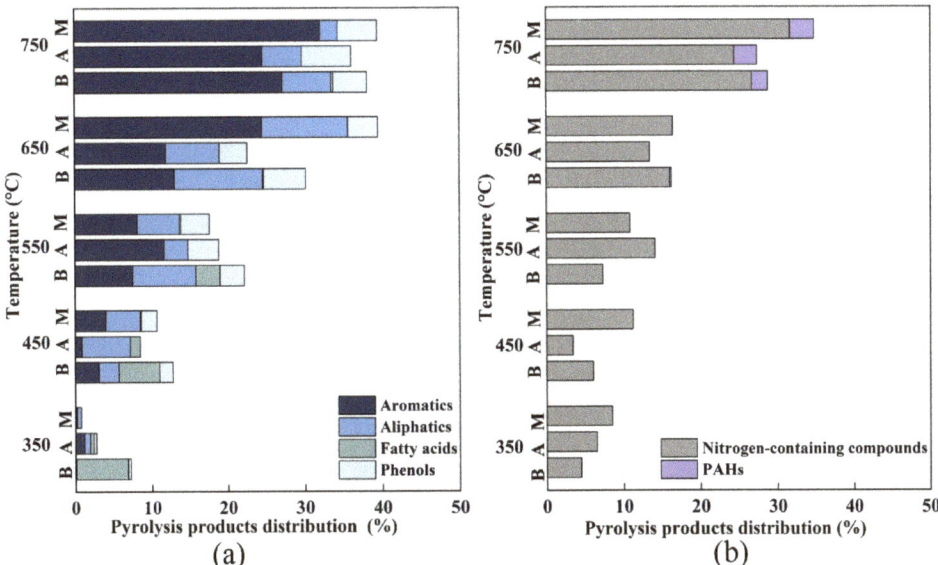

Figure 5. In the pyrolysis of microalgal leftovers grown on BG11, AW and MAW media; (**a**) useful chemicals, and (**b**) hazardous compounds were produced (A: AW; B: BG11; M: MAW). The numerical results are shown in Table A4 (Appendix D).

3.4. Life Cycle Assessment of Microalgal Residues Pyrolysis

LCA was applied to the process of pyrolysis of microalgal residues in three media at the optimum pyrolysis temperature (650 °C), and the pyrolysis of ARs in the BG11 medium required the most energy (7463.5 MJ/kg); directly using anaerobic digestion wastewater (AW) reduced the required energy by 15.93%, and in the modified anaerobic digestion wastewater (MAW) group, it was further reduced by 7.09%. This was mostly related to an increase in biomass output in wastewater that had been modified and included more plentiful and balanced nutrients for microalgal growth [31]. The pyrolysis products from MAW groups generated overall lower environmental impacts (total, 388.9 mPET$_{2000}$, Figure 6) for all four parameters selected, and led to a lower total environmental impact than the pyrolysis processes of AR obtained from AW (total, 418.1 mPET$_{2000}$) and BG11 (497.3 mPET$_{2000}$) media. The relative distribution of each environmental category during the pyrolysis operations did not differ significantly ($p > 0.05$) across the various groups (Figure 6). Eutrophication, which was primarily brought on by NO$_x$ emissions, was induced by all groups' combined largest contributions (285.3, 239.8, and 222.8 mPET$_{2000}$ by BG11, AW, and MAW, respectively) [41]. While the microalgae were given the ability to photosynthesise, the effects of pyrolysis product creation on global warming (13.6, 11.4, and 10.6 mPET$_{2000}$ by BG11, AW, and MAW, respectively) were minimal. In summary, compared with the other two cultivation methods, modified anaerobic digestion wastewater could not change the proportion of a single environmental impact, but could significantly reduce total environmental impacts.

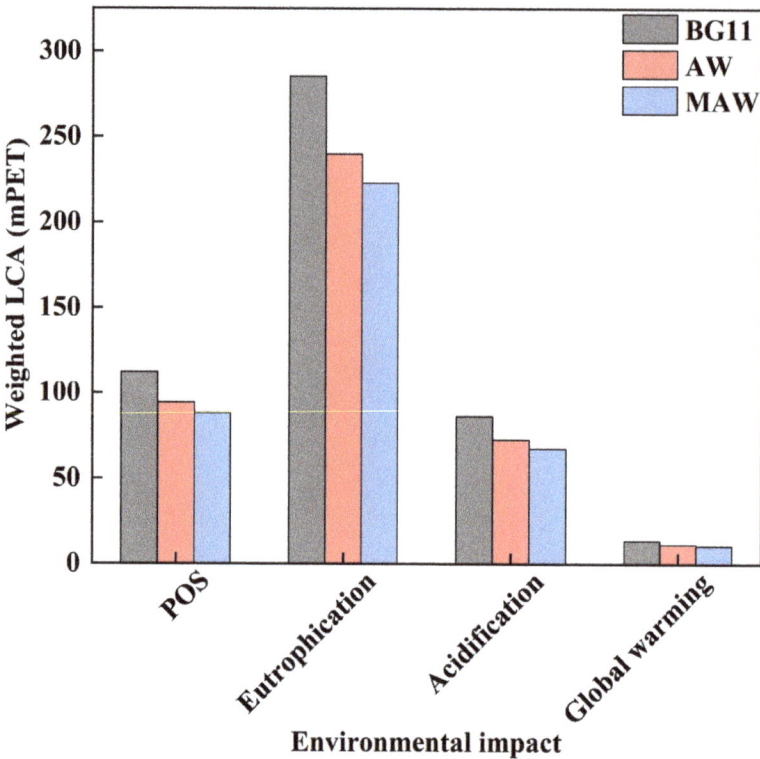

Figure 6. Weighted life cycle assessment of three kinds of medium. POS represents the impact of photochemical ozone synthesis.

3.5. Future Perspectives

In the process of microalgae growth, C source and N sources are the main nutrient sources for microalgae. The effluent of anaerobic digestion is a cheap and directly applicable nutrient source, and the cost of microalgae culture will be greatly reduced. In a real biogas project, while producing clean energy, the anaerobic digestion broth could be used to culture microalgae [43]. After extracting oil, the pyrolytic algae residues could also convert wastes into industrial available energy. This could solve problems of the high transportation cost of anaerobic digestion broth, as well as increased pollution to the surrounding environment in biogas engineering. Industries could avoid the loss and waste of beneficial elements, and produce new bioenergy at the same time, which could not only meet the goal of sustainable development, but also reduce carbon emissions. In the conversion process of biodiesel, biomass needs to be separated or purified. The transesterification reaction requires the installation of methanol recovery equipment, and due to the formation of soap, the treatment process is complex, and the purification of the product is difficult. In addition, with the increase of contents of S and N in ARs, harmful substances produced during pyrolysis will increase, such as sulphides and nitrogen-containing compounds [44]. The technological process should be further optimized to avoid the increase of harmful substances and increase the content of bio-oil at the same time.

4. Conclusions

The modified anaerobic digestion wastewater (MAW), through the extra additions of $C_6H_8FeNO_7$, K_2HPO_4, and $MgSO_4·7H_2O$, could significantly promote lipid quality in microalgal biomass, towards enhancing the quality of biofuels. The modified medium also improved the composition and properties of microalgal residues after the first step of lipid extraction, resulting in higher thermal resistance (thermal residue was 46.75 wt%) when compared to that from anaerobic digestion wastewater (AW) (17.61 wt%) and the standard BG11 (36.72 wt%) cultivation medium. MAW groups significantly reduced the pyrolysis temperature (650 °C) of optimal content and increased the relative content of valuable pyrolysis products, of which aliphatic hydrocarbons were 1.9 and 2 times as abundant as other groups, respectively, along with decreasing the contents of toxic compounds (nitrogen-containing compounds and PAHs) in products. Moreover, compared with the other two cultivation methods, MAW could not change the proportion of single environmental impacts, but could significantly reduce total environmental impacts when compared with the BG11 (reduced by 21.79%) and AW (reduced by 6.97%) groups, as indicated by the LCA.

Author Contributions: Conceptualization, G.L. and T.L.; methodology, G.L.; validation, G.L. and M.P.; formal analysis, Y.H. and M.P.; resources, G.L.; data curation, Y.H. and W.X.; writing—original draft preparation, Y.H.; writing—review and editing, T.Y., T.L., S.H. and M.P. All authors have read and agreed to the published version of the manuscript.

Funding: This research was funded by the Innovation Fund of Ministry of Education Science and Technology Development Center of China, grant number (Grant No. 2020QT04).

Institutional Review Board Statement: Not applicable.

Informed Consent Statement: Not applicable.

Data Availability Statement: Not applicable.

Acknowledgments: This research was financially supported by the Innovation Fund of Ministry of Education Science and Technology Development Center of China. Besides, we are also very grateful to Santosh Kumar for his careful revision of this article.

Conflicts of Interest: The authors declare no conflict of interest.

Appendix A

Table A1. The components of BG11 medium.

No.	Chemicals	Concentration (g/L)
1	$NaNO_3$	1.5
2	K_2HPO_4	3×10^{-2}
3	$MgSO_4·7H_2O$	7.5×10^{-2}
4	$CaCl_2·2H_2O$	36×10^{-2}
5	Iron Citrate	6×10^{-3}
6	Ammonium Citrate	6×10^{-3}
7	EDTA	1×10^{-3}
8	Na_2CO_3	6×10^{-3}
9	H_3BO_3	2.86×10^{-3}
	$MnCl_2·4H_2O$	1.81×10^{-3}
	$ZnSO_4·7H_2O$	2.22×10^{-4}
	$NaMoO_4·5H_2O$	3.9×10^{-4}
	$CuSO_4·5H_2O$	7.9×10^{-5}
	$Co(NO_2)_2·6H_2O$	4.94×10^{-4}

Appendix B

Table A2. The components of 10% (v/v) AW.

Components	Concentration (mg/L)
NH_4^+-N	42.8
PO_4^{3-}	2.09
TN	50.92
TP	2.31

Appendix C

Table A3. Compositions of fatty acids, including saturated fatty acids (SFA), monounsaturated fatty acids (MUFA) and polyunsaturated fatty acids (PUFA), of microalgae cultivated in BG11, AW and MAW medium, respectively.

Fatty Acids Sort	Fatty Acids	Fatty Acids Form	BG11		AW		MAW	
			mg/g	%	mg/g	%	mg/g	%
SFA	Tannic acid	C10:0	0.576 ± 0.007	0.35	0.599 ± 0.003	0.27	0.514 ± 0.001	0.36
	Hendecanoic acid	C11:0	0.417 ± 0.003	0.25	0.272 ± 0.001	0.12	0.211 ± 0.001	0.15
	Dodecanoic acid	C12:0	3.809 ± 0.037	2.29	2.618 ± 0.004	1.20	1.518 ± 0.006	1.07
	Tridecanoic acid	C13:0	1.017 ± 0.025	0.61	0.177 ± 0.002	0.08	0.199 ± 0.001	0.14
	Myristic acid	C14:0	0.391 ± 0.006	0.24	0.249 ± 0.003	0.11	0.171 ± 0.003	0.12
	Pentadecanoic acid	C15:0	0.655 ± 0.006	0.39	5.514 ± 0.008	2.53	2.892 ± 0.015	2.04
	Hexadecanoic acid	C16:0	0.936 ± 0.010	0.56	0.976 ± 0.002	0.45	0.462 ± 0.003	0.33
	Heptadecanoic acid	C17:0	2.537 ± 0.033	1.53	4.897 ± 0.006	2.25	2.350 ± 0.011	1.66
	Octadecanic acid	C18:0	0.363 ± 0.007	0.22	0.939 ± 0.003	0.43	0.609 ± 0.002	0.43
	Henicosanoic acid	C21:0	1.879 ± 0.052	1.13	0.262 ± 0.004	0.12	0.237 ± 0.001	0.17
MUFA	Tetradecenoic acid (cis-9)	C14:1	0.364 ± 0.004	0.22	0.437 ± 0.005	0.20	0.431 ± 0.003	0.30
	Pentadecenoic acid	C15:1	0.561 ± 0.009	0.34	0.326 ± 0.002	0.15	0.968 ± 0.005	0.68
	Hexadecanoic acid	C16:1	0.428 ± 0.005	0.26	0.460 ± 0.003	0.21	1.440 ± 0.004	1.01
	Heptadecenoic acid (cis-10)	C17:1	0.716 ± 0.010	0.43	1.310 ± 0.110	0.60	0.618 ± 0.003	0.44
	Octadecenoic acid (cis-9)	C18:1n9c	1.518 ± 0.024	0.91	1.799 ± 0.006	0.83	1.302 ± 0.006	0.92
	Eicosenoic acid	C20:1	11.501 ± 0.124	6.93	25.597 ± 0.081	11.74	9.459 ± 0.052	6.66
PUFA	Octadecadienoic acid (cis-9,12)	C18:2n6t	14.219 ± 0.326	8.57	13.026 ± 0.066	5.98	11.550 ± 0.043	8.13
	Octadecadienoic acid (anti-9,12)	C18:2n6c	1.213 ± 0.015	0.73	0.219 ± 0.001	0.10	2.168 ± 0.011	1.53
	Octadecatrienoic acid (cis-9,12,15)	C18:3n6	1.100 ± 0.020	0.66	0.239 ± 0.001	0.11	0.157 ± 0.001	0.11
	Linoleic acid	C18:3n3	0.944 ± 0.031	0.57	12.457 ± 0.041	5.71	2.126 ± 0.357	1.50
	Eicosadienoic acid (cis-11,14)	C20:2	0.211 ± 0.003	0.13	0.689 ± 0.002	0.32	0.210 ± 0.001	0.15
	Eicosatrienoic acid (cis-8,11,14)	C20:3n6	0.266 ± 0.006	0.16	0.262 ± 0.002	0.12	0.242 ± 0.002	0.17
	Arachidonic acid (cis-5,8,11,14)	C20:4n6	0.290 ± 0.001	0.17	0.502 ± 0.003	0.23	0.126 ± 0.001	0.09
	Docosadienoic acid (cis-13,16)	C20:2	0.876 ± 0.022	0.53	0.918 ± 0.036	0.42	0.442 ± 0.010	0.31

Appendix D

Table A4. The product contents of microalgae residues obtained from three kinds of medium (BG11, AW and MAW) at different pyrolysis temperatures.

Medium	Pyrolysis Products (%)	Temperature				
		350 °C	450 °C	550 °C	650 °C	750 °C
BG11	Aliphatics	0.13	2.61	8.25	11.51	6.29
	Aromatics	0.00	3.11	7.54	13.03	27.16
	Fatty acids	6.76	5.30	3.18	0.15	0.27
	Phenols	0.34	1.76	3.20	5.47	4.36
	Nitrogen-containing compounds	4.50	6.06	7.35	16.10	26.60
	PAHs	0.00	0.00	0.00	0.23	2.15

Table A4. *Cont.*

Medium	Pyrolysis Products (%)	Temperature				
		350 °C	450 °C	550 °C	650 °C	750 °C
AW	Aliphatics	0.76	6.32	3.10	7.03	5.16
	Aromatics	1.17	0.90	11.66	11.89	24.54
	Fatty acids	0.48	1.26	0.00	0.00	0.00
	Phenols	0.34	0.00	4.05	3.66	6.35
	Nitrogen-containing compounds	6.52	3.43	14.14	13.46	24.33
	PAHs	0.00	0.00	0.00	0.00	2.97
MAW	Aliphatics	0.55	4.45	5.50	11.13	2.25
	Aromatics	0.19	4.03	8.16	24.46	32.07
	Fatty acids	0.00	0.13	0.18	0.00	0.00
	Phenols	0.00	2.06	3.75	3.88	5.08
	Nitrogen-containing compounds	8.52	11.26	10.88	16.48	31.64
	PAHs	0.00	0.00	0.00	0.00	3.10

References

1. Azad, A.K.; Rasul, M.G.; Khan, M.M.K.; Sharma, S.C.; Bhuiya, M.M.K.; Mofijur, M. A review on socio-economic aspects of sustainable biofuels. *Int. J. Glob. Warm.* **2016**, *10*, 32–54. [CrossRef]
2. Hasan, K.; Yousuf, S.B.; Tushar, M.S.H.K.; Das, B.K.; Das, P.; Islam, M.S. Effects of different environmental and operational factors on the PV performance: A comprehensive review. *Energy Sci. Eng.* **2022**, *10*, 656–675. [CrossRef]
3. Moradian, J.M.; Fang, Z.; Yong, Y.C. Recent advances on biomass-fueled microbial fuel cell. *Bioresour. Bioprocess.* **2021**, *8*, 14. [CrossRef]
4. Lokke, S.; Aramendia, E.; Malskaer, J. A review of public opinion on liquid biofuels in the EU: Current knowledge and future challenges. *Biomass Bioenergy* **2021**, *150*, 106094. [CrossRef]
5. Chen, W.H.; Lin, B.J.; Huang, M.Y.; Chang, J.S. Thermochemical conversion of microalgal biomass into biofuels: A review. *Bioresour. Technol.* **2015**, *184*, 314–327. [CrossRef]
6. Ji, F.; Liu, Y.; Hao, R.; Li, G.; Zhou, Y.G.; Dong, R.J. Biomass production and nutrients removal by a new microalgae strain *Desmodesmus* sp. in anaerobic digestion wastewater. *Bioresour. Technol.* **2014**, *161*, 200–207. [CrossRef]
7. Ahmad, A.; Bhat, A.H.; Buang, A.; Shah, S.M.U.; Afzal, M. Biotechnological application of microalgae for integrated palm oil mill effluent (POME) remediation: A review. *Int. J. Environ. Sci. Technol.* **2019**, *16*, 1763–1788. [CrossRef]
8. Valev, D.; Santos, H.S.; Tyystjarvi, E. Stable wastewater treatment with *Neochloris oleoabundans* in a tubular photobioreactor. *J. Appl. Phycol.* **2020**, *32*, 399–410. [CrossRef]
9. Park, J.H.; Yoon, J.J.; Park, H.D.; Lim, D.J.; Kim, S.H. Anaerobic digestibility of algal bioethanol residue. *Bioresour. Technol.* **2012**, *113*, 78–82. [CrossRef]
10. Shahid, A.; Ishfaq, M.; Ahmad, M.S.; Malik, S.; Farooq, M.; Hul, Z.; Batawi, A.H.; Shafi, M.E.; Aloqbi, A.A. Bioenergy potential of the residual microalgal biomass produced in city wastewater assessed through pyrolysis, kinetics and thermodynamics study to design algal biorefinery. *Bioresour. Technol.* **2019**, *289*, 121701. [CrossRef]
11. Li, G.; Lu, Z.T.; Zhang, J.; Li, H.; Zhou, Y.G.; Zayan, A.M.I.; Huang, Z.G. Life cycle assessment of biofuel production from microalgae cultivated in anaerobic digested wastewater. *Int. J. Agric. Biol. Eng.* **2020**, *13*, 241–246. [CrossRef]
12. Huang, Z.G.; Zhang, J.; Pan, M.M.; Hao, Y.H.; Hu, R.C.; Xiao, W.B.; Li, G.; Lyu, T. Valorisation of microalgae residues after lipid extraction: Pyrolysis characteristics for biofuel production. *Biochem. Eng. J.* **2022**, *179*, 108330. [CrossRef]
13. Gong, Z.Q.; Fang, P.W.; Wang, Z.B.; Li, Q.; Li, X.Y.; Meng, F.Z.; Zhang, H.T.; Liu, L. Catalytic Pyrolysis of Chemical Extraction Residue from Microalgae Biomass. *Renew. Energy* **2020**, *148*, 712–719. [CrossRef]
14. Guo, F.; Wang, X.; Yang, X.Y. Potential pyrolysis pathway assessment for microalgae-based aviation fuel based on energy conversion efficiency and life cycle. *Energy Convers. Manag.* **2017**, *132*, 272–280. [CrossRef]
15. Khoo, H.H.; Sharratt, P.N.; Das, P.; Balasubramanian, R.K.; Naraharisetti, P.K.; Shaik, S. Life cycle energy and CO_2 analysis of microalgae-to-biodiesel: Preliminary results and comparisons. *Bioresour. Technol.* **2011**, *102*, 5800–5807. [CrossRef]
16. Collet, P.; Helias, A.; Lardon, L.; Steyer, J.P.; Bernard, O. Recommendations for Life Cycle Assessment of algal fuels. *Appl. Energy* **2015**, *154*, 1089–1102. [CrossRef]
17. Chen, G.Y.; Zhao, L.; Qi, Y. Enhancing the productivity of microalgae cultivated in wastewater toward biofuel production: A critical review. *Appl. Energy* **2015**, *137*, 282–291. [CrossRef]
18. Sharara, M.A.; Holeman, N.; Sadaka, S.S.; Costello, T.A. Pyrolysis kinetics of algal consortia grown using swine manure wastewater. *Bioresour. Technol.* **2014**, *169*, 658–666. [CrossRef]
19. Li, G.; Zhang, J.; Li, H.; Hu, R.C.; Yao, X.L.; Liu, Y.; Zhou, Y.G.; Lyu, T. Towards high-quality biodiesel production from microalgae using original and anaerobically-digested livestock wastewater. *Chemosphere* **2021**, *273*, 128578. [CrossRef]

20. Peccia, J.; Haznedaroglu, B.; Gutierrez, J.; Zimmerman, J.B. Nitrogen supply is an important driver of sustainable microalgae biofuel production. *Trends Biotechnol.* **2013**, *31*, 134–138. [CrossRef]
21. Li, G.; Bai, X.; Li, H.; Lu, Z.T.; Zhou, Y.G.; Wang, Y.K.; Cao, J.X.; Huang, Z.G. Nutrients removal and biomass production from anaerobic digested effluent by microalgae: A review. *Int. J. Agric. Biol. Eng.* **2019**, *12*, 8–13. [CrossRef]
22. Abou-Shanab, R.A.I.; Ji, M.; Kim, H.; Paeng, K.; Jeon, B. Microalgal species growing on piggery wastewater as a valuable candidate for nutrient removal and biodiesel production. *J. Environ. Manag.* **2013**, *115*, 257–264. [CrossRef] [PubMed]
23. Abou-Shanab, R.A.I.; Hwang, J.H.; Cho, Y.; Min, B.; Jeon, B.H. Characterization of microalgal species isolated from fresh water bodies as a potential source for biodiesel production. *Appl. Energy* **2011**, *88*, 3300–3306. [CrossRef]
24. Wang, L.; Li, Y.; Chen, P.; Min, M.; Chen, Y.; Zhu, J.; Ruan, R.R. Anaerobic digested dairy manure as a nutrient supplement for cultivation of oil-rich green microalgae *Chlorella* sp. *Bioresour. Technol.* **2010**, *101*, 2623–2628. [CrossRef]
25. Indarti, E.; Majid, M.I.A.; Hashim, R.; Chong, A. Direct FAME synthesis for rapid total lipid analysis from fish oil and cod liver oil. *J. Food Compost. Anal.* **2005**, *18*, 161–170. [CrossRef]
26. Friedl, A.; Padouvas, E.; Rotter, H.; Varmuza, K. Prediction of heating values of biomass fuel from elemental composition. *Anal. Chim. Acta* **2005**, *544*, 191–198. [CrossRef]
27. Mahinpey, N.; Murugan, P.; Mani, T.; Raina, R. Analysis of bio-oil, biogas, and biochar from pressurized pyrolysis of wheat straw using a tubular reactor. *Energy Fuels* **2009**, *23*, 2736–2742. [CrossRef]
28. Li, G.; Ji, F.; Bai, X.; Zhou, Y.G.; Dong, R.J.; Huang, Z.G. Comparative study on thermal cracking characteristics and bio-oil production from different microalgae using Py-GC/MS. *Int. J. Agric. Biol. Eng.* **2019**, *12*, 208–213. [CrossRef]
29. Marsmann, M. The ISO 14040 family. *Int. J. Life Cycle Assess.* **2000**, *5*, 317–318. [CrossRef]
30. Li, G.; Ji, F.; Zhou, Y.G.; Dong, R.J. Life cycle assessment of pyrolysis process of *Desmodesmus* sp. *Int. J. Agric. Biol. Eng.* **2015**, *8*, 105–112. [CrossRef]
31. Li, G.; Hu, R.C.; Wang, N.; Yang, T.L.; Xu, F.Z.; Li, J.L.; Wu, J.H.; Huang, Z.G.; Pan, M.M.; Lyu, T. Cultivation of microalgae in adjusted wastewater to enhance biofuel production and reduce environmental impact: Pyrolysis performances and life cycle assessment. *J. Clean. Prod.* **2022**, *355*, 131768. [CrossRef]
32. Tan, X.B.; Zhang, Y.L.; Zhao, X.C.; Yang, L.B.; Yangwang, S.C.; Zou, Y.; Lu, J.M. Anaerobic digestates grown oleaginous microalgae for pollutants removal and lipids production. *Chemosphere* **2022**, *308*, 136177. [CrossRef] [PubMed]
33. Cheng, H.H.; Narindri, B.; Chu, H.; Whang, L.M. Recent advancement on biological technologies and strategies for resource recovery from swine wastewater. *Bioresour. Technol.* **2020**, *303*, 122861. [CrossRef] [PubMed]
34. Markou, G.; Angelidaki, I.; Georgakakis, D. Microalgal carbohydrates: An overview of the factors influencing carbohydrates production, and of main bioconversion technologies for production of biofuels. *Appl. Microbiol. Biotechnol.* **2012**, *96*, 631–645. [CrossRef]
35. Chinnasamy, S.; Bhatnagar, A.; Hunt, R.W.; Das, K. Microalgae cultivation in a wastewater dominated by carpet mill effluents for biofuel applications. *Bioresour. Technol.* **2010**, *101*, 3097–3105. [CrossRef]
36. Ngoc, B.T.; Yen, T.K.N.; Jeong, Y.M.; Ediriweera, M.K.; Somi, K.C. Pentadecanoic acid, an odd-chain fatty acid, suppresses the stemness of MCF-7/SC human breast cancer stem-like cells through JAK2/STAT3 signaling. *Nutrients* **2020**, *12*, 1663. [CrossRef]
37. Bach, Q.V.; Chen, W.H. Pyrolysis characteristics and kinetics of microalgae via thermogravimetric analysis (TGA): A state-of-the-art review. *Bioresour. Technol.* **2017**, *246*, 88–100. [CrossRef]
38. Kebelmann, K.; Hornung, A.; Karsten, U.; Griffiths, G. Intermediate pyrolysis and product identification by TGA and Py-GC/MS of green microalgae and their extracted protein and lipid components. *Biomass Bioenergy* **2013**, *49*, 38–48. [CrossRef]
39. Li, G.; Bai, X.; Huo, S.H.; Huang, Z.G. Fast pyrolysis of LERDADEs for renewable biofuels. *IET Renew. Power Gener.* **2020**, *14*, 959–967. [CrossRef]
40. Li, K.; Liu, Q.; Fang, F.; Luo, R.; Lu, Q.; Zhou, W.; Huo, S.; Cheng, P.; Liu, J.; Addy, M. Microalgae-based wastewater treatment for nutrients recovery: A review. *Bioresour. Technol.* **2019**, *291*, 121934. [CrossRef]
41. Thangalazhy-Gopakumar, S.; Adhikari, S.; Chattanathan, S.A.; Gupta, R.B. Catalytic pyrolysis of green algae for hydrocarbon production using H(+)ZSM-5 catalyst. *Bioresour. Technol.* **2012**, *118*, 150–157. [CrossRef] [PubMed]
42. Gong, Z.Q.; Wang, Z.T.; Wang, Z.B.; Fang, P.W.; Meng, F.Z. Study on the migration characteristics of nitrogen and sulfur during co-combustion of oil sludge char and microalgae residue. *Fuel* **2019**, *238*, 1–9. [CrossRef]
43. Li, J.; Xiong, Z.; Zeng, K.; Zhong, D.A.; Zhang, X.; Chen, W.; Nzihou, A.; Flamant, G.; Yang, H.P.; Chen, H.P. Characteristics and Evolution of Nitrogen in the Heavy Components of Algae Pyrolysis Bio-Oil. *Environ. Sci. Technol.* **2021**, *55*, 6373–6385. [CrossRef] [PubMed]
44. Sanchez-Silva, L.; Lopez-Gonzalez, D.; Garcia-Minguillan, A.M.; Valverde, J.L. Pyrolysis, combustion and gasification characteristics of Nannochloropsis gaditana microalgae. *Bioresour. Technol.* **2013**, *130*, 321–331. [CrossRef] [PubMed]

Article

Environmentally Friendly New Catalyst Using Waste Alkaline Solution from Aluminum Production for the Synthesis of Biodiesel in Aqueous Medium

Sandro L. Barbosa [1,*], David Lee Nelson [1], Lucas Paconio [1], Moises Pedro [1], Wallans Torres Pio dos Santos [1], Alexandre P. Wentz [1], Fernando L. P. Pessoa [2], Foster A. Agblevor [3], Daniel A. Bortoleto [4], Maria B. de Freitas-Marques [5] and Lucas D. Zanatta [6]

[1] Department of Pharmacy, Federal University of Jequitinhonha and Mucuri Valleys-UFVJM, Campus JK, Rodovia MGT 367–Km 583, n° 5.000, Alto da Jacuba, Diamantina 39100-000, Brazil; dleenelson@gmail.com (D.L.N.); lucas.paconio@ufvjm.edu.br (L.P.); moises.pedro@ufvjm.edu.br (M.P.); wallanst@ufvjm.edu.br (W.T.P.d.S.); wentzap@hotmail.com (A.P.W.)

[2] University Center SENAI-CIMATEC, Av. Orlando Gomes, 1845 Piatã, Salvador 41650-010, Brazil; fernando.pessoa@fieb.org.br

[3] Utah Science Technology and Research (USTAR), Biological Engineering, Utah State University, Logan UT620 East 1600 North, Suite 130, Logan, UT 84341, USA; foster.agblevor@usu.edu

[4] Department of Geosciences, Universidade Federal do Pará, R. Augusto Corrêa, 01–Guamá, Belém 66075-110, Brazil; dabortoleto@yahoo.com.br

[5] Department of Chemistry, Instituto de Ciências Exatas, Universidade Federal de Minas Gerais, Av. Antônio Carlos, 6627, Pampulha, Belo Horizonte 31270-901, Brazil; betanialf@hotmail.com

[6] Laboratório de Química Bioinorgânica, Departamento de Química, Faculdade de Filosofia, Ciências e Letras de Ribeirão Preto, Universidade de São Paulo, Av. Bandeirantes, 3900, Ribeirão Preto 14040-901, Brazil; lucaszanatta@alumni.usp.br

* Correspondence: sandro.barbosa@ufvjm.edu.br; Tel.: +55-38-3532-1234

Abstract: Red mud (RM) is composed of a waste alkaline solution (pH = 13.3) obtained from the production of alumina. It contains high concentrations of hematite (Fe_2O_3), goethite (FeOOH), gibbsite [$Al(OH)_3$], a boehmite (AlOOH), anatase (Tetragonal–TiO_2), rutile (Ditetragonal dipyramidal–TiO_2), hydrogarnets [$Ca_3Al_2(SiO_4)_{3-x}(OH)_{4x}$], quartz ($SiO_2$), and perovskite ($CaTiO_3$). It was shown to be an excellent catalytic mixture for biodiesel production. To demonstrate the value of RM, an environmentally friendly process of transesterification in aqueous medium using waste cooking oil (WCO), MeOH, and waste alkaline solution (WAS) obtained from aluminum production was proposed. Triglycerides of WCO reacted with MeOH at 60 °C to yield mixtures of fatty acid methyl esters (FAMEs) in the presence of 0.019% (w/w) WAS/WCO using the WAS (0.204 mol L^{-1}, predetermined by potentiometric titration) from aluminum production by the Bayer process. The use of the new catalyst (WAS) resulted in a high yield of the products (greater than 99% yield).

Keywords: environmentally friendly processes; bayer residue; waste management; basic catalyst; contaminants; red mud; fatty acid methyl ester

1. Introduction

Waste alkaline solution (WAS) and red mud (RM) (bauxite residue, bauxite tailings, red sludge, or alumina refinery residues) are industrial wastes generated during the processing of bauxite into alumina using the Bayer process (over 95% of the alumina is produced globally through the Bayer process) [1–5]. With every ton of alumina produced, approximately 1 to 1.5 tons of RM is also produced. Annual production of alumina in 2020 was over 133 million tons, resulting in the generation of over 175 million tons of RM [6]. Because of this large production and the material's high alkalinity (pH of 10.5–12.5) caused by an elevated Na content and high concentrations of potentially toxic metals [7,8] and potential leaching (groundwater, surface waters, soils, and ocean ecosystems), it can pose a significant environmental hazard if not stored properly. Its storage is a critical environmental problem [9–16]. This material is typically stored in dams, which demands prior preparation of the disposal area and includes monitoring and maintenance during the storage period [17].

A small residual amount of the sodium hydroxide used in the process remains with the residue. Various stages in the solid/liquid separation process have been introduced to recycle as much hydroxide as possible from the residue back into the Bayer process to make the process as efficient as possible and to reduce production costs. These modifications also decrease the final alkalinity of the residue, making it easier and safer to handle and store [18].

The present study discusses the technical viability of RM valorization. The authors proposed the utilization of residual WAS from the aluminum industry for the first time as a catalyst in the transesterification reaction in aqueous medium for the production of fatty acid methyl esters (FAME).

FAME, or biodiesel, according to the American Society for Testing and Materials (ASTM) and European standards (EN), is a fuel consisting of "long chain fatty acids of mono-alkyl esters derived from renewable fatty raw material such as animal fats or vegetable oils" [19]. FAME content needs to be higher than 96.5 wt.%. The total glycerol, including bound glycerol [in glycerides such as monoglycerides (MGs), diglycerides (DGs), and triglycerides (TGs)] and unbound glycerol (free glycerol), needs to be limited to 0.24 and 0.25 wt.% by ASTM and EN standards, respectively.

In the search for an environmentally friendly method for biodiesel synthesis by transesterification of TGs, several alternatives for catalysis have been explored. In general, catalysts that can be used for producing biodiesel are divided into three categories: acidic, alkaline, and biocatalysts. Acidic and alkaline catalysts are classified into two groups: homogeneous and heterogeneous catalysts.

Catalysts play a vital role in the transesterification process. Both the amount and type of catalyst affect the rate of reaction and conversion efficiency. Homogeneous catalysts function in the same phase as the reactants and can be categorized into homogeneous base catalysts and homogeneous acid catalysts. Currently, most FAMEs are produced by the base-catalyzed transesterification reaction because of its high conversion rate, negligible side reactions, and short reaction time. It is a low-pressure and low-temperature process, which occurs without the formation of intermediate substances. Despite these advantages, homogeneous base catalysts have some weaknesses. The production of biodiesel from feedstocks with a high free fatty acid (FFA) content is limited. It was reported by some researchers that homogeneous base catalysts are only effective for the production of FAME via the transesterification process using the feedstocks with an FFA content of less than 2 wt.% [20]. When FFA content is >2%, the catalyst reacts with FFA to produce soap and water. The soap inhibits the separation of FAME and glycerin, and the water can hydrolyze the esters in a reaction that competes with the transesterification.

In transesterification reactions of TGs catalyzed by homogeneous bases, FAME and glycerol are produced. Zhang, Stanciulescu, and Ikura (2009) demonstrated that the use of phase transfer agents (PTA) greatly increased the rate of the base-catalyzed transesterification reaction [21]. A product containing 96.5 wt.% was obtained after only 15 min of rapid reaction at 60 °C in the presence of tetrabutylammonium hydroxide or acetate. The reaction was performed in the presence of MeOH, glycerol, refined and bleached soybean oil, and the basic catalyst, without the presence of water.

Recently, we also studied the acid-catalyzed transesterification reaction using WCO and a solid acidic catalyst (SiO_2-SO_3H) in the presence of quaternary ammonium salts as co-catalysts in toluene and DMSO [22]. We decided to study the transesterification reaction in a medium containing WAS. This study sought to utilize the industrial residues (RM; WAS) to increase their value and decrease the environmental problems resulting from the storage of these residues.

2. Experimental

2.1. Raw Materials and Chemicals

RM containing WAS was collected from ALCOA (Juruti, PA, Brazil). WCO (soybean) was donated by the university restaurant, and it was filtered through silica gel, which removed any fatty acids (FA) and polar and polymeric substances prior to use. The physical parameters determined for the yellow oil were the viscosity (41.2 mPa) and the density (0.883 g·mL^{-1}). MeOH (analytical grade) was supplied by Vetec, São Paulo, Brazil.

2.2. Typical Procedures

2.2.1. Standardization of the WAS Generated during the Processing of Bauxite into Alumina Using the Bayer Process

WAS (1.60 mL) was diluted to 100 mL. An aliquot of 1.00 mL was removed and diluted again to 100 mL. A 25 mL aliquot was collected from this solution and titrated with a 0.0017 M HCl solution using 0.48 mL of HCl solution. Titration was accomplished using an SI Analytics Titrator TitroLine® 7000 potentiometric titrator. The concentration of WAS was determined as 0.204 mol L^{-1}.

2.2.2. Elemental Analysis Based on Energy-Dispersive X-ray Fluorescence (EDXRF) of Dehydrated WAS (Dry Solid) (Chemical or Elemental Composition of WAS)

The pH of the original WAS from RM was 13.3. The WAS sample was previously dried for 24 h in a muffle furnace at 150 °C. The equipment used in this procedure was an X-ray fluorescence spectrometer, model EDX 720 (Shimadzu®; Kyoto, Japan), equipped with an X-ray tube and the use of liquid nitrogen for cooling. The software used was PCEDX, version 1.11 Shimadzu®. Energy scattering X-ray fluorescence is a non-destructive, multi-element analytical technique capable of identifying elements with an atomic number Z greater than or equal to 12. When the electrons of the innermost layer of the atom (for example, K and L) interact with photons in the X-ray region, then these electrons are ejected, creating a vacancy. To promote stability, the electronic vacancies are immediately filled by electrons from the closest layers (Kα, Kβ or Lα, Lβ), resulting in an excess of energy in the process, which is manifested in the form of the emission of X-rays characteristic of each atom present in the sample.

The EDXRF is an apparatus used for the quantitative and quali-quantitative determination of chemical elements in a wide range of samples. The analyses carried out in this work used qualitative and quantitative determination, with only the pre-calibration of the equipment (Al $\geq$ 80% and the detection of Sn and Cu), using only atmospheric air, and restricting the detection of metals included between $_{13}$Al and $_{92}$U. The samples were placed in sample holders made of polypropylene film and the analysis conditions were as follows: 10 mm collimator, scans with voltages of 0–40 KeV (Ti-U) and 0–20 KeV (Na-Sc) with a time of 100 s for each sample.

In this method, the material to be analyzed is targeted with an X-ray beam that interacts with the atoms of the sample and causes the ionization of the innermost layers of the atoms. The filling of the resulting vacancies by more peripheral electrons induces the emission of X-rays characteristic of the constituent elements of the sample. The elemental composition of WAS analyzed by EDXRF was the following (in wt.%) (see Table 1): Al, 54.58; Si, 39.54; K, 1.99; V, 1.88; Ga, 0.95; Cs, 0.33; Cr, 0.174; Fe, 0.142; Br, 0.089; Cu, 0.079; Mo, 0.069; Tl, 0.064; Ag, 0.064; and Zr, 0.047. EDXRF was used as a method for semi-quantitative elemental analysis, and the titanium derivatives were not quantified. Their concentrations were lower than the quantification limit. However, their presence was clearly observed in the powder XRD spectra.

2.2.3. Experimental X-ray Diffraction of Dehydrated WAS and Characterization of the Material by Powder XRD

The measurements were performed with a Shimadzu model XRD-6000 diffractometer using CuKα monochromatic radiation (λ = 0.15406 nm–40 kV and 30 mA) at a scan rate of 2.0 degrees·s^{-1}, covering the 2θ scale from 10–80°. WAS was dried at 150 °C before experimental X-ray diffraction. The XRD patterns of the dehydrated WAS (200 °C) are shown in Figure 1. The most common mineral phases present in the WAS were hematite (Fe_2O_3), goethite (FeOOH), gibbsite ($Al(OH)_3$), a boehmite (AlOOH), titanium mineral as anatase (Tetragonal–TiO_2), rutile (Ditetragonal dipyramidal–TiO_2), and the hydrogarnets group [$Ca_3Al_2(SiO_4)_{3-x}(OH)_{4x}$], such as $Fe_3Al_2(SiO_4)_3$, quartz (SiO_2), and perovskite ($CaTiO_3$). Iron crystalline phases were identified in the XRD diffractograms as hematite and goethite phases, and they were attributed in 2θ = 34.4°, 35.5°, 41.5°, 50.6°, 53.7°, 56.5°, 23.4°, and 33.5°, respectively. Aluminum crystalline phases were identified as gibbsite and boehmite and were attributed in 2θ = 18.6°, 21.1°, 30.2°, 38.1°, and 48.3°, respectively. The aluminosilicate crystalline phase was attributed to the hydrogarnets group with 2θ = 40°, 44.8°, 52°, 54.8°, and 67.4°. The silicon crystalline phase was attributed to quartz with 2θ = 60.8° and 63.9°. Finally, titanium crystalline phases were the last type of material identified as titanium dioxide and perovskite by XRD diffractograms, with signals in 2θ = 26.1°, 27.8°, 62.1°, 46.7°, 59°, and 69.3°, respectively. The WAS composition agrees with that of the EDXRF identification, in which aluminosilicate was confirmed to be the principal component, with traces of iron, chromium, and copper that can act as active species in the catalysis or in a synergistic catalysis process with potassium [23,24].

Table 1. The elemental composition of WAS determined by EDXRF.

Elemental Composition	wt.%
Al	54.58
Si	39.54
K	1.99
V	1.88
Cs	0.330
Cr	0.174
Fe	0.142
Br	0.089
Cu	0.079
Mo	0.069
Tl	0.064
Ag	0.064
Zr	0.047

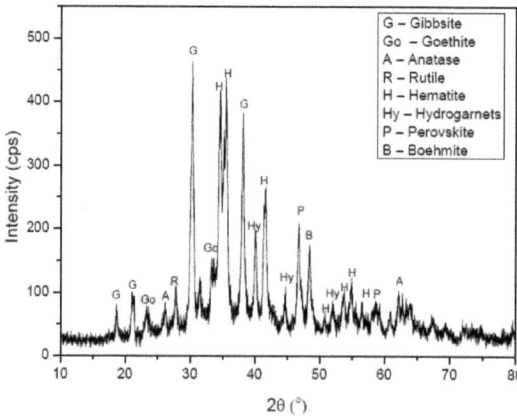

Figure 1. PXRD (X-ray diffraction patterns) of dehydrated WAS (dry solid).

2.2.4. Thermal Gravimetric (TG) Analysis Technology of Dehydrated WAS

WAS had been dried at 150 °C before thermogravimetric analysis. The TG curve (Figure 2, red) of the WAS indicated an intense mass loss (i.e., 76%) between the initial heating phase, 30 °C up to 253 °C, with a corresponding endotherm displayed by the simultaneous DTA curve (Figure 2, black), which is equivalent to the removal of moisture from the sample. A small endothermic peak around 300 °C (T_{onset} 274.4 °C) was observed in the DTA curve without mass loss in the same temperature range. It corresponds to the melting of the residue obtained at a temperature greater than 253 °C. The highlighted photos show the crucible containing the sample before (upper left) and after (upper right) the analysis. Note the presence of a white solid residue representing 24% of the sample.

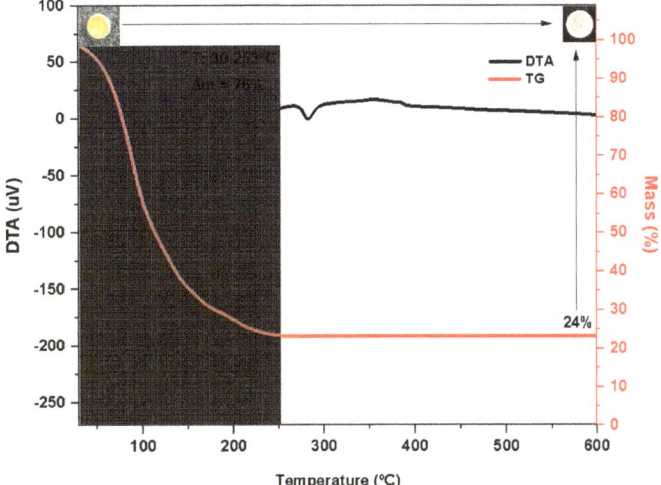

Figure 2. Thermal analysis WAS.

2.2.5. Reacting the TGs from WCO with MeOH Using WAS as Catalyst

The procedure utilized for the transesterification reaction was based on various trials to determine the optimum conditions for this reaction. A 150-mL round bottom flask, equipped with a reflux condenser, containing WAS (0.204 mol L^{-1}; or 0.019% w/w of WCO) with MeOH (3.75 mL, 2.9738 g, 0.0928 mol; or a 1:18 molar ratio of WCO/MeOH) and WCO (4.4170 g, 5.0 mL; 5.0510^{-3} mol) were mixed, and the mixture was refluxed for 30 min at 60 °C. The mixture was cooled and transferred to a separatory funnel where the biofuel-containing upper phase was separated from the lower phase containing glycerol by decantation. The MeOH was removed from the biodiesel phase on a rotary evaporator, purified by distillation, and used in new reaction processes within this study. The recovered glycerol was stored for future treatments. The biofuel phase was dissolved in hexane (20 mL), extracted with 20 mL of a saturated solution of NaCl, dried over MgSO$_4$, and concentrated.

2.3. WCO and Biodiesel Analysis

The official methods proposed by ISO 12966 were used to determine the compositional profile by gas chromatography using a flame ionization detector (GC-FID) (Shimadzu GC-2010). The chromatographic system used to separate and identify FFAs (wt.%) included a cross-bound polyethylene glycol capillary column (Supelco SP 2560, 100 m × 0.25 mm × 20 μm). The initial temperature was 60 °C for 2 min; the temperature increased to 220 °C at 10 °C·min^{-1}, and finally, to 240 °C at 5 °C·min^{-1}, where it was held for 7 min. The injector and detector temperatures were 350 °C, and the sample (0.5 μL injected) was dissolved in 99% isooctane.

The EN 14103 and the Brazilian Technical Standards Association (ABNT NBR 15908) were used to quantify FAMEs and remaining mono-, di-, and triglycerides (MG, DG, TG) in the FAME (biodiesel). For quantification of FAMEs, a Thermo Trace GC-Ultra chromatograph equipped with a flame ionization detector and a Thermo Scientific TR-BD (FAME) Capillary GC Column (L × I.D. 30 m × 0.25 mm, df 0.25 μm) containing a polyethylene glycol stationary phase was used, according to the EN 14103 analytical procedure. Pure methyl nonadecanoate (C19:0, Sigma-Aldrich—Sao Paulo, Brazil) was used as an internal standard to normalize the peak areas of the chromatograms. The integration was achieved from the methyl hexanoate (C6:0) peak to that of the methyl nervonate (C24:1), including all the peaks identified as FAMEs. To analyze the FAME samples, approximately 100 mg (accuracy ± 0.1 mg) of homogenized sample and approximately 100 mg (accuracy ± 0.1 mg) of nonadecanoic acid methyl ester were weighed in a 10 mL vial and diluted with 10 mL of toluene before injection into the equipment. All the samples were prepared in duplicate. Chromatographic conditions are described as follows: (a) column temperature: 60 °C held for 2 min, programmed at 10 °C·min^{-1} to 200 °C, and then programmed at 5 °C·min^{-1} to 240 °C; the final temperature was held for 7 min; (b) injector and detector temperature: 250 °C; (c) helium carrier gas flow rate: 1–2 mL·min^{-1}; a minimum flow rate of 1 mL·min^{-1} was warranted when operating at the maximum temperature; (d) injected volume: 1 μL; and (e) split flow: 100 mL·min^{-1}.

For the quantification of the glycerides (MG, DG, and TG), a Shimadzu GC2010 chromatograph equipped with a flame ionization detector was used according to the ASTM D6584 analytical procedure. The chromatographic system was configured to separate and identify MG, DG, and TG with a CrossbondTM 5% Phenyl/95% dimethylpolysiloxane capillary column (Zebron ZB-5HT, 30 m × 0.32 mm × 0.1 mm–Phenomenex, Torrence, CA, USA) with on-column injection. The initial temperature in the capillary column was 50 °C (1 min); the temperature increased to 180 °C at 15 °C min$^-$ to 230 °C at 7 °C·min^{-1}, and finally, to 380 °C at 20 °C·min^{-1}, where it was held for 10 min. The injector and detector temperatures were 380 °C, and the sample (0.5 mL injected) was prepared using heptane 99%. ^{1}H- and ^{13}C-NMR spectra were recorded on Bruker *Avance* 400 and *Avance* 500 spectrometers. These data are included in the Supplementary Material.

Thermogravimetry Analysis

The thermal behavior of FAME was evaluated by thermogravimetry (TG), and the data were treated simultaneously in the first derivative, thermogravimetry derivative (DTG), and differential thermal analysis (DTA). TG/DTA curves were obtained on a DTG60H Shimadzu thermobalance; the heating rate was 10 °C·min^{-1} and the temperature range was 25–350 °C under a controlled nitrogen atmosphere at 50 mL·min^{-1} and under oxidizing conditions (synthetic air). An open alumina crucible had an accurately weighed sample mass of about 20 mg. The derivative curve was obtained by T.A. data software.

3. Results and Discussion

The compositional profile analysis of the WCO used in this work is described in Table 2. The main fatty acids (FAs) in that WCO were linoleic (C18:2) and oleic (C18:1) acids; accordingly, the mean molecular weight (MW) of the FAs was determined to be 277.41 g·mol^{-1}, and the mean molecular mass (MM) of the TGs was 873.22 g·mol^{-1}. The composition of the WCO was very similar to that of soybean oil (SO) described in the literature [25]. The following profile was considered for calculating the molar ratio of WCO: MeOH for the transesterification reaction.

Lit [17]: palmitic acid, 11.6%; stearic acid, 3.22%; oleic acid, 25.09%; linoleic acid, 52.93%; linolenic acid, 5.95%; others, 1.08%.

Sodium methoxide (MeONa) is formed by the deprotonation of MeOH and is frequently prepared by treating MeOH with metallic sodium [26]:

$$Na + CH_3OH \rightarrow CH_3ONa + 1/2H_2 \tag{1}$$

The resulting solution, MeONa/MeOH, which is colorless, is often used as a source of MeONa. The MeONa absorbs CO_2 from the air to form MeOH and Na_2CO_3, thereby diminishing the alkalinity of the base.

$$CH_3ONa + CO_2 + H_2O \rightarrow 2\ CH_3OH + Na_2CO_3 \qquad (2)$$

Table 2. FAs composition of the WCO used in this study.

FA	MW (g·mol^{-1})	wt.%
Linoleic acid (C18:2)	280.45	51.66
Oleic acid (C18:1)	282.46	26.52
Palmitic acid (C16:0)	256.43	10.41
Linolenic acid (C18:3)	278.43	5.55
Stearic acid (C18:0)	284.48	3.91
Others	-	1.95
Average MW of FAs (g·mol^{-1})	277.41	
MM of TGs (g·mol^{-1})	873.22	

MeONa is highly basic and reacts with water, resulting in MeOH and NaOH. In this work, we used a solution of WAS (0.204 mol/L, determined by potentiometric titration) which leads to a very low concentration of MeONa. Thus, we believe that the reaction did not occur as a result of a nucleophilic attack by MeONa on the ester carbonyl groups of TGs, DGs, and MGs.

In an unprecedented process, biodiesel was produced by catalysis using WAS obtained from the aluminum industry for the transesterification reaction of WCO. The WAS-catalyzed transesterification is a rapid reaction at both initial and final reaction stages because it is not limited by mass transfer between the polar water/MeOH/glycerol phase and the non-polar MGs, DGs, and TGs phase.

Zhang, Stanciulescu, and Ikura (2009) used a total OH^-/oil molar ratio of 0.22 [27]. In this study, we believe that the low molar concentration of hydroxide (0.019% w/w of WCO or 0.0040 mol/mol of WCO) is related to gibbsite ($Al(OH)_3$) and to the fact that the hematite (Fe_2O_3) contained in WAS forms an anionic iron species in the presence of hydroxide ion. According to Ishikawa, Yoshioka, Sato, and Okuwaki (1997), hematite in the presence of sodium hydroxide leads to the formation of $NaFeO_2$ (FeO_2^-) [28].

$$Fe_2O_3(s) + 2NaOH(aq) \rightarrow 2FeO_2^- + 2Na^+ + H_2O \qquad (3)$$

Species like FeO_2^- might facilitate the formation of the nucleophilic species MeO^- in a medium containing MeOH. Wang et al. (2022) have shown that the $NaFeO_2$-Fe_3O_4 composite from blast furnace dust acts as an efficient catalyst for the production of biodiesel [29]. They obtained the catalyst by treating the blast furnace dust with different proportions of sodium carbonate and calcining the mixture. The catalyst was recycled, and high yields were still obtained ($Na_2CO_3 \cdot H_2O$@BFD300): 93.00 wt.% at the eighth use ($NaHCO_3$@BFD300) and 96.16 wt.% at the seventh use ($Na_2CO_3 \cdot 10H_2O$@BFDun). The highest yields for sustainable biodiesel production were obtained with $Na_2CO_3 \cdot H_2O$@BFD300 catalyst as a result of the reaction of impregnated Na_2CO_3 with Fe_2O_3 in the blast furnace dust to produce stable and active nanocomponents of $NaFeO_2$ (32.42 nm) and a magnetic nanocomponent of Fe_3O_4 (3.14 nm and Ms of 6.16 Am_2/kg). Blast furnace dust was a suitable raw material for catalyst synthesis to produce soybean biodiesel by the transesterification reaction in a non-aqueous medium. This composite would also be formed in the residue from aluminum production and would explain the high yield of biodiesel obtained, even though the free hydroxide concentration was very low.

Liu et al. [30] and Wang et al. [31] reported that the alkaline content of the material from the Bayer process includes soluble NaOH, Na_2CO_3, $NaAlO_2$, Na_2SiO_3, and sodium aluminosilicate hydrate ($Na_2O \cdot Al_2O_3 \cdot xSiO_2 \cdot mH_2O$; zeolite) generated from the reaction between the bauxite and the highly alkaline solution. Pamparano and Debecker [32] observed that sodium aluminate ($NaAlO_2$) was an effective catalyst for the preparation of biodiesel from refined sunflower oil in a moisture-free process. The catalyst was calcined to remove any moisture because the material was hygroscopic. The oil was also heated to remove any moisture. They showed that the activity of the catalyst was a function of its basicity. The waste material utilized in our study contained 54.581% aluminum [gibbsite ($Al(OH)_3$), a boehmite (AlOOH), and hydrogarnets $Ca_3Al_2(SiO_4)_{3-x}(OH)_{4x}$], and one would expect that this component would be the principal agent responsible for the catalysis of the transesterification reaction, either directly or indirectly via the equilibrium with the hematite (Fe_2O_3), goethite (FeOOH), anatase (tetragonal–TiO_2), rutile (ditetragonal dipyramidal–TiO_2), and perovskite ($CaTiO_3$). The solution was used directly in the form that it was produced by in the industry. No calcination was performed. The activity of WAS was slightly higher than that of the $NaFeO_2$-Fe_3O_4, $Na_2CO_3 \cdot H_2O$@BFD300, and $Na_2CO_3 \cdot 10H_2O$@BFDun described above.

3.1. Transesterification of TGs from WCO with MeOH Using WAS

In this study, an excess of MeOH was mixed with standardized WAS and WCO and refluxed at 60 °C. The progress of the reaction was monitored using thin layer chromatography (TLC). The total consumption of TGs occurred after 60 min. The FAMEs and glycerides (MG, DG, and TG) contained in the biodiesel phase were confirmed by GC-FID using the methods defined in EN 14103 and ASTM D6584. The composition of the products (FAMEs and glycerides) is presented in Table 2; they represent the average values of five different measurements. More than 30 experiments were performed in which the proportion of WAS catalyst to WCO was varied. The greatest efficiency was observed when the mass ratio of WAS to WCO was 1:4700.

The composition of the products (FAMEs and glycerides) after 10.0 min of reaction is presented in Table 3; they represent the average values of five different measurements. The yields for completed transesterification reactions are presented in Table 4.

Table 3. Composition of the WCO feedstock and the product mixture using refluxing MeOH.

Products	WCO (wt.%)	WAS Catalyst Mixture *
Triacylglycerides (%)	96.2 ± 2.84	5.9 ± 0.17
Diacylglycerides (%)	2.80 ± 1.80	11.0 ± 0.32
Monoacylglycerides (%)	1.0 ± 0.92	22.9 ± 0.68
FAME (%)	0	60.2 ± 1.78

* The composition of the products (FAMEs and glycerides) after 10.0 min of reaction.

A comparison of the results obtained using the WAS catalyst and a traditional catalyst (KOH) for FAME (biodiesel) production from WCO as feedstock is presented in Table 5. Refaat et al. [33] stated that FAME production is worthy of continued study and optimization of production procedures because of its environmentally beneficial attributes and its renewable nature. Their study was intended to consider aspects related to the feasibility of the production of biodiesel from WCO in an attempt to help reduce the cost of biodiesel and reduce waste and pollution resulting from WCO. The variables affecting the yield and characteristics of the biodiesel produced from WCO were studied. The best yield was obtained using a 6:1 MeOH/WCO with KOH as the catalyst (1%) at 65 °C for 60 min. The yield obtained from WCO reached 96.15% under optimum conditions.

Xiangmei Meng, Guanyi Chen, and Yonghong Wang [34] used WCO, which contained large amounts of free fatty acids produced in restaurants and was collected by the Environmental Protection Agency in the main cities of China. The optimum experimental conditions, which were obtained from the orthogonal test, were MeOH/WCO molar ratio 9:1 with 1.0 wt.% NaOH, a temperature of 50 °C, and 90 min. The 6:1 MeOH:WCO molar ratio was most suitable in the process, and WCO conversion efficiency was 89.8%.

Table 4. Composition of the product of transesterification of TGs from WCO with MeOH using WAS catalyst.

Assay	WAS Catalyst Mixture
	Yield * (% w/w)
FAME	99.5 ± 2.94
Free glycerol (FG)	0.01 ± 0.01
Total glycerol (TG)	0.07 ± 0.03
MG	0.26 ± 0.24
DG	0.06 ± 0.02
TG	0.01 ± 0.01

* The composition of the products (FAMEs and glycerides) after 30.0 min of reaction at 60 °C.

Table 5. Comparison of the yield of the transesterification with WAS catalyst and traditional catalysts (NaOH and KOH) for biodiesel production using WCO as feedstock.

	WAS Catalyst	NaOH Catalyst	KOH Catalyst
MeOH/WCO molar ratio	18:1	9:1	6:1
Temperature °C	60	50	65
Time (min)	30	90	60
FAME (%)	99.5%	89.8%	96.15%

3.2. Determination of Thermophysical Properties of FAME

The significant properties of FAME are determined by the various tests and methods as per the ASTM specifications. Table 6 gives the standard test methods used for the determination of various properties of FAME.

Table 6. ASTM standards for FAME properties and the experimental values.

Property	FAME	Reference	Test Standard
Kinematic Viscosity (40 °C; mm^2/s)	5.03	[35]	ASTM D 445-04e
Density	0.87	[35]	Density ASTM D7371-12
Cloud Point	−1	[35]	Cloud Point ASTM-D 2500-05
High heating value	41.28	[35]	ASTM D-240-02
Cloud Filter Plugging Point	−7	[35]	ASTM D6377-05
Cetane Number	61	[35]	ASTM D 613-05
Pour Point	−16	[36]	ASTM-D97
Flash Point	164	[36]	EN ISO 2719

3.3. Thermogravimetry Analysis of FAME

Under inert conditions (N_2 atmosphere), FAME ignites at 140 °C, as is seen in the TG curve (top Figure 3, red), involving two overlapping thermal decomposition mechanisms illustrated by the dotted highlight of the DTG curve (top Figure 3, pink). In the thermal decomposition range, 140–295 °C, a broad endothermic event characteristic of this phenomenon occurs, as is seen in the DTA curve (top Figure 3, black). The biodiesel was almost completely consumed, with 98.6% loss in mass. The photos in the detail of Figure 3, top, show the alumina crucible containing the sample before ignition and after firing up to 350 °C.

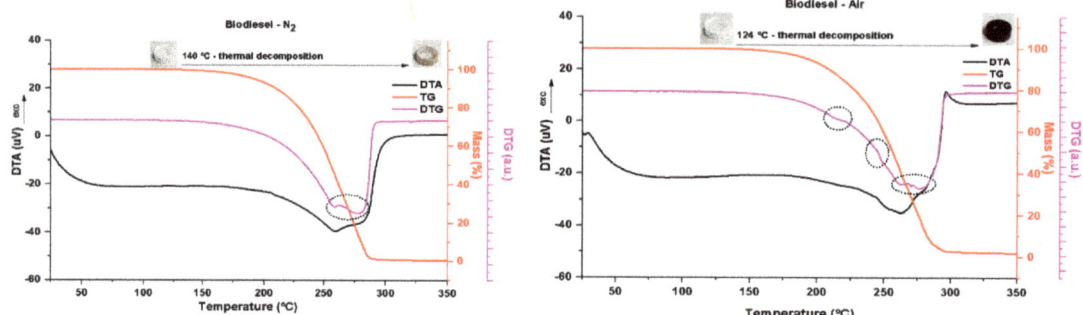

Figure 3. Thermal behavior of biodiesel under inert (N_2) conditions (**left graph**) and under oxidant (air) conditions (**right graph**).

Biodiesel underwent ignition from 124 °C under oxidizing conditions (under an atmosphere of synthetic air), as observed in the TG curve (Figure 1, bottom, red); that is, at a lower temperature than that observed under inert conditions because the supply of oxygen favors the combustion process. Some inflections of the TG curve occur, as is seen in dotted details of the DTG curve (Figure 3, bottom, pink), showing the complexity of the mechanisms of thermal decomposition of biodiesel under oxidizing conditions. This phenomenon is also marked by a broad endothermic curve (DTA curve, Figure 3, bottom, black) over the full range of decomposition (124–307 °C). Biodiesel burned almost completely, with 98.6% mass loss. The photos in the detail of Figure 3, top, show the alumina crucible containing the sample before ignition and after firing up to 350 °C. Note the intensity of carbonization of the sample, consistent with the oxidizing condition.

4. Conclusions

The transesterification reaction of TGs from WCO was catalyzed by WAS {containing hematite (Fe_2O_3), goethite (FeOOH), gibbsite (Al(OH)$_3$), a boehmite (AlOOH), anatase (Tetragonal–TiO_2), rutile (Ditetragonal dipyramidal–TiO_2), hydrogarnets [$Ca_3Al_2(SiO_4)_{3-x}(OH)_{4x}$], SiO_2 (quartz), and perovskite ($CaTiO_3$)}. The WAS (catalytic red mud solution) was used successfully for the transesterification of TGs from WCO with MeOH. Analysis of the quantities of TGs, DGs, MGs, and FAMEs in the products indicated excellent catalytic behavior of the WAS. The WAS-catalyzed transesterification rate was indicated by the high FAME yield obtained after only 30 min of reaction. The rapid transesterification observed can be explained by the fact that WAS can facilitate ion transfer between the polar water/MeOH/glycerol phase and non-polar WCO phase, overcome mass transfer limitations, and speed up reaction rates. Product analyses showed that a FAME content greater than 99.5 wt.% was achieved after only 30 min of rapid transesterification. Free and total glycerol contents in the final products after 30 min of transesterification were lower than the maximum legal limits in standard specifications for FAME. The catalyst was suitable for the synthesis of FAME from WCO by the transesterification reaction in aqueous medium. This method employing WAS as a catalyst could enable recycling of the waste alkaline solution from the bauxite process, minimize contaminants, and reduce the cost of the catalyst. In addition, it would be of great environmental value through the decontamination of groundwater and soils, as well as the elimination of areas intended for the disposal of these alkaline solutions that result from the Bayer process in the production of alumina. The use of this industrial residue as a catalyst by biodiesel-producing industries could lead, in the short term, to a total replacement of traditional catalysts by the use of WAS and zero investment in traditionally used catalysts. This highly efficient and low-cost WAS catalyst could make the process of FAME production more economical. The global sodium methoxide solution as a biodiesel catalyst market is estimated to value at around US \$0.3 bn in 2021 and is expected to register a CAGR of 3.1% [37]. In addition to FAME production, such environmentally friendly WAS catalysts should find application in a wide range of other important base-catalyzed organic reactions, such as the Michael reaction or Michael 1,4-addition.

Supplementary Materials: The following supporting information can be downloaded at: https://www.mdpi.com/article/10.3390/bioengineering10060692/s1, Figure S1: ^{1}H NMR WCO. Figure S2: ^{13}C NMR WCO. Figure S3: ^{1}H NMR FAME using WAS. Figure S4: ^{13}C NMR FAME using WAS as catalyst.

Author Contributions: Conceptualization, S.L.B., D.A.B. and D.L.N.; methodology, S.L.B., L.P. and M.P.; validation, S.L.B. and W.T.P.d.S., L.P. and M.P.; formal analysis, S.L.B.; investigation, S.L.B., L.P. and M.P.; resources, A.P.W.; data curation, S.L.B., L.D.Z., M.B.d.F.-M.; writing—original draft preparation, S.L.B.; writing—review and editing, S.L.B. and D.L.N.; visualization, S.L.B.; supervision, S.L.B.; project administration, F.A.A.; funding acquisition, F.L.P.P. All authors have read and agreed to the published version of the manuscript.

Funding: SENAI CIMATEC and PRPPG/UFVJM in response to Resolução 15/2019.

Institutional Review Board Statement: Not applicable.

Informed Consent Statement: Not applicable.

Acknowledgments: The authors acknowledge the support by the SENAI CIMATEC and PRPPG/UFVJM in response to Resolução 15/2019 and the Fundação de Apoio à Pesquisa do Estado de Minas Gerais–FAPEMIG (Chamada Universal), 0004022 code.

Conflicts of Interest: Everyone involved in the peer review process carefully considers and declares that there are no conflict of interest when participating in the review, decision-making process, and publication of an article.

References

1. Habashi, F. A Hundred Years of the Bayer Process for Alumina Production. In *Essential Readings in Light Metals*; Donaldson, D., Raahauge, B.E., Eds.; Springer: Cham, Switzerland, 2016. [CrossRef]
2. Đurić, I.; Mihajlović, I.; Živković, Ž.; Kešelj, D. Artificial neural network prediction of aluminum extraction from bauxite in the Bayer process. *J. Serb. Chem. Soc.* **2012**, *76*, 1259–1271. [CrossRef]
3. Hind, A.R.; Bhargava, S.K.; Grocott, S.C. The surface chemistry of Bayer process solids: A review. *Colloids Surf. A Physicochem. Eng. Asp.* **1999**, *146*, 359–374. [CrossRef]
4. Scarsella, A.A.; Noack, S.; Gasafi, E.; Klett, C.; Koschnick, A. Energy in Alumina Refining: Setting New Limits. In *Light Metals*; Hyland, M., Ed.; Springer: Cham, Switzerland, 2015. [CrossRef]
5. Whittington, B.I. The chemistry of CaO and Ca(OH)$_2$ relating to the Bayer process. *Hydrometallurgy* **1996**, *43*, 13–35. [CrossRef]
6. Evans, K. The History, Challenges and new developments in the management and use of Bauxite Residue. *J. Sustain Metall.* **2016**, *2*, 316–331. [CrossRef]
7. Wang, L.; Sun, N.; Tang, H.; Sun, W. A review on comprehensive utilization of red mud and prospect analysis. *Minerals* **2019**, *9*, 362. [CrossRef]
8. Chao, X.; Zhang, T.A.; Lv, G.; Chen, Y.; Li, X.; Yang, X. Comprehensive application technology of bauxite residue treatment in the ecological environment: A review. *Bull. Environ. Contam. Toxicol.* **2022**, *109*, 209–214. [CrossRef]
9. Bray, A.W.; Stewart, D.I.; Courtney, R.; Rout, S.P.; Humphreys, N.; Mayes, W.M.; Burke, I.T. Sustained Bauxite Residue Rehabilitation with Gypsum and Organic Matter 16 years after Initial Treatment. *Environ. Sci. Technol.* **2018**, *52*, 152–161. [CrossRef]
10. Burke, I.T.; Mayes, W.M.; Peacock, C.L.; Brown, A.P.; Jarvis, A.P.; Gruiz, K. Speciation of Arsenic, Chromium, and Vanadium in Red Mud Samples from the Ajka Spill Site, Hungary. *Environ. Sci. Technol.* **2012**, *46*, 3085–3092. [CrossRef]
11. Burke, I.T.; Peacock, C.L.; Lockwood, C.L.; Stewart, D.I.; Mortimer, R.J.G.; Ward, M.B.; Renforth, P.; Gruiz, K.; Mayes, W.M. Behavior of Aluminum, Arsenic, and Vanadium during the Neutralization of Red Mud Leachate by HCl, Gypsum, or Seawater. *Environ. Sci. Technol.* **2013**, *47*, 6527–6535. [CrossRef]
12. Gelencsér, A.; Kováts, N.; Turóczi, B.; Rostási, Á.; Hoffer, Á.; Imre, K.; Nyirő-Kósa, I.; Csákberényi-Malasics, D.; Tóth, Á.; Czitrovszky, A.; et al. The Red Mud Accident in Ajka (Hungary): Characterization and Potential Health Effects of Fugitive Dust. *Environ. Sci. Technol.* **2011**, *45*, 1608–1615. [CrossRef]
13. Gupta, V.K.; Sharma, S. Removal of Cadmium and Zinc from Aqueous Solutions Using Red Mud. *Environ. Sci. Technol.* **2002**, *36*, 3612–3617. [CrossRef]

14. Ruyters, S.; Mertens, J.; Vassilieva, E.; Dehandschutter, B.; Poffijn, A.; Smolders, E. The Red Mud Accident in Ajka (Hungary): Plant Toxicity and Trace Metal Bioavailability in Red Mud Contaminated Soil. *Environ. Sci. Technol.* **2011**, *45*, 1616–1622. [CrossRef] [PubMed]
15. Santini, T.C.; Fey, M.V. Spontaneous Vegetation Encroachment upon Bauxite Residue (Red Mud) As an Indicator and Facilitator of In Situ Remediation Processes. *Environ. Sci. Technol.* **2013**, *47*, 12089–12096. [CrossRef]
16. De Resende, E.C.; Gissane, G.; Nicol, R.; Heck, R.J.; Guerreiro, M.C.; Coelho, J.V.; de Oliveira, L.C.A.; Palmisano, P.; Berruti, F.; Briens, C.; et al. Synergistic co-processing of Red Mud waste from the Bayer process and a crude untreated waste stream from bio-diesel production. *Green Chem.* **2013**, *15*, 496–510. [CrossRef]
17. Silveira, N.C.G.; Martins, M.L.F.; Bezerra, A.C.S.; Araújo, F.G.S. Red Mud from the Aluminium Industry: Production, Characteristics, and Alternative Applications in Construction Materials-A Review. *Sustainability* **2021**, *13*, 12741. [CrossRef]
18. Chandra, S. Red Mud Utilization. In *Waste Materials Used in Concrete Manufacturing*; Elsevier: Amsterdam, The Netherlands, 1996; pp. 292–295, ISBN 978-0-8155-1393-3.
19. *ASTM D6751-20a*; License Agreement, Standard Specification for Biodiesel Fuel Blend Stock (B100) for Middle Distillate Fuels. ASTM: Philadelphia, PA, USA, 2020.
20. Patel, R.L.; Sankhavara, C. Biodiesel production from Karanja oil and its use in diesel engine: A review. *Renew. Sustain. Energy Rev.* **2017**, *71*, 464–474. [CrossRef]
21. Zhang, Y.; Stanciulescu, M.; Ikura, M. Rapid transesterification of soybean oil with phase transfer catalysts. *Appl. Catal. A Gen.* **2009**, *366*, 176–183. [CrossRef]
22. Barbosa, S.L.; Rocha, A.C.P.; Nelson, D.L.; Freitas, M.S.; Mestre, A.A.P.F.; Klein, S.I.; Clososki, G.C.; Caires, F.J.; Flumignan, D.L.; Santos, L.K.; et al. Catalytic Transformation of Triglycerides to Biodiesel with SiO_2-SO_3H and Quaternary Ammonium Salts in Toluene or DMSO. *Molecules* **2022**, *27*, 953. [CrossRef] [PubMed]
23. Thangaraj, B.; Solomon, P.R.; Muniyandi, B.; Ranganathan, S.; Lin, L. Catalysis in biodiesel production-a review. *Clean Energy* **2020**, *3*, 2–23. [CrossRef]
24. Botti, R.F.; Innocentini, M.D.M.; Faleiros, T.A.; Mello, M.F.; Flumignan, D.L.; Santos, L.K.; Franchin, G.; Colombo, P. Biodiesel Processing Using Sodium and Potassium Geopolymer Powders as Heterogeneous Catalysts. *Molecules* **2020**, *25*, 2839. [CrossRef]
25. Martínez, G.; Sánchez, N.; Encinar, J.M.; González, J.F. Fuel properties of biodiesel from vegetable oils and oil mixtures. Influence of methyl esters distribution. *Biomass Bioenergy* **2014**, *63*, 22–32. [CrossRef]
26. Charoensuppanimit, P.; Kitsahawong, K.; Kim-Lohsoontorn, P.; Assabumrungrat, S. Incorporation of hydrogen by-product from $NaOCH_3$ production for methanol synthesis via CO_2 hydrogenation: Process analysis and economic evaluation. *J. Clean. Prod.* **2019**, *212*, 893–909. [CrossRef]
27. Chandran, K.; Kamruddin, M.; Ajikumar, P.K.; Gopalan, A.; Ganesan, V. Kinetics of thermal decomposition of sodium methoxide and ethoxide. *J. Nucl. Mater.* **2006**, *358*, 111–128. [CrossRef]
28. Ishikawa, K.; Yoshioka, T.; Sato, T.; Okuwaki, A. Solubility of hematite in LiOH, NaOH and KOH solutions. *Hydrometallurgy* **1997**, *45*, 129–135. [CrossRef]
29. Wang, X.-M.; Zeng, Y.-N.; Jiang, L.-Q.; Wang, Y.-T.; Li, J.-G.; Kang, L.-L.; Ji, R.; Gao, D.; Wang, F.-P.; Yu, Q.; et al. Highly stable $NaFeO_2$-Fe_3O_4 composite catalyst from blast furnace dust for efficient production of biodiesel at low temperature. *Ind. Crops Prod.* **2022**, *182*, 114937. [CrossRef]
30. Liu, Y.; Lin, C.X.; Wu, Y.G. Characterization of red mud derived from a combined Bayer process and bauxite calcinations method. *J. Hazard Mater.* **2007**, *146*, 255–261. [CrossRef] [PubMed]
31. Wang, Z.; Han, M.F.; Zhang, Y.H.; Zhou, F.S. Study on the dealkalization technics of Bayer process red mud with CO_2 by carbonation. *Bull. Chin. Ceramic. Soc.* **2013**, *32*, 1851–1855. (In Chinese)
32. Pampararo, G.; Debecker, D. Sodium aluminate-catalyzed synthesis of biodiesel. In *ChemRxiv*; Cambridge Open Engage: Cambridge, UK, 2023. [CrossRef]
33. Refaat, A.A.; Attia, N.K.; Sibak, H.A.; El Sheltawy, S.T.; ElDiwani, G.I. Production optimization and quality assessment of biodiesel from waste vegetable oil. *Int. J. Environ. Sci. Technol.* **2008**, *5*, 75–82. [CrossRef]
34. Meng, X.; Chen, G.; Wang, Y. Biodiesel production from waste cooking oil via alkali catalyst and its engine test. *Fuel Process. Technol.* **2008**, *89*, 851–857. [CrossRef]
35. Ramirez-Verduzco, L.F.; Rodriguez-Rodriguez, J.E.; Jaramillo-Jacob, A.R. Predicting cetane number, kinematic viscosity, density and higher heating value of biodiesel from its fatty acid methyl ester composition. *Fuel* **2012**, *91*, 102–111. [CrossRef]
36. Singh, S.P.; Singh, D. Biodiesel Production through the use of different sources and characterization of oils and their esters as the substitute of diesel. *Renew. Sustain. Energy Rev.* **2010**, *14*, 200–216. [CrossRef]
37. Available online: https://marketresearch.biz/report/sodium-methoxide-solution-as-a-biodiesel-catalyst-market/ (accessed on 16 May 2023).

Disclaimer/Publisher's Note: The statements, opinions and data contained in all publications are solely those of the individual author(s) and contributor(s) and not of MDPI and/or the editor(s). MDPI and/or the editor(s) disclaim responsibility for any injury to people or property resulting from any ideas, methods, instructions or products referred to in the content.

Review

Biofuels from Renewable Sources, a Potential Option for Biodiesel Production

Dhurba Neupane

Department of Biochemistry and Molecular Biology, University of Nevada, Reno, NV 89557, USA; dneupane@unr.edu; Tel.: +1-775-409-2265

Citation: Neupane, D. Biofuels from Renewable Sources, a Potential Option for Biodiesel Production. *Bioengineering* **2023**, *10*, 29. https://doi.org/10.3390/bioengineering10010029

Academic Editors: Zhijian Pei and Liang Luo

Received: 21 October 2022
Revised: 9 December 2022
Accepted: 16 December 2022
Published: 25 December 2022

Copyright: © 2022 by the author. Licensee MDPI, Basel, Switzerland. This article is an open access article distributed under the terms and conditions of the Creative Commons Attribution (CC BY) license (https://creativecommons.org/licenses/by/4.0/).

Abstract: Ever-increasing population growth that demands more energy produces tremendous pressure on natural energy reserves such as coal and petroleum, causing their depletion. Climate prediction models predict that drought events will be more intense during the 21st century affecting agricultural productivity. The renewable energy needs in the global energy supply must stabilize surface temperature rise to 1.5 °C compared to pre-industrial values. To address the global climate issue and higher energy demand without depleting fossil reserves, growing bioenergy feedstock as the potential resource for biodiesel production could be a viable alternative. The interest in growing biofuels for biodiesel production has increased due to its potential benefits over fossil fuels and the flexibility of feedstocks. Therefore, this review article focuses on different biofuels and biomass resources for biodiesel production, their properties, procedure, factors affecting biodiesel production, different catalysts used, and greenhouse gas emissions from biodiesel production.

Keywords: generation of biofuels; biodiesel; renewable resources; fossil fuel; population growth; greenhouse gas emissions

1. Introduction

The rising world population is predicted to reach over 9 billion by 2050 [1]. Increasing global prices and higher energy demand have put tremendous pressure on natural energy reserves, causing their depletion [2–4]. The burning of fossil fuels has several environmental implications, including an increase in greenhouse gas (GHG) emissions, particularly carbon dioxide (CO_2) [5,6]. Over the last few decades, global primary energy consumption has increased dramatically due to rapid industrialization and higher living standards [2,7]. Developing countries such as Brazil, the South Asian region, and South Africa require 12–24 gigajoules (GJ)/cap of energy annually to have a decent standard of living [8]. Currently, over 80% of the world's energy comes from fossil fuels, including natural gas, oil, and coal, and about 98% of it is generated via carbon emissions from fossil fuels [8,9]. The duration and intensity of drought are expected to become more severe, thus reducing water reserves by five-fold throughout the 21st century [1].

An increased share of renewable energy in the global energy supply will help to stabilize surface temperature rise to 1.5 °C compared to pre-industrial levels [10]. The temperature increase could be as much as 3–5 °C depending on certain regions [11]. Further, a shift in rainfall was found, ranging from 19.2 to 37.2 mm over different growing seasons [12]. With the inadequate pool of sources, particularly water, and an ever-increasing need for global energy, alternative fuels are the most practical way to meet the rising demand [13]. Researchers have already figured out alternatives to address this demand [14]. Further, the potential options to mitigate the effect of climate change and reduce dependence on fossil fuels are urgently needed and are already in development. There is an increasing interest in growing biofuels at a global and national level as a low-carbon alternative to fossil fuels due to their potential to reduce GHG emissions and the associated climate change impact from transport [15]. The use of bioenergy/biofuels is one of the promising renewable energy alternatives [16] because these are cheaper in synthesis [4]. Biofuels, generally biodiesel, have attracted researchers' attention due to their potential benefits over fossil fuels and the flexibility of feedstocks. For example, sulfur-free, adequate oxygen content, an easy manufacturing process, and reduced GHG emissions are critical advantages of biodiesel [17]. Biodiesel as a diesel fuel is bio-degradable [18], non-toxic [19], portable, environmentally sustainable [20], efficient, and has low sulfur as well as aromatic content [21]. Additionally, due to the higher flask point of biodiesel, transportation and storage of biodiesel are safer than diesel fuels. However, it has some disadvantages; biodiesel is more expensive and emits more NO gas than diesel [9].

Due to its crucial characteristics and usage of versatile feedstock, for example, from waste frying oil to cheap non-edible resources, biodiesel has tremendous potential to use as an alternative fuel [22]. It is a promising and economical alternative to diesel that can reduce the global reliance on imported petroleum fuels. This article provides a comprehensive review of the types and generation of biofuels, biomass sources, properties, and factors affecting biodiesel production. This article also highlights various catalysts in biodiesel production, greenhouse gas emissions from several literatures, and finally, the conclusion and future perspective.

2. Types and Generation of Biofuels

Biofuels are classified into four generations, namely first, second, third, and fourth based on their sources and production of various biomaterials. A brief description of each of the generations is highlighted below.

2.1. First-Generation Biofuels

First-generation biofuels are conventional biofuels, mainly generated from two types of edible feedstock, namely starch-based (e.g., potato, corn, barley, and wheat) and sugar-based (e.g., sugarcane and sugar beet) feedstocks [23,24]. The main advantages of first-generation raw materials are the availability of crops and comparative simple conversion processes. However, using edible food crops for biodiesel production, there is a reduced food supply, thus potentially increasing food prices [25]. Another concern is the diverting of agricultural land to fuel production. Using a significantly large amount of fertilizer and pesticides for agricultural production could negatively impact the environment [15]. There are several types of conventional biofuels based on the technological approach they use to generate (Figure 1).

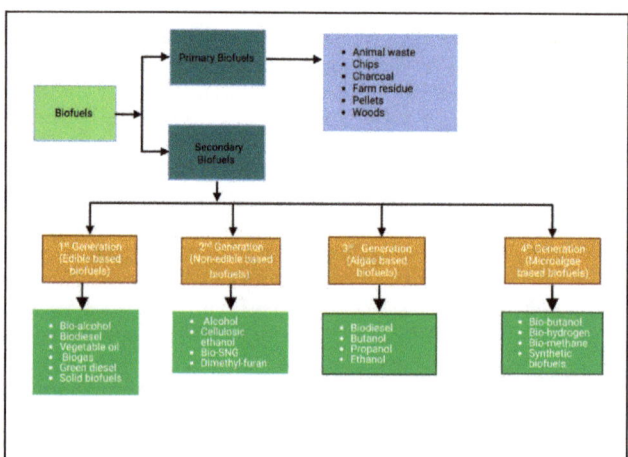

Figure 1. Types and generations of Biofuel. Adopted from [23].

2.1.1. Bio Alcohols

Bio alcohols are extracted with the help of enzymes and microorganisms by alcohol fermentation of cellulose, glucose, starches, carbohydrates, and other sugars. Bio alcohols are further categorized into bioethanol, biopropanol, and biobutanol [26].

2.1.2. Biodiesels

Biodiesels are the forms of diesel extracted from renewable feedstocks, including lignocellulosic biomass, which consists of long-chain fatty acid esters. Biodiesels are produced chemically by reacting lipids, such as animal fat (tallow), soybean oil, or other vegetable oils with alcohol and produce methyl, ethyl, or propyl ester [27]. The commonly used catalyst used during biodiesel production includes NaOH or KOH [28].

2.1.3. Vegetable Oil

Vegetable oils are produced from fat, olive oil, castor oil, and sunflower oil. The fuels produced from vegetable oil are economical and environmentally friendly. Recent studies reported that waste cooking and vegetable oils are considered alternative fuels for diesel engines in some precise applications [29].

2.1.4. Green Diesel

The hydrotreating of triglycerides produces green diesel in vegetable oils with hydrogen. Three main reactions during the process are hydrodeoxygenation (HDO), decarbonylation (DCO), and decarboxylation (DCO_2) [30].

2.1.5. Biogas

Biogas is produced by anaerobic digestion with the help of microbial consortium without oxygen, and digestate as a nutrient-rich byproduct is also produced [31,32]. Biogas produced during the process contains about 60% CH_4, 35% CO_2, and 5% a mixture of H_2, N_2, CO, NH_3, O_2, and volatile amines [33]. Biogas can be used for industrial energy, cooking in rural areas [33,34], and combined heat and power production [10].

2.1.6. Solid Biofuels

Raw materials, including wood, wood chips, leaves, sawdust, charcoal, and animal dung, are commonly used as solid biofuels. The use of solid biofuels in the energy sector is limited to particular markets [23]. For example, firewood is the most common strategy to generate bioenergy, which can be used for cooking food [28].

2.2. Second-Generation Biofuels

The controversy of using first-generation biofuel feedstock due to the food vs. energy debate has forced us to move to second-generation biofuels, such as lignocellulosic or carbohydrate biomass, as the potential alternative source for biofuels and chemical production [24]. These feedstocks do not rely on edible plants and do not require agricultural land [35]. Cellulosic biomass comprises various chemical compositions such as cellulose, lignin, and polyose. Lignocellulosic biomass is composed of cellulose (35–50%), lignin (15–20%), hemicellulose (20–35%), and other components (15–20%). The lignocellulosic-based biofuel production process has the potential to lower GHG emissions, boost the economy, and aid energy security. The biotechnological approach in the United States has been estimated to produce 1.3 billion tons of dry biomass annually without compromising food security [36]. Second-generation biofuels are advanced biofuels obtained from several trees, grass, bushes, and agricultural residues [23]. Based on the technologies used to produce them, second-generation biofuels include the following (Figure 1).

2.2.1. Cellulosic Ethanol

The fermented sugars obtained from polyose and cellulose compounds of lignocellulose are used for making cellulosic ethanol [23]. Cellulosic biofuels can contribute to rural economic development and enhance the sustainability of agricultural landscapes [37,38].

2.2.2. Algae-Based Biofuels

Algae is the fastest-growing raw material for biofuel production and an essential substitute for biofuel extraction. Techniques of extraction and concentration of biomass from algae include processes such as centrifugation, aggregation, floatation, purification, and flocculation [39,40]. Biofuels such as biodiesel, biogas, and hydrogen can be produced from algae using the advanced feature [41].

2.2.3. Alcohol

Alcohol is obtained from syngas by fermenting biomass with the help of specific microorganisms [42].

2.2.4. Dimethylfuran

Dimethylfuran is an oxygenated hydrocarbon with an oxygen content of 17%. It is an additive in diesel fuels. This is highly competitive in reducing emissions from engines [43].

2.2.5. Biosynthetic Natural Gas (Bio-SNG)

Biogas can be produced from anaerobic digestion with the help of microbes. Bio-SNG is used in the form of CNG and LNG in vehicles and for refilling a natural gas cylinder [44].

2.3. Third-Generation Biofuels

Third-generation biofuels are produced from algal biomass and waste oil. The advantages of using third-generation biofuels include higher growth and productivity, no agricultural land required, higher oil content, and less impact on food supply. Microalgae, fish oil, animal fat, and waste cooking oil are the primary sources of third-generation biodiesel feedstocks [45]. Because of the cost involved during harvesting, drying, and extraction processes, using algal biomass as biodiesel feedstock is expensive. However, it produces about 10–100 times more biofuel or oil per unit area. Seaweed or macro-algae is third-generation biomass that can be used in bio-energy production and has many advantages such as short cultivation time, high carbohydrate, proteins, and lipids content, and low or no lignin content [46]. Algal-based biofuel includes bioethanol, biodiesel, and biohydrogen (by the process of bio photolysis, photo fermentation, and dark fermentation) [47]. A study showed that the lipid in algae could be converted to biodiesel by the conventional approach, such as the conversion method used for vegetable oil. The conversion process of algal biodiesel production involves transesterifications, enzymatic, wet extraction, alcoholysis and acidolysis, and finally, biodiesel [48]. Algal oil blended with diesel fuel in a 20% ratio reduced hydrocarbon exhaust and better emission characteristics [49,50]; however, the complete combustion of algae releases a higher % of NO_x into the atmosphere due to the significant presence of nitrogen in algae (5–8%) [51].

In the case of waste oil or waste cooking oil, the variation in using different feedstocks and their chemical composition, and impurities, limit their productivity at large scale [52]. Waste coffee ground oils and bardawil lagoon are an example of third-generation feedstock used in recent years [53].

2.4. Fourth-Generation Biofuels

With the application of molecular biology, genetic engineering, and interdisciplinary physicochemical approaches, which include the use of CRISPR/Cas9 with guided RNA for genetic modification in algae [54] to optimize and enhance the yield of biofuel production, the biofuel generated by such process is considered a fourth-generation biofuel. The fourth-generation biofuel production employs genetically modified algae that accumulate high lipid and carbohydrate content to improve biofuel yield [55]. The raw materials used for biofuel production are microalgae, macroalgae, and cyno-bacteria. Cyno-bacteria are non-photosynthetic prokaryotes, and micro and macro algae are eukaryotes [56]. The inactivation of ADP-glucose phosphorylase in a *Chlamydomonas starchless* mutant led to a 10-fold increase in TAG [57]. Similarly, a modification in the CoA-dependent 1-butanol production pathway into a cyanobacterium, *Synechococcus elongatus*, can produce butanol from CO_2 directly [58].

3. Biomass Sources for Biodiesel Production

Biodiesel or fatty acid methyl ester (FAME) is a processed diesel fuel from different biological sources, including edible, non-edible, animal fats, and waste cooking oils. FAME combines long-chain fatty acid monoalkyl esters of fatty acids [2]. It is a green biological ester-based oxygenated oil that comprises organic fats and oils [3]. The world's biodiesel production is projected to reach 10.3 billion gallons by 2024, which reached 8.5 billion gallons in 2016 [59]. It is also estimated that food-based feedstocks (first-generation biofuel) will dominate the world's market [60].

Biodiesel is intended to be used in standard diesel engines as a standalone fuel or blended with petroleum. In 2021, the total volume of biodiesel production in the United States amounted to over 1.6 billion gallons, compared to 9 million gallons in 2001 and 991 million in 2012. After 2012, there were fluctuations in biodiesel production volume in different years, with the highest quantity attained in 2018 (Figure 2a). Similarly, total biomass production in the United States was 1375.56 billion kW hours in 2021, which is expected to increase gradually in the coming decades. It is estimated to reach 1630.73 billion kW hours by 2050 (Figure 2b).

Various feedstock sources can be used for biodiesel production, including various vegetable oils, animal fats, microbial oil, algal oils, and waste oils [63,64]. Palm oil, stearic oil, lauric oil, oleic oil, soybean oil, sunflower oil, palmitic oil, rapeseed oil, canola oil, and vegetable derivates are included under vegetable oils. With the use of catalyst, animal fats or vegetable with alcohols also produces biodiesel and glycerin [3]. Feedstock selection is a crucial step in biodiesel production, which is impacted by different factors, such as yield, cost, composition, and purity of the produced biodiesel. Another significant factor affecting biodiesel production is availability and the types of sources (non-edible, edible, or waste) [65].

Further, the choice of materials used for its production depends on the geographical regions; for example, soybean is the primary source of biodiesel in the United States, whereas, in Europe and the tropical parts of the world, rapeseed (canola) and palm oil serve as the primary sources [66–68]. Different feedstocks produce biodiesel with distinct qualities that must be considered when blending biodiesel with petroleum diesel for their use in transportation. Biodiesel is blended with petroleum diesel from 5% to 20% biodiesel, or B5-B20. However, the Renewable Fuel Standard (RFS), a federal program that mandates the blending of biofuels into the nation's fuel supply, has suggested including higher biodiesel blends. Soybean and canola oil are the most common biodiesel in the United States. Soybean accounted for about 50% of biodiesel feedstock input between 2014 and 2017. The soybean oil used for biodiesel production increased by 30% in 2017 compared to 2014 (https://www.eia.gov/todayinenergy/detail.php?id=36052. Accessed on 3 May 2018).

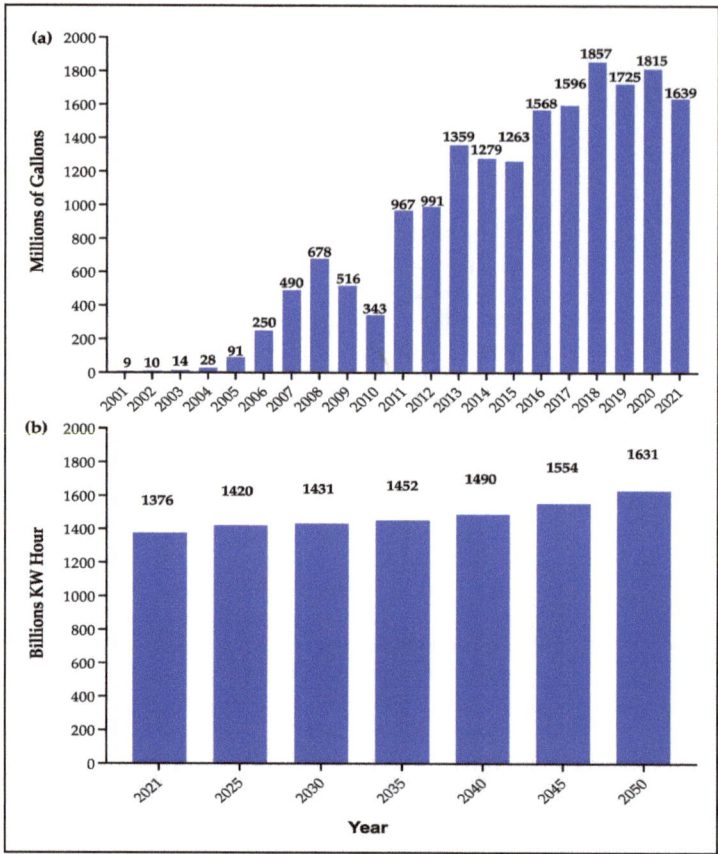

Figure 2. (a) US biodiesel production change for the past 20 years since 2001 [61] (Source: U.S. Energy Information Administration (EIA), Monthly Energy Review, Table, 10.4. Release date: April 2022. Available at: https://www.eia.gov/totalenergy/data/monthly/pdf/mer.pdf), and (b) US biomass energy production forecast from 2021 to 2050 [62] (Source: EIA, Annual Energy Outlook 2022, Table 1. Available online: https://www.statista.com/statistics/264029/us-biomass-energy-production/. Accessed on 21 June 2022).

In 2020, approximately 71.7 % of the biodiesel feedstock came from soybean, while other small amounts of vegetable oils and animal fats (AF) such as canola oil (10.7 %), corn oil (13.0%), tallow beef fat (3.1%), poultry fat (1.5%), and other (0.1%) were used (Table 1). Based on 2012–2019 data (Table 2), rapeseed oil is still the dominant biodiesel feedstock in Europe and worldwide. In 2016, rapeseed (canola) input to global contribution for biodiesel production was 68%, followed by soybean (15%), animal fat and yellow grease (5% each), palm oil (6%), and sunflower (1%) [68]. However, rapeseed share in the feedstock mix in Europe has significantly decreased; for example, its share was 62.3% in 2012 compared to only 37.9% in 2019 (Table 2). This decrease in the share of rapeseed oil in Europe is primarily because of recycled vegetable oil/used cooking oil (UCO) and palm oil. UCO, or yellow grease, has become the second-most important feedstock for Europe since 2015. In the USA, biodiesel production from yellow grease (13%) dominated both rapeseed-based biodiesel (10%), corn-based biodiesel (12%), and animal fats-based biodiesel (10%) (based on 2016 data reported by Kim et al., 2018 [68]).

Table 1. US inputs to biodiesel production (million kilograms).

Period	Vegetable Oil (Million kg)					Animal Fats (Million kg)	
	Canola Oil	Corn Oil	Cottonseed Oil	Soybean Oil	Other	Poultry	Tallow
January	49.4	80.3	-	236.3	W	5.0	W
February	42.4	60.7	-	260.7	W	5.4	9.4
March	59.6	65.7	-	297.6	W	10.7	W
April	62.8	38.0	-	304.7	S	W	10.9
May	58.9	38.3	-	365.3	W	3.9	5.3
June	50.0	42.7	W	338.9	5.9	W	9.7
July	W	60.5	W	351.5	W	W	24.6
August	W	67.3	W	338.0	W	W	20.0
September	W	61.7	-	334.0	W	10.4	12.4
October	W	45.8	-	328.0	W	9.5	23.6
November	W	60.3	-	309.8	-	6.4	15.0
December	W	66.7	-	337.5	-	3.2	17.2
Total	565.2	687.6	0.3	3802.5	W	78.5	166.9
% of total	10.7	13.0	0.0	71.7	0.1	1.5	3.1

Table with -, W, S indicates no data, withheld to avoid disclosure of individual company data, and value is less than 0.5 of the table metrics. However, the value is included in any associated total. Source: U.S. Energy Information Administration (EIA), Form EIA-22M "Monthly Biodiesel Production Survey." U.S. EIA I Monthly Biodiesel Production Report (2020).

Table 2. The feedstock was used for biodiesel + renewable diesel (HVO; hydrotreated vegetable oil) in Europe from 2012 to 2019.

Feedstocks	2012	2013	2014	2015	2016	2017	2018	2019
Rapeseed oil	6500	5710	6200	6400	6060	6300	5200	5000
Used cooking oil (UCO)	800	1150	1890	2400	2620	2770	2860	2750
Palm oil	1535	2340	2240	2340	2315	2650	2570	2640
Soybean oil	720	870	840	540	610	930	1000	1100
Animal fats	360	420	920	1030	795	795	800	800
Sunflower oil	300	290	310	210	250	180	185	190
other, pine/tall oils, fatty acid	220	335	370	560	615	635	680	700
Share of rapeseed oil (%)	62.3	51.4	48.6	47.5	45.7	44.2	39.1	37.9

The original data were collected in a metric ton (MT) and then converted to kilogram (kg) using a conversion rate of 1 MT = 1000 kg (Source: EU-28. Available online: https://apps.fas.usda.gov/newgainapi/api/report/downloadreportbyfilename?filename=Biofuels%20Annual_The%20Hague_EU-28_7-15-2019.pdf. Accessed on 15 July 2019).

The types of raw materials/feedstocks for biodiesel production rapidly diversified for economic and environmental reasons [69]. A market survey reported that biodiesel's feedstock market is transitioning from first-generation feedstock such as soybean, rapeseed, and palm oil to non-food and lower-cost feedstock such as jatropha, castor, UCO, and AF [70]. In countries such as Brazil, effective programs are underway to promote jatropha and castor production for biodiesel production. Similarly, another emerging feedstock for Biodiesel is HVO, which is produced through hydrotreating [69]. Production and use of biofuel generate emissions such as particulate matter (PM), carbon monoxide (CO), carbon dioxide (CO_2), nitrogen oxides (NOx), hydrocarbons, and volatile organic compounds (VOCs). The VOCs, unburnt hydrocarbon (UBHC), and NOx are the precursors for forming smog and ground-level ozone, which are associated with increased morbidity and mortality from cardiovascular and respiratory diseases and certain cancers [15]. Compared to fossil diesel, biodiesel produces lower PM, CO, VOCs, and NO_X emissions [71]. Among NOx, nitrous oxide (N_2O) is only the greenhouse gas of great environmental concern. It is a substantial anthropogenic greenhouse gas, and agriculture represents its most significant source. The global warming potential of N_2O is 298 times that of CO_2 [72]. Previous studies on biofuel production systems revealed that emissions of N_2O may counterbalance a substantial part of the global warming reduction by fossil fuel displacement [73]. Using optimized crop management, which involves state-of-the-art agricultural technologies coupled with an optimized fertilization regime, and nitrification inhibitors, N_2O emissions can significantly be reduced by −135% points (pp) compared to conventional management. However, uncertainties in using statistical N_2O emission models and data on non-land use GHG emissions due to biofuel production are significant, which can change the GHG emission reduction by between −152 and 87 pp [74].

While selecting the raw materials for biodiesel production, various parameters are considered, including oil content, suitability, chemical composition, and physical properties [75] (Table 3).

A brief description of various feedstocks used for biodiesel production with their oil content is summarized in Table 3.

Different studies were conducted to investigate the suitability of various feedstocks, for example, edible, non-edible oils, animal fats, and algal oils, for biodiesel production. The transformation of edible oil is biodiesel was considered the most feasible approach. As reported in Table 3, the biodiesel feedstocks such as olive oil and microalgae oil have the highest oil content, up to 70%, followed by rubber seed oil (up to 68.4%) and coconut oil (up to 65%). The lowest oil content was reported for soybean oil (15–20%). Edible oils such as sunflower, soybean, and rapeseed (Table 3) served as important substrates for biodiesel production. However, a vast disparity in food use affects the use of these first-generation feedstocks as fuel [76]. This will create a significant conflict with food vs. fuel, and competition with the food market can also adversely affect the price of biodiesel. The shift for non-edible oil such as castor oil, jatropha oil, and rubber seed oil was associated with the higher price of biofuel from edible oils because of their higher demand for food. Using raw materials from non-edible oils, animal fats, and waste oils has several advantages, including reducing the price of raw materials and avoiding competition with the food market [25,64].

In recent years, there has been significant interest in renewable and sustainable oils, and the life cycle assessment of raw materials plays a vital role in biodiesel production, it is essential to consider the oil content (%) and oil yield to determine the quality of biodiesel [65]. Additionally, microalgae are a great source of biodiesel production. These organisms can produce well-graded bioactive compounds by converting carbon dioxide (CO_2) with the help of sunlight [77,78]. With the increase in the price of petroleum and the concern with greenhouse gas emissions, microalgae have become an environmentally friendly alternative for biodiesel production. Though it is challenging for commercial-scale production, several companies have already started algal-based fuel production [77,78].

Similarly, animal fats, the byproducts of meat processing and cooking, are also important sources for biodiesel production. These include mutton or beef tallow, yellow grease, and lard, the residues after producing omega-3 fatty acids from fish oil [78]. Commercial-scale biodiesel production has been attained from animal fat-based feedstocks such as tallow, lard, and chicken fats. Unlike edible oils, animal fats-based biodiesel feedstocks have economic, environmental, and food security advantages. However, higher amounts of saturated fatty acids and free fatty animal fats demand complex production techniques. On the other hand, animal waste fats with lower saturated fatty acids have good oxidative stability, elevated calorific value, and shorter ignition [78,79]. Another important source of biodiesel feedstock is waste cooking oil. The waste cooking or frying oils include yellow and brown grease that does not directly conflict with food security. Yellow grease has < 15% fatty acid and can be used as a potential low-cost raw material for biodiesel production compared to brown grease (>15% fatty acid), which has an adverse effect on biodiesel production [79].

Feedstocks' chemical composition and physical properties are essential when selecting raw materials for biodiesel production. The chemical composition of different fatty acids from different sources is highlighted in Table 4. The differences in the degree of saturation and the carbon chain length are mainly due to the fatty acids of different architecture in the oil [66]. The degree of saturation from different sources is 14.7 % (soybean oil), 49.6% (palm oil, 6.1% (rapeseed oil), 21.6% (jatropha oil), 28.7% (used cooking oil), 46.9 % (animal fats), and 36.1% (algal oil) [68]. The percentage of carbon found at higher concentrations with $C \geq 18$ in most of the feedstock oils except for algal oil, which has only 33.1% compared to 85% (soybean oil), 55% (palm oil), 87.4% (rapeseed oil), 85.7% (jatropha oil), 73.1% (used cooking oil), and 68.9% (animal fats) [68]. This study compiled the fatty acid profile of different fatty acids from various sources, including edible and non-edible oil, animal fats, and other sources. The predominant fatty acids were monosaturated fatty acids, saturated fatty acids and polyunsaturated fatty acids, oleic acid (C18:1; 2.9–72.2), palmitic acid (C16:0; 1.3–48), and linoleic acid (C18:2; 1–70) (Table 4).

Table 3. Different sources of feedstocks/raw materials are used for the production of biodiesel [75].

Edible Oils	Oil Content (%)	Non-Edible Oils	Oil Content (%)	Animal Fats and Other Sources	Oil Contents (%)
Sunflower oil	25–35	[1] Jatropha oil	30–60	Mutton fat	-
Soybean oil	15–20	Stillingia oil	44.15	Broiler chicken waste	41 [80]
Rapeseed oil	38–46	[1] Karanja oil	27–40	Algae oil	20–60 [81]
Peanut oil	45–55	Neem oil	20–30	Waste cooking oil	33–53 [82]
Palm oil	30–60	[1] Castor oil	45–60	Microbial oil	23–70 [83]
Olive oil	45–70	Rubber seed oil	53.7–68.4	Waste fish oil	40–65 [84]
Mustard oil	40–42 [85]	[1] Mahua	35–40	Microalgae	30–70, 15–77
[1] Linseed oil	35–45	-	-	Pine and Kapok oil	-
Coconut oil	63–65	-	-	-	-
Canola oil	40–45	-	-	-	-

[1] represents feedstocks for biodiesel production reported by [86–88]; Ambat et al., 2018 [75] gathered information on sources of biodiesel feedstocks from different studies and reported them in their review paper.

Table 4. Fatty acid composition (%) of different biodiesel feedstocks.

Fatty Acid	Octanoic C8:0	Decanoic C10:0	Lauric C12:0	Myristic C14:0	Palmitic C16:0	Palmitoleic C16:1	Stearic C18:0	Oleic C18:1	Linoleic C18:2	Linolenic C18:3	Arachidic C20:0	Eicosenoic C20:1	Eicosapentaenoic C20:5	Behenaic C22:0	Erucic C22:1	others
Edible																
Soybean				0.1 [a]	6–11 [abc]	11 [a]	2–5 [abc]	20–30 [abc]	50–60 [abc]	5–11 [abc]						
Rapeseed					1–3.5 [bc]	9.1 [c]	0–1 [bc]	10–15 [b], 64.1 [c]	12–15 [b], 22.3 [c]	8–12 [b], 0.1 [c]		7–10 [b]			45–60 [b]	
Sunflower					5–8 [b]		2–6 [ab]	15–40 [ab]	30–70 [ab]	3–5 [b]						
Peanut					8–9 [b]		2–3 [b]	50–65 [b]	20–30 [b]							
Olive					9–10 [b]		2–3 [b]	72–85 [b]	10–12 [b]	0–1 [b]	0.3 [a]					
Palm				16.3 [a], 0.5–2 [b]	8.4 [a], 39–48 [bc]		2.4–6 [abc]	15.4 [a], 36–44 [bc]	2.4 [a], 9–12 [bc]		0.1 [a]					
Mustard							1–2 [b]	8–23 [b]	10–24 [b]	8–18 [b]					20–50 [b]	
Coconut			45–53 [b]	16–21 [b]	7–10 [b]		2–4 [b]	5–10 [b]	1–2.5 [b]			5–13 [b]				
Almond kernel					6.5 [e]		1.4 [e]	70.7 [e]	20 [e]	0.9 [e]						
Walnut kernel					7.2 [e]		1.9 [e]	18.5 [e]	56 [e]	16.2 [e]						
Sesame					13 [e]		4 [e]	53 [e]	30 [e]							
Non-edible																
Linseed					4–7 [b] 13.6–16.2 [b]		2–4 [b]	25–40 [b] 49.1–61.9 [b]	35–40 [b]	25–60 [b]						
Neem				0–0.1 [a], 14.1–15.3 [b]	14.1–15.3 [ac], 0–13 [b]											
Jatropha						0–1.3 [a]	3.7–9.8 [ac]	34.3–45.8 [abc]	14.1–15.3 [b], 29–44.2 [ac]	0–0.3 [ab]	0–0.3 [a]			0–0.2 [a]		1.4
Cotton seed					23–28.3 [b]		0.8–0.9 [b]	13.3–18.3 [b]		0.2 [b]						
Rubber				2.2 [f]	10.2 [f]		8.7 [f]	24.6 [f]	39.6 [f]	16.3 [f]						
Karanja					9.8 [a], 3.7–7.9 [f]		2.4–8.6 [af]	44.5–72.2 [af]	10.8–18.3 [af]							

Table 4. Cont.

Fatty Acid	Octanoic C8:0	Decanoic C10:0	Lauric C12:0	Myristic C14:0	Palmitic C16:0	Palmitoleic C16:1	Stearic C18:0	Oleic C18:1	Linoleic C18:2	Linolenic C18:3	Arachidic C20:0	Eicosenoic C20:1	Eicosapentaenoic C20:5	Behenic C22:0	Erucic C22:1	others
Pongamia				11.65 [f]				51.5 [f]	11.65 [f]	41.5 [f]						
Stillingia			0.4 [f]	0.1 [f]	7.5 [f]		2.3 [f]	16.7 [f]	31.5 [f]							
Animal fat and other sources																
Animal fats				2.52 [c]	28.4 [c]		15.7 [c]	42.2 [c]	9.4 [c]	0.6 [c]	0.16 [c]	0.86 [c]		0.01 [c]	0.01 [c]	
Chicken fats				3.1 [g]	19.82 [g]		3.06 [g]	37.62 [g]								
Used/waste cooking oil				0.9 [c]	20.4 [c], 8.5 [g]	4.6 [c]	4.8 [c], 3.1 [g]	52.9 [c], 21.2 [g]	13.5 [c], 55.2 [g]	0.8 [c], 5.9 [g]	0.12 [c]	0.84 [c]		0.03 [c]	0.07 [c]	0.04 [c]
Tallow				23.3 [f]	19.3 [f]		42.4 [f]	2.9 [f]	0.9 [f]	2.9 [f]						
Brown grease				1.66 [f]	22.83 [f]		12.54 [f]	42.36 [f]	12.09 [f]	0.82 [f]						
Microalgal	0.2 [d]			12–15 [g]	34.8 [d], 10–20 [g]	32 [d]	1.1 [d]	21.7 [d]	1.4 [d]				8.9 [d]			
Yellow grease				2.43 [fh]	23.24 [fh]		12.96 [fh]	44.32 [fh]	6.97 [fh]	0.67 [fh]						

The values of different fatty acids reported by different studies are represented by superscripts [a] [88], [b] [89], [c] [68,90–94]; microalgae species (*Nannochlopsis oculate*) [d] [95], [e] [66], [f] [96], [g] [75] and [h] [97].

4. Biodiesel and Its Properties

Biodiesel, also known as FAME, is produced by mixing methanol with vegetable oil, animal fat, or other triacylglycerol-carrying material. Differences in feedstocks significantly fluctuate the value of characteristics of FAME, including cloud point, Cetane number (CN), oxidative stability, saponification value, iodine value, and acid value [88]. The main physicochemical properties of biodiesel obtained from various feedstock/raw materials are discussed below (Table 5).

Table 5. Physicochemical properties of different biofuel feedstocks.

Sources	CP (°C)	CN	OS (mg/100 mL)	SV	IN	AV (mg KOH/g oil)
Soybean oil	0.9	47	16.0	189–195	117–143	0.1–0.2
Canola oil	−3.3	55	44.9	188–193	109–126	0.6–0.8
Olive	-	-	-	184–196	75–94	0.94–2.11
Corn	-	-	-	187–198	103–140	0.1–5.75
Jatropha curcas	5.66	55.43	-	177–189	92–112	15.6–43
Palm oil	14.24	60.21	-	186–209	35–61	6.9–50.8
Rapeseed	-	-	-	168–187	94–129	0.2
Sunflower	-	-	-	186–194	110–143	0.2–0.5
Camelina	2.5	48.91	-	-	146.5	0.2
Poultry fat	-	-	-	-	78.8	0.55
Choice white grease	7.0	64	72.0	-	-	-
Inedible tallow	16.0	62	6.2	-	-	-
Yellow grease	6.0	58	2.3	-	-	-
Ultra-low sulfur diesel (ULSD)	−45 to −7	47	-	-	-	-

Cloud point (CP), cetane number (CN), oxidative stability (OS), saponification value (SN), iodine number (IN), and Acid value (AV). Values shaded with green are adopted from [88], blue from a study by [75], and not highlighted text black are from the U.S. Energy Information Administration (EIA), compiled from the U.S. Department of Energy, National Renewable Energy Laboratory, and Renewable Energy Group.

4.1. Cloud Point

The cloud point (CP) is the minimum temperature below which wax begins to form crystals in fuels, resulting in a cloudy appearance [98]. Solidified waxes can clog engine fuel filters and injectors. Biodiesel has higher CP due to the high melting points of saturated fatty acids compared to unsaturated fatty acids [88]. Biodiesel produced from feedstocks such as inedible tallow and waste frying oil may require additives or blend at higher levels with lower cloud point ULSD to mitigate cold weather concerns.

4.2. Cetane Number

The cetane number (CN) represents the ignition behavior and quality of the fuel. Higher cetane is often associated with improved performance and a cleaner burning fuel [99]. Most biodiesel feedstocks have slightly higher cetane numbers than ultra-low sulfur diesel (ULSD), which usually has a minimum allowable cetane value of 40 (https://www.eia.gov/todayinenergy/detail.php?id=36052. Accessed on 3 May 2018). The CN value of biodiesel increases with the length of the fatty-acid chain and the degree of saturation; hence, a higher CN means a higher oxygen concentration in the biodiesel and a better combustion efficiency [98]. Studies reported the highest CN value of 70 for Spirulina platensis [100] vs. the lowest CN value of 34.6 for biodiesel obtained from linseed oil [101,102]. The raw materials and feedstocks reported in this study have a CN value range between 47 for soybean oil and 64 for choice white grease (Table 5).

4.3. Oxidative Stability

Oxidative stability is the ability of the fuel to resist oxidation during storage and use. This essential factor significantly influences the storage duration and condition [103]. Fuels with lower oxidative stability are more likely to form peroxides, acids, and deposits that adversely affect the engine performance. Because it generally has lower oxidative stability, petroleum diesel can be stored longer than biodiesel feedstocks such as white grease and tallow. Biodiesel producers may use additives to extend the storage and usage timelines of Biodiesel (Source: EIA). Biodiesel with high oxidative stability is highly susceptible to oxidation deterioration. Oxidative stability varies according to fatty acid composition [104]. The fuel's oxidative stability is greatly affected by polyunsaturated FAME. For example, Camelina-oil-based Biodiesel has low oxidative stability because it has approximately 35% polyunsaturated FAME occurrence (i.e., α-linolenic [C18:3]) [105] compared to coconut-oil-based biodiesel, which has better oxidative stability due to 2% polyunsaturated FAME in its oil [88]. Oxidative stability reported in this study ranges from 2.3 mg/100 mL (yellow grease) to 44.9 mg/100 mL (Canola oil) (Table 5).

4.4. Saponification Value

Saponification value (SV) is an index of the molecular weights of triglycerides in the oil. It is inversely proportional to the average molecular weight or the chain length of the fatty acids [106]. Thus, the shorter the chain length, the higher the SV of the oil. The expected SV should range between 195 and 205 mg/KOH/g of oil [107,108]. Any value below that value needs refining to meet the required standard and would be better fitted for an industrial purpose [109]. The SV reported in this study is comparable and lies close to the required range of 195–205 (Table 5).

4.5. Iodine Number

Iodine number (IN) represents the amount of iodine absorbed by double bonds of the FAME molecules in 100 g of the fuel sample. A higher iodine value indicates higher fats and oils [110,111]. In the case of biodiesel fuels, linseed methyl ester showed the highest IN of 178 compared to the lowest IN of 37.59 reported for Kusum-oil-based Biodiesel [102,112]. This study reported the lowest IN for Palm oil (35–61) vs. the highest value of IN for Camelina oil (146.5) (Table 5).

4.6. Acid Value

The acid value represents the fuel sample's quantity of free fatty acids. A high acid number causes corrosion problems in the engine's fuel delivery system [112]. A high acid value of 6.9–50.8 mg KOH/mg of oil is reported for biodiesel from palm oil compared to the lowest acid value of 0.1–0.2 mg KOH/mg of oil from soybean oil (Table 5). Further, descriptions of additional fuel properties of biodiesel from different generation oil feedstocks are reported in our previous study [88].

5. Procedures for Biodiesel Production

Different physicochemical processes could produce biodiesel, and the primary methods include pyrolysis, micro-emulsion, and transesterification [113]. Each method has its merits and demerits. For example, micro-emulsion is a simple and environmentally safer method that generates fewer pollutants. Biodiesel synthesized using this method has a good cetane number (CN). Similarly, alcohol in the micro-emulsion process improves the CN of Biodiesel [114]. Microemulsion-based fuel systems reduce the combustion temperature, which leads to lower emissions of thermal NO_x, CO, black smoke, and particulate matter. However, one major problem of using ethanol to formulate a microemulsion system is its lower miscibility with diesel. The immiscibility can be visualized for a wide range of temperatures, particularly at lower temperatures [115,116]. Furthermore, environmentally benign bio-based non-ionic surfactants and cosurfactant without N and S are of environmental concern [117,118]. Biodiesel production from the pyrolysis method (also known as thermal cracking) has low CN, volatility, and high viscosity [21]. By comparing these methods, the transesterification method is reliable and effective because the transesterification method demands low temperature, low pressure, and less processing time. The transesterification method is simple and highly efficient [119]. A description of various procedures to generate biodiesel is highlighted below.

5.1. Micro-Emulsion

This method uses isotropic fluid to form a colloidal dispersion of dimensions ranging from 1 to 150 nm. A study using soybean oil has already demonstrated that by using this method, maximum viscosity was achieved that involves both ionic and non-ionic aqueous solutions [120,121]. A study revealed that using a ternary phase system (a clear and thermodynamically stable, isotropic liquid mixture of oil, water, and surfactant) counters the viscosity problems of vegetable oils by forming micro-emulsions with different solvents (ethanol, methanol, propanol, n-butanol, and hexanol). These alcohols act as emulsifying agents, dispersing the oil into tiny droplets, usually with diameters ranging from 100 to 1000 Å [122].

5.2. Pyrolysis

Thermal cracking or pyrolysis converts organic materials to fuels without oxygen using thermal decomposition (temperature: 300–1300 °C) [122]. Chemically, pyrolysis reaction cleaves the bonds in a substance, converting it into many smaller compounds. The process is similar to the process used to synthesize petroleum-diesel; therefore, it yields a product with similar combustion characteristics and results in less waste formation and no pollution [123,124].

The substrate used for pyrolysis includes vegetable oils, animal fats, natural fatty acids, or methyl esters of fatty acids. It sometimes produces a higher yield than the transesterification reaction, which is the most widely used [122]. The pyrolysis of organic feedstock for the manufacture of synthetic diesel has yet to be viable on an economic scale [124]. Based on operating parameters, pyrolysis can be divided into three types, namely conventional pyrolysis (550–900 K), fast pyrolysis (850–1250 K), and flash pyrolysis (1050–1300 K) [124]. The pyrolysis of biomass for bio-oil generation can be performed using both conventional and flash pyrolysis. In conventional pyrolysis, the vapor residence time ranges from 5 to 30 min, and thus this contributes to overall reaction time. Depending upon residence time, the vapors can be removed continuously. Whereas in flash pyrolysis, the heating rate is predominantly high. Some of the prerequisites for flash pyrolysis include a high heat transfer rate, finely grounded materials, and short vapor residence times (<2 s) [125]. The product obtained from pyrolysis has desired characteristics of biodiesel, such as low viscosity, less amount of sulfur and water, and high cetene number; however, it has less ash and residual carbon content than the desirable amount [123,124,126,127].

5.3. Transesterification

Transesterification is a standard and widely used procedure for high-quality biodiesel production [128]. This procedure involves the transformation of fats or oils using alcohol, particularly methanol or ethanol, with the help of catalysts (e.g., heterogeneous, homogeneous, or enzyme) [129,130]. Compared to the transesterification process facilitated by enzymes, the process is energy-consuming because of the presence of soap byproducts, and separation and purification of the chemically produced biodiesel require more complex steps than enzymatically produced biodiesel [131]. Ethanol is cost-effective and abundant commodity obtained from the fermentation of sucrose from sugarcane. Propanol or butanol could be a better option because these two alcohols promote better miscibility between the alcohol and the oil phases [132]. Transesterification can be combined with ultrasound-assisted member technology [25,66,120]. There are merits and demerits of using various biodiesel production technologies based on several studies (Table 6). However, these production technologies were centered on reducing problems during biodiesel production, such as oil's high viscosity, acid value, and fatty acid content [97,133]. Among those technologies, transesterification using a homogeneous catalyst was the most typical and commercially used technology [134]. From an environmental point of view, enzyme catalysts and heterogeneous catalysts are suitable options for the future [75].

Table 6. Merits and demerits of using various biodiesel production technologies [27,97,133,135].

Production Technologies	Merits	Demerits
Micro-emulsion	Micro-emulsion is a simple process, a potential solution for solving the problem of vegetable oil viscosity [136]. It is the dispersion of water, oil, and surfactant. Alcohols such as methanol and ethanol are used to lower viscosity, higher alcohols are used as surfactants, and alkyl nitrates are used as cetane improvers [137]. Micro-emulsion is an alternative method that produces biofuel with suitable properties with low energy consumption [138].	Some of the disadvantages of micro-emulsion include high viscosity, poor stability, and volatility. Therefore, pre-treatment technology such as cracking, blending, and hydrodeoxygenation is required to minimize the viscosity and FFAs content before producing biodiesel [138].
Pyrolysis	Pyrolysis is a simple and pollution-free process. The product from pyrolysis has a lower viscosity, flash point, and pour point than petroleum diesel; however, it has equivalent calorific values and a lower value of cetane number. Thus, pyrolyzed vegetable oil has an acceptable amount of sulfur, water, sediment, and copper corrosion values [139]. A study suggested that pyrolytic oil, also known as bio-oil, derived from non-edible feedstock such as Jatropha, Castor, Kusum, Mahua, Neem, and Polanga, has drawn interest to be used as an alternative biofuel. The advantages of using pyrolytic bio-oil are that it is easy to handle, store, and transport and has a high cetane number, low viscosity, and low sulfur quantities [138,140].	The bio-oils derived from edible and non-edible plant seeds are acidic. They are denser than petroleum diesel fuel and thus require a pre-treatment process to remove moisture and neutralize prior to use as an alternative biofuel [138,141]. The disadvantages of pyrolysis include high temperature, expensive apparatus, and low purity due to intolerable amounts of carbon residue and clinker [138,141].
Transesterification	The transesterification process has several advantages over the biodiesel synthesis methods, which include eco-friendly, mild chemical reactions, and are suitable for biodiesel feedstock. It effectively reduces moisture, FFAs, and viscosity during producing biodiesel from non-edible oil [138,142].	The type of catalyst used will determine the conversion efficiency, reusability, cost, and applicability of feedstocks with water and high fatty acid content. The enzymes used during the process are costly, and the reaction is time-consuming [4].
Catalytic distillation	Catalytic distillation is a green reactor technology that integrates chemical reactions and product separation into a single operation. This method simultaneously carries out the chemical reaction and product separation within a single-stage operation. The continuous removal of the product from the reactive section via distillation action can lead to increased product yield and enhanced productivity. Catalytic distillation has several advantages, such as mitigating catalyst hot spots, better temperature control, and improved energy integration due to the conduction of an exothermic chemical reaction in a boiling medium. Recent studies show that catalytic distillation is a novel approach to biodiesel production, which is more efficient and cost-effective [143].	The conversion process and solvent usage for post-treatment depend on catalyst recovery.

Table 6. *Cont.*

Production Technologies	Merits	Demerits
Dilution	Dilution is a simple process that results in a reduction in the viscosity and density of vegetable oils. A study revealed that adding 4% ethanol to diesel fuel increases the brake thermal efficiency, brake torque, and power [144]. Another study reported that blending non-edible oil with diesel fuel increases the storability, potential improvement of physical properties, and engine performance. Additionally, dilution reduces poor atomization and difficulty handling by conventional fuel injection systems of compression ignition engines [55].	The issues with blending include the formation of carbon in the engine and incomplete combustion.
Microwave technology	The electromagnetic waves generated in the microwave through electric energy transfer energy directly at the molecular level, allowing quick reaction activity and better energy transfer [135]. The catalyst (homogeneous or heterogeneous) in microwave radiation lowers microwave power usage while keeping the reaction equilibrium and achieving transesterification at very low input power with a very fast conversion rate [53]. The high input power can directly degrade oils into different byproducts. Thus, controlling the radiation level is vital to achieving a complete transesterification reaction.	Removal of the catalyst after the process is needed, and process conversion depends on catalyst activity and is not appropriate for solid feedstocks.
Reactive distillation	Reactive distillation offers new and exciting opportunities for manufacturing fatty acid alkyl esters in the industrial production of biodiesel and specialty chemicals. The processes can be enhanced by heat integration and powered by heterogeneous catalysts to eliminate all conventional catalyst-related operations by efficiently using raw materials and reaction volume. At the same time, reactive distillation offers higher conversion, selectivity, and high energy savings [145]. This method combines the reaction and separation stages in a single unit, thereby reducing the capital cost and increasing heat integration [25]. Overall, this method is applicable with feedstock with high FFAs content, simple process, less use of methanol, and easy to separate product.	However, it requires high energy, and process conversion depends on catalyst efficiency.
Supercritical fluid method	In the supercritical fluid method, the reaction is carried out at supercritical conditions. The mixture becomes homogeneous, where both the esterification of free fatty acids and the transesterification of triglycerides occur without needing a catalyst, making the process suitable for all types of raw materials. The combination of two stages has attracted research interest recently, where simultaneous extraction and reaction from solid matrices are carried out using methanol with supercritical CO_2 as a co-solvent [25]. This method involves less reaction time, high conversion, and no catalyst required.	This method demands a high cost of apparatus and energy consumption.

6. Factors Affecting Biodiesel Production

Biodiesel production using biomass feedstock is influenced by several factors described below.

6.1. Free Fatty Acids

Free fatty acids affect biodiesel production. The higher amount of free fatty acid leads to soap and water formation [146]. The slow rate of acid-catalyzed reaction requires low-temperature conditions [147]. Base-catalyzed transesterification reactions demand raw materials with low acid value (<1) and free from water [148]. With 3% free fatty acids, there is no need to use a homogeneous base catalyst during the transesterification reaction [149].

6.2. Water Content

The amount of water content in the feedstock accelerates the hydrolysis and lowers the formation of ester [150]. A study has revealed that for a 90% biodiesel yield for an acid-catalyzed reaction, the water content should be less than 0.5% [151]. Additionally, water obtained as a byproduct inhibits the reaction and decreases engine performance. However, water in oil can be removed by preheating it up to 120 °C or by using anhydrous sodium sulfate or anhydrous magnesium sulfate [152].

6.3. Types of Alcohol

Methanol is used for biodiesel production for a higher conversion rate from waste cooking oil with lower viscosity and is cheaper than other alcohol-based biofuels [153]. However, it is more toxic [154] and causes enzyme deactivation, denaturation, or inhibition at higher concentrations [155]. In order to address these issues, ethanol is used in most enzymatic reactions [153].

6.4. Alcohol to Oil Ratio

In order to obtain one mole of alkyl ester, 3 mol of alcohol and 1 mol of triglyceride are needed [156]. The rate of biodiesel production increases with higher alcohol concentration, i.e., increasing the alcohol-to-oil ratio [157]. The maximum conversion with 99% biodiesel production was achieved from waste sunflower oil transesterification using methanol and NaOH as the catalyst, with an alcohol-to-oil ratio of 6:1 [158,159], compared to 49.5% yield in waste canola petroleum using 1:1 methanol to oil [158].

6.5. Reaction Time

Reaction time plays a significant role in product conversion. Suppose more time is needed to give to the reaction. In that case, some parts of the oil may remain unreacted and ultimately reduce ester yield and exceed reaction time than usual, affecting the end product and leading to soap formation [160]. The reaction time for lipase-catalyzed reactions differs from 7 to 48 h [161]. Studies also suggested that reaction time also controls production costs. A study found no significant change in the conversion of biodiesel with the reaction time of 1 h (96.10%) versus 3 h (96.35%) [162]. However, a longer reaction time may lead to the reduction in biodiesel due to reversible transesterification reaction resulting in loss of esters and soil formation. Thus, reaction time needs to be optimized to bring the production cost down to a minimum. Maximum ester conversion can be achieved within <90 min.

6.6. Reaction Temperature

High temperatures lead to lower oil viscosity, resulting in a high reaction rate and reduced reaction time. However, if the temperature increases beyond the desirable range, the biodiesel yield is lowered due to the saponification of triglycerides accelerated by high temperature [163]. Biodiesel viscosity improves as the reaction temperature falls below 50 °C. For waste cooking oil, it is necessary to pre-heat up to 120 °C and cool down to 60 °C [164]. Higher reaction temperature increased the reaction rate and shortened the reaction time due to the reduction in the viscosity of oils. For the esterification reaction, the temperature should be below the boiling point of alcohol to prevent alcohol evaporation [165,166]. The highest conversion was achieved for cottonseed oil at 50 °C and Jatropha oil at 55 °C using lipase as a catalyst [167]. Further, the maximum yield of biodiesel was reported at 65 °C for domestic and commercial (waste and fresh) oils using KOH as a catalyst [162].

6.7. pH

Though pH is not crucial for acid/base catalysts, for lipase catalysts, pH plays an important role; for example, the enzyme may decompose at higher or lower pH. For example, a study found that a pH of 7 is optimal for biodiesel production using Jatropha oil-immobilized *Pseudomonas fluorescence* [168].

6.8. Catalyst Concentration

The most commonly used catalyst for biodiesel production is sodium hydroxide (NaOH) or potassium hydroxide (KOH) [165], and other catalysts used are sodium methoxy and potassium methoxide [169]. Increasing the catalyst concentration with oil samples also increases the conversion of triglycerides into biodiesel. However, it also increased soap formation. Lowering the amount of catalyst leads to incomplete conversion into fatty acid ester, resulting in lower methyl esters yield [166]. Optimum biodiesel production is achieved when the concentration of NaOH reaches 1.5% weight [93]. Again, using an excess amount of catalyst can have a negative impact on biodiesel yield [93,170]. For soybean oil biodiesel, a 1.5 % copper vanadium phosphate (CuVOP) concentration was found to be the most effective [171].

6.9. Agitation Speed

Agitation is mandatory for the reaction, and its speed is essential for product formation. Lower agitation speed cause less product formation. Lower agitation speed cause less product formation. However, higher agitation speed favors soap formation [166]. There should be an optimum stirrer speed, which varies with our feedstocks. A study revealed a stirrer speed of 200 mm found to be optimum for biodiesel production using enzymatic reactions [172]. However, another study reported that at 400 rpm, there was a higher conversion of end product compared to 200, 600, and 800 rpm for an hour [165].

7. Catalyst Use for Biodiesel Production

Biodiesel is fatty acid methyl esters (FAME) with lower alkyl esters and long-chain fatty acids. It is synthesized by two procedures: esterification of fatty acids and transesterification with lower alcohol. Even without a catalyst, transesterification reactions can happen. However, they demand high temperatures, pressure, and reaction time. It also increases the overall cost of the reaction process [173]. The biodiesel thus produced has high purity of ester and glycerol (soap-free); however, from a commercial scale standpoint, it is imperative to use catalysts. Hence, there are three different catalysts: acidic, alkaline, and enzyme [174].

7.1. Acidic Catalysts

Acidic catalysts support higher efficiency for the esterification of FFAs over alkaline catalysts, with up to 90% conversion [175]. These catalysts favor feed oil with high acid value (including edible waste oil) and have good potential for transesterifying low-quality feeds [3]. Transesterification is performed at high temperatures (100 °C), pressure (~5 bar), and a high amount of alcohol. However, the process is slower compared to alkaline catalysts [3]. The most commonly used acid catalysts are sulfuric acid, hydrochloric acid, organic sulfonic acid, sulfonic acid, and ferric sulfate [75].

7.2. Alkaline Catalysts

Alkaline catalysts for biodiesel production include NaOH, KOH, alkaline metal carbonate, sodium and potassium carbonates, sodium methoxide, and sodium ethoxide. These catalysts are appropriate for oil with low FFAs due to the sensitivity as oils with higher FFAs, are converted to soap rather than biodiesel. This process restricts the separation of glycerin, biodiesel, and water. In order to cope with the issue, a deacidification step is necessary before the transesterification of vegetable oil [3].

7.3. Enzyme Catalysts

Enzymes such as lipases from microorganisms act as catalysts during transesterification reactions [176]. Lipase enzymes are abundant in nature and are synthesized by microorganisms (fungi, bacteria, and yeast), plants (rapeseed, oat, papaya, latex, and caster seeds), and animals (cattle, pigs, hogs, and pancreases of sheep) [177]. During biodiesel production, no or little residual or soap is formed at the end, resulting in high-quality glycerol production. This is also useful for feedstocks with high acidic values. Some limitations of using enzyme catalysts are high concentration and long reaction time. Separating the final product from the reaction results in a high cost of biodiesel production [3]. Further, applying metagenomics in enzyme technology opens the door for developing stable and solvent-tolerant biocatalysts for biodiesel production [178].

7.4. Homogeneous Catalysts

Homogeneous catalysis involves a series of reactions involving a catalyst from the same phase as the reactants, whether in the liquid or gaseous state. A homogeneous catalyst is dissolved or co-dissolved in the solvent with all the reactants [166]. Sodium hydroxide (NaOH) or potassium hydroxide (KOH) is the most popular homogeneous catalyst for biodiesel production [179]. Homogeneous catalysts are acidic and basic and widely used for biodiesel production. Acid catalysts are less active than base catalysts (i.e., lower reaction time). Therefore, a base catalyst involves high temperature and pressure. When FFAs exceed 1% in the oil, acid catalysts become effective. Acid catalysts prevent soap from forming. These catalysts catalyze the esterification of FFAs to form FAME and thus enhance biodiesel production [75,180]. The deep eutectic solvents (DESs) with acidic nature were evaluated for biodiesel production and found to have over 90% conversion efficiency [3]. Alkaline catalysts react with alcohol to form alkoxide and protonated catalysts. The carbonyl atom of the triglyceride molecule is attacked by nucleophilic alkoxide to form a tetrahedral intermediate, which reacts with alcohol to revive the anion. Further, the tetrahedral structure undergoes structural reorganization to form a fatty acid ester and diglyceride [66,181]. The higher conversion rate is obtained at low temperatures and pressure, resulting in lower production costs of biodiesel [3]. The alkaline catalysts are less efficient than acidic catalysts for converting oils containing high FFAs, producing soap, and inhibiting the separation of ester and glycerin. Thus, acid catalysts are recommended for biodiesel production [75].

7.5. Heterogeneous Catalysts

Catalysts with a state or phase different from reactants are heterogeneous catalysts. Most of the heterogeneous catalysts are solid. However, reactants are either in liquid or gaseous forms [166]. The separation process in heterogenous catalysts is easy and aids faster recycling and reuse than homogeneous catalysts. Therefore, it resolves problems related to homogeneous catalysis while lowering the material and processing costs [25,120,182–185]. Heterogenous catalysts can also tolerate high FFA and moisture content [186]. These catalysts, even at severe reactions conditions, can recover from a reaction mixture, stand up to aqueous treatment steps, and can be easily modified to achieve a high level of activity, selectivity, and long lifetime. Solid base heterogeneous catalysts include hydrotalcite, metal oxides (CaO, MgO, or SrO), oxides of mixed metals (Ca/Mg, Ca/Zn), alkali metal oxides (Na/NaOH/γ-Al$_2$O$_3$, K$_2$CO$_3$/Al$_2$O$_3$, magnetic composites, and alkali-doped metal oxides (MgO/Al$_2$O$_3$, CaO/Al$_2$O$_3$, Li/CaO) [187]. However, some limitations of using heterogenous catalysts include diffusion due to phase separation between alcohol and oil, low surface area, and leaching. Strategies to resolve these issues include using n-hexane and tetrahydrofuran as co-solvents, increasing the area of specific activities, and providing more pores for reactive components. This can be possible with the help of supporters for catalysts as well as promoters for its structure [25,120]. The study also suggested that through immobilization or in the liquid phase, higher biodiesel yield can be obtained with robust lipase enzymes (how lipase technology contributes to the evolution of biodiesel production using multiple feedstocks).

A clear distinction between acid versus alkali and homogeneous versus heterogeneous catalyzed transesterification reactions is shown in Tables 7 and 8.

Table 7. Comparison of acid versus alkali-catalyzed transesterification process of biodiesel production, reported by [27].

Transesterification Process	Merits	Demerits
Acid-based catalyzed reaction	Suitable in the presence of high levels of FFA and water. No need for pretreatment. Fewer environmental problems and less toxic effect. Few main processing units.	Slow reaction. High temperature, pressure, and alcohol/oil ratio. Environmental contamination. Required costly equipment.
Alkali-based catalyzed reaction	Low temperature, pressure, and alcohol/oil ratio. High reaction rate. Smaller equipment, good corrosion resistance properties. Low cost of catalyst.	Need of pretreatment. Low ester yields and byproducts without pretreatment. Saponification occurs.

Table 8. Comparison of homogeneous versus heterogeneous-catalyzed transesterification process of biodiesel production, reported by [27].

Factors	Homogeneous Catalysis	Heterogenous Catalysis
Reaction rate	Fast and high conversion	Moderate conversion
Post-treatment	No recovery of catalyst	Catalysts can be recovered
Processing methodology	Mild reaction and less energy consumption	Continuous operation possible
Process of water and FFA	Sensitive and not suitable	Not sensitive and suitable
Reuse of catalyst	Not possible	Possible
Cost	Comparatively cost-effective than the currently available heterogeneous catalyzed transesterification	Potentially cheaper, high conversion efficiency, and technologically available

Overall, the reusability and recyclability are complex in homogeneous catalysts, whereas heterogeneous catalysts offer efficient, yielding results and can be reused again. Nanocatalysts that come under the heterogeneous catalyst group reveal better yield due to the large surface area at the Nanoscale and are preferable for biofuel reproduction with the help of transesterification [4]. Homogeneous catalysts are also considered fuel performance catalysts due to their ability to improve fuel efficiency and reduce smoke emissions, unburned hydrocarbons, and carbon monoxide.

8. Evaluation of Greenhouse Gas Emissions from Biodiesel Production

Overall, biodiesel reduces GHG emissions of carbon monoxide (CO), carbon dioxide (CO_2), unburnt hydrocarbon (UBHC), and particulate matter (PM) to a significant extent, except nitrogen oxide (NO_X), compared to diesel. These emission gases are the primary causes of atmospheric pollution and human health [71].

Carbon monoxide reduction from different biodiesel feedstock range from 9.4% (microalgae) to 63% (palm oil) compared to CO emissions from diesel [188]. Several studies have shown the different proportions of CO emission reduction as engine speed increases. For example, CO emission reduction for soybean biodiesel was reported at 14% at 1400 rpm, 27% at 2000 rpm [189], and 37% at 3600 rpm engine speed [90]. Carbon monoxide reduction using rapeseed oil was 29.7% [91] and 26% [90]. Similarly, CO emission from jatropha oil ranged from 14 to 30% [190–192], waste cooking oil ranged from 17.8 to 20% [91,193], animal fats from 26% [90], and microalgae ranged from 9.4 to 32% [194–196].

Carbon dioxide emissions in some biodiesels are almost the same or even higher than the regular diesel [68]. For example, CO_2 emissions from soybean oil (SO) and used/waste cooking oil (UCO)-based biodiesel were 60% and 33% more compared to regular diesel [90]. However, another study reported an increase in CO_2 emission from SO and UCO by 1.8% and 1.2%, respectively. Similarly, palm, rapeseed oil [90], and jatropha oil [191] generated 41%, 32%, and 3% more CO_2 emissions than regular diesel. Animal fats and microalgae biodiesels emitted 3% and 2.6% more CO_2 emissions than regular diesel [197].

Nitrogen oxide emissions from biodiesels are more than diesel, except for palm oil and microalgae-based biodiesel. Studies have shown that biodiesel with long-chain fatty acids produced fewer NOx emissions than short-chain fatty acids. On the contrary, NOx emissions increased as the number of double bonds, i.e., the degree of saturation of fatty acids, increased [68]. NOx emissions from soybean biodiesel increased due to its highest degree of saturation (14.7%) and 85.3% of the chain. NOx emission was lowered for palm-based biodiesel due to more short chains and a high degree of saturation than other biodiesels [68].

PM emissions from biodiesel are lowered compared to diesel. PM emissions from SO biodiesel decreased from 56% [90] to 69% [198], palm oil by 50% [91], rapeseed oil by 36% [90] to 70% [198], jatropha oil by 11% [191] to 15% [192], and used/waste cooking oil by 17% [90]. Similarly, PM emissions from animal fats-based biodiesel decreased by 61% [99] to 77% [198] and microalgae by 31% [196].

Overall, soybean and animal fats-based biodiesel produced the lowest PM, jatropha oil-, animal fats-, and microalgae-based biodiesel produced the lowest CO_2 emissions. Palm oil-based biodiesel produced the lowest CO and NOx emissions [68].

9. Conclusions and Recommendations

The demands for fossil fuels are gradually increasing due to the improvement in technology (for example, urbanization and improved life standard), which also requires more fuels. This increasing use of energy reserves will decrease fossil fuels in the future. A rapid increase in population and associated energy demand cannot be fulfilled by using fossil fuels alone. Using first-generation crops such as soybean and corn as bioenergy creates conflict in the food versus energy debate. Likewise, second-generation crops, particularly grasses, are unsuitable for biodiesel production. One of the significant problems in using second-generation vegetable oil is that it lessens engine life if the oil is not refined correctly. These issues of using first- and second-generation biofuels, such as economic, social, and food insecurity [48], can be resolved using third and fourth-generation biofuels. Third and fourth-generation biofuels are generated from various types of algae, which is highly efficient, and algal-based biofuels have great potential and no competition for food or land. In recent times, fourth-generation biofuels have great promise to overcome the inherent flaws and meet the world's growing energy demands. Though algal cultivation is simple, feedstock production is complex due to high lipid content, and harvesting needs should be addressed. Detailed work on the parameters for fuel compatibility is required. Many things need to be worked out to make an algal biofuel a commercially viable option to fossil fuel, as the production of biofuels from microalgae is an energy-intensive process [199]. Further, greenhouse gas emissions are much lower; mainly, there is no emission of CO or CO_2 using this generation of biofuels. Thus, these fuels could be potential options to replace fossil fuels. It is also recommended to consider the potential benefits of using other resources for energy sources that are more cost-effective, climate resilient, and sustainable. This could reduce the burden on fossil fuels in the future.

Author Contributions: Conceptualization and writing, D.N., writing—review and editing, compilation, and supervision, D.N. All authors have read and agreed to the published version of the manuscript.

Funding: There is no external funding (except the reviewers' vouchers) for this publication.

Institutional Review Board Statement: Not applicable.

Informed Consent Statement: Not applicable.

Data Availability Statement: Not applicable. There is no primary data used in this study. However, secondary data that supports this study are available from the corresponding author (D.N.) upon reasonable request.

Acknowledgments: The author would like to acknowledge Eric Olson, an Associate Professor of the Department of Mathematics and Statistics at the University of Nevada, Reno, for critical reading of the manuscript. The author would also like to acknowledge all individuals, including anonymous reviewers and editors, for their valuable comments and suggestions to improve the quality of this article.

Conflicts of Interest: The authors declare no conflict of interest. The funders had no roles in the design of the study; in the collection, analysis, or interpretation of data; in the writing of the manuscript, or in the decision to publish the results.

Abbreviations

GJ: gigajoules; GHGs: Greenhouse gases; CO_2: carbon dioxide; NO_X: nitrogen oxides; CH_4: methane; FAME: fatty acid methyl ester; EIA: energy information administration; AF: animal fats; UCO: used cooking oil; UBHC: unburnt hydrocarbon; FFA: free fatty acid.

References

1. Neupane, D.; Adhikari, P.; Bhattarai, D.; Rana, B.; Ahmed, Z.; Sharma, U.; Adhikari, D. Does Climate Change Affect the Yield of the Top Three Cereals and Food Security in the World? *Earth* **2022**, *3*, 45–71. [CrossRef]
2. Mehmood, M.A.; Ibrahim, M.; Rashid, U.; Nawaz, M.; Ali, S.; Hussain, A.; Gull, M. Biomass production for bioenergy using marginal lands. *Sustain. Prod. Consum.* **2017**, *9*, 3–21. [CrossRef]
3. Maheshwari, P.; BelalHaider, M.; Yusuf, M.; Klemeš, J.J.; Bokhari, A.; Beg, M.; Al-Othman, A.; Kumar, R.; Jaiswal, A.K. A review on latest trends in cleaner biodiesel production: Role of feedstock, production methods, and catalysts. *J. Clean. Prod.* **2022**, *335*, 131588. [CrossRef]
4. Singh, A.R.; Singh, S.K.; Jain, S. A review on bioenergy and biofuel production. *Mater. Today Proc.* **2022**, *49*, 510–516.
5. Alengebawy, A.; Mohamed, B.A.; Ghimire, N.; Jin, K.; Liu, T.; Samer, M.; Ai, P. Understanding the environmental impacts of biogas utilization for energy production through life cycle assessment: An action towards reducing emissions. *Environ. Res.* **2022**, *213*, 113632. [CrossRef]
6. Edrisi, S.A.; Abhilash, P. Exploring marginal and degraded lands for biomass and bioenergy production: An Indian scenario. *Renew. Sustain. Energy Rev.* **2016**, *54*, 1537–1551. [CrossRef]
7. BP. BP, 2022. Statistical Review of World Energy-August 2022. Available online: https://www.bp.com/en/global/corporate/energy-economics/statistical-review-of-world-energy/primary-energy.html (accessed on 21 July 2022).
8. Yusuf, M.; Beg, M.; Ubaidullah, M.; Shaikh, S.F.; Keong, L.K.; Hellgardt, K.; Abdullah, B. Kinetic studies for DRM over high-performance Ni–W/Al2O3–MgO catalyst. *Int. J. Hydrog. Energy* **2021**, *47*, 42150–42159. [CrossRef]
9. Seffati, K.; Honarvar, B.; Esmaeili, H.; Esfandiari, N. Enhanced biodiesel production from chicken fat using $CaO/CuFe_2O_4$ nanocatalyst and its combination with diesel to improve fuel properties. *Fuel* **2019**, *235*, 1238–1244. [CrossRef]
10. Bacenetti, J.; Negri, M.; Fiala, M.; González-García, S. Anaerobic digestion of different feedstocks: Impact on energetic and environmental balances of biogas process. *Sci. Total Environ.* **2013**, *463*, 541–551. [CrossRef]
11. Fatima, Z.; Ahmed, M.; Hussain, M.; Abbas, G.; Ul-Allah, S.; Ahmad, S.; Ahmed, N.; Ali, M.A.; Sarwar, G.; Iqbal, P. The fingerprints of climate warming on cereal crops phenology and adaptation options. *Sci. Rep.* **2020**, *10*, 1–21. [CrossRef]
12. Sharma, R.K.; Kumar, S.; Vatta, K.; Bheemanahalli, R.; Dhillon, J.; Reddy, K.N. Impact of recent climate change on corn, rice, and wheat in southeastern USA. *Sci. Rep.* **2022**, *12*, 1–14. [CrossRef] [PubMed]
13. Iqbal, F.; Abdullah, B.; Oladipo, H.; Yusuf, M.; Alenazey, F.; Nguyen, T.D.; Ayoub, M. Chapter 18—Recent Developments in Photocatalytic Irradiation from CO_2 to Methanol. In *Nanostructured Photocatalysts*; Nguyen, V.-H., Vo, D.-V.N., Nanda, S., Eds.; Elsevier: Amsterdam, The Netherlands, 2021; pp. 519–540. [CrossRef]
14. Rosdin, R.D.b.; Yusuf, M.; Abdullah, B. Dry reforming of methane over Ni-based catalysts: Effect of ZrO_2 and MgO addition as support. *Mater. Lett. X* **2021**, *12*, 100095. [CrossRef]
15. Jeswani, H.K.; Chilvers, A.; Azapagic, A. Environmental sustainability of biofuels: A review. *Proc. R. Soc. A* **2020**, *476*, 20200351. [CrossRef] [PubMed]

16. Cherubini, F.; Strømman, A.H. Life cycle assessment of bioenergy systems: State of the art and future challenges. *Bioresour. Technol.* **2011**, *102*, 437–451. [CrossRef]
17. Tayari, S.; Abedi, R.; Rahi, A. Comparative assessment of engine performance and emissions fueled with three different biodiesel generations. *Renew. Energy* **2020**, *147*, 1058–1069. [CrossRef]
18. Okoye, P.U.; Longoria, A.; Sebastian, P.J.; Wang, S.; Li, S.; Hameed, B.H. A review on recent trends in reactor systems and azeotrope separation strategies for catalytic conversion of biodiesel-derived glycerol. *Sci. Total Environ.* **2020**, *719*, 134595. [CrossRef]
19. Garcia, R.; Figueiredo, F.; Brandão, M.; Hegg, M.; Castanheira, É.; Malça, J.; Nilsson, A.; Freire, F. A meta-analysis of the life cycle greenhouse gas balances of microalgae biodiesel. *Int. J. Life Cycle Assess.* **2020**, *25*, 1737–1748. [CrossRef]
20. Sun, S.; Li, K. Biodiesel production from phoenix tree seed oil catalyzed by liquid lipozyme TL100L. *Renew. Energy* **2020**, *151*, 152–160. [CrossRef]
21. Esmaeili, H. A critical review on the economic aspects and life cycle assessment of biodiesel production using heterogeneous nanocatalysts. *Fuel Process. Technol.* **2022**, *230*, 107224. [CrossRef]
22. Singh, R.; Singh, S.; Kumar, M. Impact of n-butanol as an additive with eucalyptus biodiesel-diesel blends on the performance and emission parameters of the diesel engine. *Fuel* **2020**, *277*, 118178. [CrossRef]
23. Deora, P.S.; Verma, Y.; Muhal, R.A.; Goswami, C.; Singh, T. Biofuels: An alternative to conventional fuel and energy source. *Mater. Today Proc.* **2022**, *48*, 1178–1184.
24. Mat Aron, N.S.; Khoo, K.S.; Chew, K.W.; Show, P.L.; Chen, W.H.; Nguyen, T.H.P. Sustainability of the four generations of biofuels–a review. *Int. J. Energy Res.* **2020**, *44*, 9266–9282. [CrossRef]
25. Aransiola, E.F.; Ojumu, T.V.; Oyekola, O.; Madzimbamuto, T.; Ikhu-Omoregbe, D. A review of current technology for biodiesel production: State of the art. *Biomass Bioenergy* **2014**, *61*, 276–297. [CrossRef]
26. Obergruber, M.; Hönig, V.; Procházka, P.; Kučerová, V.; Kotek, M.; Bouček, J.; Mařík, J. Physicochemical properties of biobutanol as an advanced biofuel. *Materials* **2021**, *14*, 914. [CrossRef]
27. Hajjari, M.; Tabatabaei, M.; Aghbashlo, M.; Ghanavati, H. A review on the prospects of sustainable biodiesel production: A global scenario with an emphasis on waste-oil biodiesel utilization. *Renew. Sustain. Energy Rev.* **2017**, *72*, 445–464. [CrossRef]
28. Salehi Jouzani, G.; Aghbashlo, M.; Tabatabaei, M. Biofuels: Types, Promises, Challenges, and Role of Fungi. In *Fungi in Fuel Biotechnology*; Springer: Berlin/Heidelberg, Germany, 2020; pp. 1–14.
29. Corsini, A.; Marchegiani, A.; Rispoli, F.; Sciulli, F.; Venturini, P. Vegetable Oils as Fuels in Diesel Engine. Engine Performance and Emissions. *Energy Procedia* **2015**, *81*, 942–949. [CrossRef]
30. Faungnawakij, K.; Suriye, K. *New and Future Developments in Catalysis: Chapter 4. Current Catalytic Processes with Hybrid Materials and Composites for Heterogeneous Catalysis*; Elsevier Inc. Chapters: Amsterdam, The Netherlands, 2013.
31. Ai, P.; Jin, K.; Alengebawy, A.; Elsayed, M.; Meng, L.; Chen, M.; Ran, Y. Effect of application of different biogas fertilizer on eggplant production: Analysis of fertilizer value and risk assessment. *Environ. Technol. Innov.* **2020**, *19*, 101019. [CrossRef]
32. Aryal, N.; Ghimire, N.; Bajracharya, S. Coupling of microbial electrosynthesis with anaerobic digestion for waste valorization. In *Advances in Bioenergy*; Elsevier: Amsterdam, The Netherlands, 2020; Volume 5, pp. 101–127.
33. do Nascimento Santos, T.; Dutra, E.D.; do Prado, A.G.; Leite, F.C.B.; de Souza, R.d.F.R.; dos Santos, D.C.; de Abreu, C.A.M.; Simões, D.A.; de Morais Jr, M.A.; Menezes, R.S.C. Potential for biofuels from the biomass of prickly pear cladodes: Challenges for bioethanol and biogas production in dry areas. *Biomass Bioenergy* **2016**, *85*, 215–222. [CrossRef]
34. Wasajja, H.; Lindeboom, R.E.; van Lier, J.B.; Aravind, P. Techno-economic review of biogas cleaning technologies for small scale off-grid solid oxide fuel cell applications. *Fuel Process. Technol.* **2020**, *197*, 106215. [CrossRef]
35. Bhuiya, M.; Rasul, M.; Khan, M.M.K.; Ashwath, N.; Azad, A.K.; Hazrat, M. Second generation biodiesel: Potential alternative to-edible oil-derived biodiesel. *Energy Procedia* **2014**, *61*, 1969–1972. [CrossRef]
36. Perlack, R.D. *Biomass as Feedstock for a Bioenergy and Bioproducts Industry: The Technical Feasibility of a Billion-Ton Annual Supply*; Oak Ridge National Laboratory: Oak Ridge, Tennessee, 2005.
37. Werling, B.P.; Dickson, T.L.; Isaacs, R.; Gaines, H.; Gratton, C.; Gross, K.L.; Liere, H.; Malmstrom, C.M.; Meehan, T.D.; Ruan, L. Perennial grasslands enhance biodiversity and multiple ecosystem services in bioenergy landscapes. *Proc. Natl. Acad. Sci. USA* **2014**, *111*, 1652–1657. [CrossRef] [PubMed]
38. Lynd, L.R.; Sow, M.; Chimphango, A.F.; Cortez, L.A.; Brito Cruz, C.H.; Elmissiry, M.; Laser, M.; Mayaki, I.A.; Moraes, M.A.; Nogueira, L.A. Bioenergy and African transformation. *Biotechnol. Biofuels* **2015**, *8*, 1–18. [CrossRef] [PubMed]
39. Demirbas, A. Use of algae as biofuel sources. *Energy Convers. Manag.* **2010**, *51*, 2738–2749. [CrossRef]
40. Ho, S.-H.; Chen, C.-Y.; Lee, D.-J.; Chang, J.-S. Perspectives on microalgal CO_2-emission mitigation systems—A review. *Biotechnol. Adv.* **2011**, *29*, 189–198. [CrossRef]
41. Posten, C.; Schaub, G. Microalgae and terrestrial biomass as source for fuels—A process view. *J. Biotechnol.* **2009**, *142*, 64–69. [CrossRef]
42. Hong, M.; Zhukareva, V.; Vogelsberg-Ragaglia, V.; Wszolek, Z.; Reed, L.; Miller, B.I.; Geschwind, D.H.; Bird, T.D.; McKeel, D.; Goate, A. Mutation-specific functional impairments in distinct tau isoforms of hereditary FTDP-17. *Science* **1998**, *282*, 1914–1917. [CrossRef]
43. Xu, H.; Wang, C. A Comprehensive Review of 2,5-Dimethylfuran as a Biofuel Candidate. *Biofuels Lignocellul. Biomass Innov. Beyond Bioethanol.* **2016**, 105–129.

44. Zhang, W.; He, J.; Engstrand, P.; Björkqvist, O. Economic evaluation on bio-synthetic natural gas production integrated in a thermomechanical pulp mill. *Energies* **2015**, *8*, 12795–12809. [CrossRef]
45. Alaswad, A.; Dassisti, M.; Prescott, T.; Olabi, A.G. Technologies and developments of third generation biofuel production. *Renew. Sustain. Energy Rev.* **2015**, *51*, 1446–1460. [CrossRef]
46. Wirth, R.; Lakatos, G.; Böjti, T.; Maróti, G.; Bagi, Z.; Rákhely, G.; Kovács, K.L. Anaerobic gaseous biofuel production using microalgal biomass–a review. *Anaerobe* **2018**, *52*, 1–8. [CrossRef]
47. Anto, S.; Mukherjee, S.S.; Muthappa, R.; Mathimani, T.; Deviram, G.; Kumar, S.S.; Verma, T.N.; Pugazhendhi, A. Algae as green energy reserve: Technological outlook on biofuel production. *Chemosphere* **2020**, *242*, 125079. [CrossRef] [PubMed]
48. Ganesan, R.; Manigandan, S.; Samuel, M.S.; Shanmuganathan, R.; Brindhadevi, K.; Chi, N.T.L.; Duc, P.A.; Pugazhendhi, A. A review on prospective production of biofuel from microalgae. *Biotechnol. Rep.* **2020**, *27*, e00509. [CrossRef] [PubMed]
49. Subramani, L.; Venu, H. Evaluation of methyl ester derived from novel Chlorella emersonii as an alternative feedstock for DI diesel engine & its combustion, performance and tailpipe emissions. *Heat Mass Transf.* **2019**, *55*, 1513–1534.
50. Popovich, C.A.; Pistonesi, M.; Hegel, P.; Constenla, D.; Bielsa, G.B.; Martin, L.A.; Damiani, M.C.; Leonardi, P.I. Unconventional alternative biofuels: Quality assessment of biodiesel and its blends from marine diatom Navicula cincta. *Algal Res.* **2019**, *39*, 101438. [CrossRef]
51. Wang, W.; Xu, Y.; Wang, X.; Zhang, B.; Tian, W.; Zhang, J. Hydrothermal liquefaction of microalgae over transition metal supported TiO_2 catalyst. *Bioresour. Technol.* **2018**, *250*, 474–480. [CrossRef]
52. Talebian-Kiakalaieh, A.; Amin, N.A.S.; Mazaheri, H. A review on novel processes of biodiesel production from waste cooking oil. *Appl. Energy* **2013**, *104*, 683–710. [CrossRef]
53. Bashir, M.A.; Wu, S.; Zhu, J.; Krosuri, A.; Khan, M.U.; Aka, R.J.N. Recent development of advanced processing technologies for biodiesel production: A critical review. *Fuel Process. Technol.* **2022**, *227*, 107120. [CrossRef]
54. Godbole, V.; Pal, M.K.; Gautam, P. A critical perspective on the scope of interdisciplinary approaches used in fourth-generation biofuel production. *Algal Res.* **2021**, *58*, 102436. [CrossRef]
55. Abdullah, B.; Muhammad, S.A.F.a.S.; Shokravi, Z.; Ismail, S.; Kassim, K.A.; Mahmood, A.N.; Aziz, M.M.A. Fourth generation biofuel: A review on risks and mitigation strategies. *Renew. Sustain. Energy Rev.* **2019**, *107*, 37–50. [CrossRef]
56. Brennan, L.; Owende, P. Biofuels from microalgae—A review of technologies for production, processing, and extractions of biofuels and co-products. *Renew. Sustain. Energy Rev.* **2010**, *14*, 557. [CrossRef]
57. Li, Y.; Han, D.; Hu, G.; Dauvillee, D.; Sommerfeld, M.; Ball, S.; Hu, Q. Chlamydomonas starchless mutant defective in ADP-glucose pyrophosphorylase hyper-accumulates triacylglycerol. *Metab. Eng.* **2010**, *12*, 387–391. [CrossRef] [PubMed]
58. Lan, E.I.; Liao, J.C. Metabolic engineering of cyanobacteria for 1-butanol production from carbon dioxide. *Metab. Eng.* **2011**, *13*, 353–363. [CrossRef] [PubMed]
59. USDA. Agricultural Projections to 2025. Available online: https://www.usda.gov/oce/commodity/projections/USDA_Agricultural_Projections_to_2025.pdf (accessed on 2 June 2018).
60. OECD-FAO. Agricultural Outlook. OECD Agriculture Statistics (Database). Available online: http://www.fao.org/fileadmin/templates/est/COMM_MARKETS_MONITORING/Oilcrops/Documents/ECD_Reports/OECD_biofuels2015_2024.pdf (accessed on 21 June 2022).
61. EIA. Monthly Energy Review. In *U.S. Energy Information Administration (EIA), Monthly Energy Review*. Available online: https://www.eia.gov/totalenergy/data/monthly/pdf/mer.pdf (accessed on 21 June 2022).
62. Jaganmohan, M. US Biomass Energy Production Forecast from 2021 to 2050. In *Energy Information Administration (EIA). Annual Energy Outlook 2022*; Available online: https://www.statista.com/statistics/264029/us-biomass-energy-production/ (accessed on 3 March 2022).
63. Kaur, M.; Ali, A. Lithium ion impregnated calcium oxide as nano catalyst for the biodiesel production from karanja and jatropha oils. *Renew. Energy* **2011**, *36*, 2866–2871. [CrossRef]
64. Mahdavi, M.; Abedini, E.; hosein Darabi, A. Biodiesel synthesis from oleic acid by nano-catalyst (ZrO_2/Al_2O_3) under high voltage conditions. *RSC Adv.* **2015**, *5*, 55027–55032. [CrossRef]
65. Demirbas, A. Production of biodiesel fuels from linseed oil using methanol and ethanol in non-catalytic SCF conditions. *Biomass Bioenergy* **2009**, *33*, 113–118. [CrossRef]
66. Singh, S.P.; Singh, D. Biodiesel production through the use of different sources and characterization of oils and their esters as the substitute of diesel: A review. *Renew. Sustain. Energy Rev.* **2010**, *14*, 200–216. [CrossRef]
67. Umana, U.S.; Ebong, M.S.; Godwin, E.O. Biomass production from oil palm and its value chain. *J. Hum. Earth Future* **2020**, *1*, 30–38. [CrossRef]
68. Kim, D.S.; Hanifzadeh, M.; Kumar, A. Trend of biodiesel feedstock and its impact on biodiesel emission characteristics. *Environ. Prog. Sustain. Energy* **2018**, *37*, 7–19. [CrossRef]
69. Gülşen, E.; Olivetti, E.; Freire, F.; Dias, L.; Kirchain, R. Impact of feedstock diversification on the cost-effectiveness of biodiesel. *Appl. Energy* **2014**, *126*, 281–296. [CrossRef]
70. Thurmond, W. Global Biodiesel Market Trends, Outlook and Opportunities. *Emerg. Mark. Online Glob. Energy Biofuels Intell.* **2008**, *2*.
71. Schumacher, L.G.; Borgelt, S.C.; Fosseen, D.; Goetz, W.; Hires, W. Heavy-duty engine exhaust emission tests using methyl ester soybean oil/diesel fuel blends. *Bioresour. Technol.* **1996**, *57*, 31–36. [CrossRef]

72. Forster, P.; Ramaswamy, V.; Artaxo, P.; Berntsen, T.; Betts, R.; Fahey, D.; Haywood, J.; Lean, J.; Lowe, D.; Myhre, G. *Climate Change 2007: The Physical Science Basis. Contribution of Working Group I to the Fourth Assessment Report of the Intergovernmental Panel on Climate Change*; Solomon, S., Qin, D., Manning, M., Chen, Z., Marquis, M., Averyt, K.B., Tingor, M., Miller, H.L., Eds.; Cambridge University Press: Cambridge, UK, 2007; pp. 129–234.
73. Carter, M.S.; Hauggaard-Nielsen, H.; Heiske, S.; Jensen, M.; Thomsen, S.T.; Schmidt, J.E.; Johansen, A.; Ambus, P. Consequences of field N_2O emissions for the environmental sustainability of plant-based biofuels produced within an organic farming system. *GCB Bioenergy* **2012**, *4*, 435–452. [CrossRef]
74. Smeets, E.M.; Bouwman, L.F.; Stehfest, E.; Van Vuuren, D.P.; Posthuma, A. Contribution of N_2O to the greenhouse gas balance of first-generation biofuels. *Glob. Change Biol.* **2009**, *15*, 1–23. [CrossRef]
75. Ambat, I.; Srivastava, V.; Sillanpää, M. Recent advancement in biodiesel production methodologies using various feedstock: A review. *Renew. Sustain. Energy Rev.* **2018**, *90*, 356–369. [CrossRef]
76. Gashaw, A.; Lakachew, A. Production of biodiesel from non edible oil and its properties. *Int. J. Sci. Environ. Technol.* **2014**, *3*, 1544–1562.
77. Hannon, M.; Gimpel, J.; Tran, M.; Rasala, B.; Mayfield, S. Biofuels from algae: Challenges and potential. *Biofuels* **2010**, *1*, 763–784. [CrossRef]
78. Paul Abishek, M.; Patel, J.; Prem Rajan, A. Algae oil: A sustainable renewable fuel of future. *Biotechnol. Res. Int.* **2014**, *2014*, 272814. [CrossRef]
79. Adewale, P.; Dumont, M.-J.; Ngadi, M. Recent trends of biodiesel production from animal fat wastes and associated production techniques. *Renew. Sustain. Energy Rev.* **2015**, *45*, 574–588. [CrossRef]
80. Sangkharak, K.; Mhaisawat, S.; Rakkan, T.; Paichid, N.; Yunu, T. Utilization of mixed chicken waste for biodiesel production using single and combination of immobilized lipase as a catalyst. *Biomass Convers. Biorefinery* **2020**, *12*, 1–14. [CrossRef]
81. Zhao, C.; Brück, T.; Lercher, J.A. Catalytic deoxygenation of microalgae oil to green hydrocarbons. *Green Chem.* **2013**, *15*, 1720–1739. [CrossRef]
82. Li, Y.; Jin, Y.; Borrion, A.; Li, J. Influence of feed/inoculum ratios and waste cooking oil content on the mesophilic anaerobic digestion of food waste. *Waste Manag.* **2018**, *73*, 156–164. [CrossRef] [PubMed]
83. Ma, Y.; Gao, Z.; Wang, Q.; Liu, Y. Biodiesels from microbial oils: Opportunity and challenges. *Bioresour. Technol.* **2018**, *263*, 631–641. [CrossRef] [PubMed]
84. Arruda, L.F.d.; Borghesi, R.; Oetterer, M. Use of fish waste as silage: A review. *Braz. Arch. Biol. Technol.* **2007**, *50*, 879–886. [CrossRef]
85. Sharma, A.K.; Samuchia, D.; Sharma, P.; Mandeewal, R.L.; Nitharwal, P.K.; Meena, M. Effect of Nitrogen and Sulphur in the Production of the Mustard Crop [Brassica juncea (L.)]. *Int. J. Environ. Clim. Change* **2022**, *12*, 132–137. [CrossRef]
86. Koh, M.Y.; Mohd; Ghazi, T.I. A review of biodiesel production from *Jatropha curcas* L. oil. *Renew. Sustain. Energy Rev.* **2011**, *15*, 2240–2251. [CrossRef]
87. Ho, D.P.; Ngo, H.H.; Guo, W. A mini review on renewable sources for biofuel. *Bioresour. Technol.* **2014**, *169*, 742–749. [CrossRef] [PubMed]
88. Neupane, D.; Bhattarai, D.; Ahmed, Z.; Das, B.; Pandey, S.; Solomon, J.K.; Qin, R.; Adhikari, P. Growing Jatropha (*Jatropha curcas* L.) as a potential second-generation biodiesel feedstock. *Inventions* **2021**, *6*, 60. [CrossRef]
89. Li, Y.; Khanal, S.K. *Bioenergy: Principles and Applications*; John Wiley & Sons: Hoboken, NJ, USA, 2016.
90. Cecrle, E.; Depcik, C.; Duncan, A.; Guo, J.; Mangus, M.; Peltier, E.; Stagg-Williams, S.; Zhong, Y. Investigation of the effects of biodiesel feedstock on the performance and emissions of a single-cylinder diesel engine. *Energy Fuels* **2012**, *26*, 2331–2341. [CrossRef]
91. Behcet, R.; Aydın, H.; Ilkılıç, C.; İşcan, B.; Aydın, S. Diesel engine applications for evaluation of performance and emission behavior of biodiesel from different oil stocks. *Environ. Prog. Sustain. Energy* **2015**, *34*, 890–896. [CrossRef]
92. Akbar, E.; Yaakob, Z.; Kamarudin, S.K.; Ismail, M.; Salimon, J. Characteristic and composition of Jatropha curcas oil seed from Malaysia and its potential as biodiesel feedstock feedstock. *Eur. J. Sci. Res.* **2009**, *29*, 396–403.
93. Leung, D.; Guo, Y. Transesterification of neat and used frying oil: Optimization for biodiesel production. *Fuel Process. Technol.* **2006**, *87*, 883–890. [CrossRef]
94. Canoira, L.; Rodriguez-Gamero, M.; Querol, E.; Alcántara, R.n.; Lapuerta, M.n.; Oliva, F.n. Biodiesel from low-grade animal fat: Production process assessment and biodiesel properties characterization. *Ind. Eng. Chem. Res.* **2008**, *47*, 7997–8004. [CrossRef]
95. Islam, M.A.; Magnusson, M.; Brown, R.J.; Ayoko, G.A.; Nabi, M.N.; Heimann, K. Microalgal species selection for biodiesel production based on fuel properties derived from fatty acid profiles. *Energies* **2013**, *6*, 5676–5702. [CrossRef]
96. Wan Ghazali, W.N.M.; Mamat, R.; Masjuki, H.H.; Najafi, G. Effects of biodiesel from different feedstocks on engine performance and emissions: A review. *Renew. Sustain. Energy Rev.* **2015**, *51*, 585–602. [CrossRef]
97. Verma, P.; Sharma, M.P. Review of process parameters for biodiesel production from different feedstocks. *Renew. Sustain. Energy Rev.* **2016**, *62*, 1063–1071. [CrossRef]
98. Sakthivel, R.; Ramesh, K.; Purnachandran, R.; Mohamed Shameer, P. A review on the properties, performance and emission aspects of the third generation biodiesels. *Renew. Sustain. Energy Rev.* **2018**, *82*, 2970–2992. [CrossRef]
99. Lapuerta, M.; Armas, O.; Rodríguez-Fernández, J. Effect of biodiesel fuels on diesel engine emissions. *Prog. Energy Combust. Sci.* **2008**, *34*, 198–223. [CrossRef]

100. Mostafa, S.S.M.; El-Gendy, N.S. Evaluation of fuel properties for microalgae Spirulina platensis bio-diesel and its blends with Egyptian petro-diesel. *Arab. J. Chem.* **2017**, *10*, S2040–S2050. [CrossRef]
101. Dixit, S.; kanakraj, S.; Rehman, A. Linseed oil as a potential resource for bio-diesel: A review. *Renew. Sustain. Energy Rev.* **2012**, *16*, 4415–4421. [CrossRef]
102. Kumar, R.; Tiwari, P.; Garg, S. Alkali transesterification of linseed oil for biodiesel production. *Fuel* **2013**, *104*, 553–560. [CrossRef]
103. Rizwanul Fattah, I.M.; Masjuki, H.H.; Kalam, M.A.; Wakil, M.A.; Rashedul, H.K.; Abedin, M.J. Performance and emission characteristics of a CI engine fueled with Cocos nucifera and Jatropha curcas B20 blends accompanying antioxidants. *Ind. Crops Prod.* **2014**, *57*, 132–140. [CrossRef]
104. Kumar, K.; Sharma, M. Performance and emission characteristics of a diesel engine fuelled with biodiesel blends. *Int. J. Renew. Energy Res.* **2016**, *6*, 658–662.
105. Neupane, D.; Lohaus, R.H.; Solomon, J.K.; Cushman, J.C. Realizing the Potential of Camelina sativa as a Bioenergy Crop for a Changing Global Climate. *Plants* **2022**, *11*, 772. [CrossRef] [PubMed]
106. Muhammad, N.; Bamishaiye, E.; Bamishaiye, O.; Usman, L.; Salawu, M.; Nafiu, M.; Oloyede, O. Physicochemical properties and fatty acid composition of cyperus esculentus (Tiger Nut) Tuber Oil. *Biores Bull* **2011**, *5*, 51–54.
107. Standard Organization of Nigeria. *Standards for Edible Refined Palm Oil and Its Processed Form*; Scientific Research Publishing Inc.: Irvine, CA, USA, 2000; pp. 2–5.
108. NIS. *Nigerian Industrial Standards. Standard for Edible Vegetable Oil*; Scientific Research Publishing Inc.: Irvine, CA, USA, 1992; pp. 5–12.
109. Nduka, J.K.C.; Omozuwa, P.O.; Imanah, O.E. Effect of heating time on the physicochemical properties of selected vegetable oils. *Arab. J. Chem.* **2021**, *14*, 103063. [CrossRef]
110. Knothe, G. Analyzing biodiesel: Standards and other methods. *J. Am. Oil Chem. Soc.* **2006**, *83*, 823–833. [CrossRef]
111. Kyriakidis, N.B.; Katsiloulis, T. Calculation of iodine value from measurements of fatty acid methyl esters of some oils: Comparison with the relevant American oil chemists society method. *J. Am. Oil Chem. Soc.* **2000**, *77*, 1235–1238. [CrossRef]
112. Atabani, A.E.; Silitonga, A.S.; Ong, H.C.; Mahlia, T.M.I.; Masjuki, H.H.; Badruddin, I.A.; Fayaz, H. Non-edible vegetable oils: A critical evaluation of oil extraction, fatty acid compositions, biodiesel production, characteristics, engine performance and emissions production. *Renew. Sustain. Energy Rev.* **2013**, *18*, 211–245. [CrossRef]
113. Rasouli, H.; Esmaeili, H. Characterization of MgO nanocatalyst to produce biodiesel from goat fat using transesterification process. *3 Biotech.* **2019**, *9*, 1–11. [CrossRef]
114. Deshmukh, S.; Kumar, R.; Bala, K. Microalgae biodiesel: A review on oil extraction, fatty acid composition, properties and effect on engine performance and emissions. *Fuel Process. Technol.* **2019**, *191*, 232–247. [CrossRef]
115. Attaphong, C.; Do, L.; Sabatini, D.A. Vegetable oil-based microemulsions using carboxylate-based extended surfactants and their potential as an alternative renewable biofuel. *Fuel* **2012**, *94*, 606–613. [CrossRef]
116. Chang, Y.-C.; Lee, W.-J.; Lin, S.-L.; Wang, L.-C. Green energy: Water-containing acetone–butanol–ethanol diesel blends fueled in diesel engines. *Appl. Energy* **2013**, *109*, 182–191. [CrossRef]
117. Attaphong, C.; Sabatini, D.A. Phase behaviors of vegetable oil-based microemulsion fuels: The effects of temperatures, surfactants, oils, and water in ethanol. *Energy Fuels* **2013**, *27*, 6773–6780. [CrossRef]
118. Lif, A.; Holmberg, K. Water-in-diesel emulsions and related systems. *Adv. Colloid Interface Sci.* **2006**, *123*, 231–239. [CrossRef] [PubMed]
119. Grebemariam, S.; Marchetti, J.M. Biodiesel Production Technologies: Review. *AIMS Energy* **2017**, *5*, 425–457.
120. Baskar, G.; Aiswarya, R. Trends in catalytic production of biodiesel from various feedstocks. *Renew. Sustain. Energy Rev.* **2016**, *57*, 496–504. [CrossRef]
121. Agarwal, A.K. Biofuels (alcohols and biodiesel) applications as fuels for internal combustion engines. *Prog. Energy Combust. Sci.* **2007**, *33*, 233–271. [CrossRef]
122. Akram, F.; ul Haq, I.; Raja, S.I.; Mir, A.S.; Qureshi, S.S.; Aqeel, A.; Shah, F.I. Current trends in biodiesel production technologies and future progressions: A possible displacement of the petro-diesel. *J. Clean. Prod.* **2022**, 133479. [CrossRef]
123. Singh, D.; Sharma, D.; Soni, S.; Sharma, S.; Sharma, P.K.; Jhalani, A. A review on feedstocks, production processes, and yield for different generations of biodiesel. *Fuel* **2020**, *262*, 116553. [CrossRef]
124. Karmakar, B.; Halder, G. Progress and future of biodiesel synthesis: Advancements in oil extraction and conversion technologies. *Energy Convers. Manag.* **2019**, *182*, 307–339. [CrossRef]
125. Mohan, D.; Pittman, C.U., Jr.; Steele, P.H. Pyrolysis of wood/biomass for bio-oil: A critical review. *Energy Fuels* **2006**, *20*, 848–889. [CrossRef]
126. Avhad, M.; Marchetti, J. A review on recent advancement in catalytic materials for biodiesel production. *Renew. Sustain. Energy Rev.* **2015**, *50*, 696–718. [CrossRef]
127. Sani, Y.; Daud, W.; Aziz, A.A. Biodiesel feedstock and production technologies: Successes, challenges and prospects. *Biodiesel-Feedstocks Prod. Appl.* **2012**, *10*, 52790.
128. Vyas, A.P.; Verma, J.L.; Subrahmanyam, N. A review on FAME production processes. *Fuel* **2010**, *89*, 1–9. [CrossRef]
129. Demirbas, A. Biofuels securing the planet's future energy needs. *Energy Convers. Manag.* **2009**, *50*, 2239–2249. [CrossRef]
130. Hashmi, S.; Gohar, S.; Mahmood, T.; Nawaz, U.; Farooqi, H. Biodiesel production by using CaO-Al_2O_3 Nano catalyst. *Int. J. Eng. Res. Sci.* **2016**, *2*, 43–49.

131. Rodrigues, R.C.; Volpato, G.; Wada, K.; Ayub, M.A.Z. Enzymatic synthesis of biodiesel from transesterification reactions of vegetable oils and short chain alcohols. *J. Am. Oil Chem. Soc.* **2008**, *85*, 925–930. [CrossRef]
132. Iso, M.; Chen, B.; Eguchi, M.; Kudo, T.; Shrestha, S. Production of biodiesel fuel from triglycerides and alcohol using immobilized lipase. *J. Mol. Catal. B Enzym.* **2001**, *16*, 53–58. [CrossRef]
133. Mahmudul, H.M.; Hagos, F.Y.; Mamat, R.; Adam, A.A.; Ishak, W.F.W.; Alenezi, R. Production, characterization and performance of biodiesel as an alternative fuel in diesel engines—A review. *Renew. Sustain. Energy Rev.* **2017**, *72*, 497–509. [CrossRef]
134. Abbaszaadeh, A.; Ghobadian, B.; Omidkhah, M.R.; Najafi, G. Current biodiesel production technologies: A comparative review. *Energy Convers. Manag.* **2012**, *63*, 138–148. [CrossRef]
135. Gude, V.G.; Patil, P.; Martinez-Guerra, E.; Deng, S.; Nirmalakhandan, N. Microwave energy potential for biodiesel production. *Sustain. Chem. Process.* **2013**, *1*. [CrossRef]
136. Ma, F.; Hanna, M.A. Biodiesel production: A review. *Bioresour. Technol.* **1999**, *70*, 1–15. [CrossRef]
137. Chiaramonti, D.; Bonini, M.; Fratini, E.; Tondi, G.; Gartner, K.; Bridgwater, A.; Grimm, H.; Soldaini, I.; Webster, A.; Baglioni, P. Development of emulsions from biomass pyrolysis liquid and diesel and their use in engines—Part 1: Emulsion production. *Biomass Bioenergy* **2003**, *25*, 85–99. [CrossRef]
138. Shaah, M.A.H.; Hossain, M.S.; Allafi, F.A.S.; Alsaedi, A.; Ismail, N.; Ab Kadir, M.O.; Ahmad, M.I. A review on non-edible oil as a potential feedstock for biodiesel: Physicochemical properties and production technologies. *RSC Adv.* **2021**, *11*, 25018–25037. [CrossRef]
139. Mahanta, P.; Shrivastava, A. Technology development of bio-diesel as an energy alternative. *Dep. Mech. Eng. Indian Inst. Technol.* **2004**, *1*, 1–19.
140. Yan, B.; Zhang, S.; Chen, W.; Cai, Q. Pyrolysis of tobacco wastes for bio-oil with aroma compounds. *J. Anal. Appl. Pyrolysis* **2018**, *136*, 248–254. [CrossRef]
141. Shadangi, K.P.; Mohanty, K. Production and characterization of pyrolytic oil by catalytic pyrolysis of Niger seed. *Fuel* **2014**, *126*, 109–115. [CrossRef]
142. Naik, M.; Meher, L.; Naik, S.; Das, L. Production of biodiesel from high free fatty acid Karanja (Pongamia pinnata) oil. *Biomass Bioenergy* **2008**, *32*, 354–357. [CrossRef]
143. Gaurav, A.; Ng, F.T.; Rempel, G.L. A new green process for biodiesel production from waste oils via catalytic distillation using a solid acid catalyst–Modeling, economic and environmental analysis. *Green Energy Environ.* **2016**, *1*, 62–74. [CrossRef]
144. Bilgin, A.; Durgun, O.; Sahin, Z. The effects of diesel-ethanol blends on diesel engine performance. *Energy Sources* **2002**, *24*, 431–440. [CrossRef]
145. Kiss, A.A.; Bildea, C.S. A review of biodiesel production by integrated reactive separation technologies. *J. Chem. Technol. Biotechnol.* **2012**, *87*, 861–879. [CrossRef]
146. Adenuga, A.A.; Oyekunle, J.A.O.; Idowu, O.O. Pathway to reduce free fatty acid formation in Calophyllum inophyllum kernel oil: A renewable feedstock for biodiesel production. *J. Clean. Prod.* **2021**, *316*, 128222. [CrossRef]
147. Elgharbawy, A.S.; Sadik, W.A.; Sadek, O.M.; Kasaby, M.A. Maximizing biodiesel production from high free fatty acids feedstocks through glycerolysis treatment. *Biomass Bioenergy* **2021**, *146*, 105997. [CrossRef]
148. Demirbas, A. Progress and recent trends in biodiesel fuels. *Energy Convers. Manag.* **2009**, *50*, 14–34. [CrossRef]
149. Chang, M.Y.; Chan, E.-S.; Song, C.P. Biodiesel production catalysed by low-cost liquid enzyme Eversa®Transform 2.0: Effect of free fatty acid content on lipase methanol tolerance and kinetic model. *Fuel* **2021**, *283*, 119266. [CrossRef]
150. Felizardo, P.; Baptista, P.; Menezes, J.C.; Correia, M.J.N. Multivariate near infrared spectroscopy models for predicting methanol and water content in biodiesel. *Anal. Chim. Acta* **2007**, *595*, 107–113. [CrossRef]
151. Fregolente, P.B.L.; Fregolente, L.V.; Wolf Maciel, M.R. Water Content in Biodiesel, Diesel, and Biodiesel–Diesel Blends. *J. Chem. Eng. Data* **2012**, *57*, 1817–1821. [CrossRef]
152. Marchetti, J.M.; Miguel, V.U.; Errazu, A.F. Possible methods for biodiesel production. *Renew. Sustain. Energy Rev.* **2007**, *11*, 1300–1311. [CrossRef]
153. Parthiban, K.S.; Pandian, S.; Subramanian, D. Conventional and in-situ transesterification of Annona squamosa seed oil for biodiesel production: Performance and emission analysis. *Environ. Technol. Innov.* **2021**, *23*, 101593. [CrossRef]
154. Musa, I.A. The effects of alcohol to oil molar ratios and the type of alcohol on biodiesel production using transesterification process. *Egypt. J. Pet.* **2016**, *25*, 21–31. [CrossRef]
155. Norjannah, B.; Ong, H.C.; Masjuki, H.; Juan, J.; Chong, W. Enzymatic transesterification for biodiesel production: A comprehensive review. *RSC Adv.* **2016**, *6*, 60034–60055. [CrossRef]
156. IŞIk, M.Z. Comparative experimental investigation on the effects of heavy alcohols- safflower biodiesel blends on combustion, performance and emissions in a power generator diesel engine. *Appl. Therm. Eng.* **2021**, *184*, 116142. [CrossRef]
157. Razak, N.H.; Hashim, H.; Yunus, N.A.; Klemeš, J.J. Reducing diesel exhaust emissions by optimisation of alcohol oxygenates blend with diesel/biodiesel. *J. Clean. Prod.* **2021**, *316*, 128090. [CrossRef]
158. Binhweel, F.; Bahadi, M.; Pyar, H.; Alsaedi, A.; Hossain, S.; Ahmad, M.I. A comparative review of some physicochemical properties of biodiesels synthesized from different generations of vegetative oils. *J. Phys. Conf. Ser.* **2021**, *1900*, 012009. [CrossRef]
159. Hossain, A.; Boyce, A.; Salleh, A.; Chandran, S. Impacts of alcohol type, ratio and stirring time on the biodiesel production from waste canola oil. *Afr. J. Agric. Res.* **2010**, *5*, 1851–1859.

160. Freedman, B.; Butterfield, R.O.; Pryde, E.H. Transesterification kinetics of soybean oil 1. *J. Am. Oil Chem. Soc.* **1986**, *63*, 1375–1380. [CrossRef]
161. Silitonga, A.S.; Mahlia, T.M.I.; Ong, H.C.; Riayatsyah, T.M.I.; Kusumo, F.; Ibrahim, H.; Dharma, S.; Gumilang, D. A comparative study of biodiesel production methods for Reutealis trisperma biodiesel. *Energy Sources Part A Recovery Util. Environ. Eff.* **2017**, *39*, 2006–2014. [CrossRef]
162. Refaat, A.; Attia, N.; Sibak, H.A.; El Sheltawy, S.; ElDiwani, G. Production optimization and quality assessment of Biodiesel from waste vegetable oil. *Int. J. Environ. Sci. Technol.* **2008**, *5*, 75–82. [CrossRef]
163. Kusdiana, D.; Saka, S. Kinetics of transesterification in rapeseed oil to biodiesel fuel as treated in supercritical methanol. *Fuel* **2001**, *80*, 693–698. [CrossRef]
164. Daramola, M.; Mtshali, K.; Senokoane, L.; Fayemiwo, O. Influence of operating variables on the transesterification of waste cooking oil to biodiesel over sodium silicate catalyst: A statistical approach. *J. Taibah Univ. Sci.* **2016**, *10*, 675–684. [CrossRef]
165. Mathiyazhagan, M.; Ganapathi, A. Factors affecting biodiesel production. *Res. Plant Biol.* **2011**, *1*.
166. Mathew, G.M.; Raina, D.; Narisetty, V.; Kumar, V.; Saran, S.; Pugazhendi, A.; Sindhu, R.; Pandey, A.; Binod, P. Recent advances in biodiesel production: Challenges and solutions. *Sci. Total Environ.* **2021**, *794*, 148751. [CrossRef] [PubMed]
167. Shah, S.; Gupta, M.N. Lipase catalyzed preparation of biodiesel from Jatropha oil in a solvent free system. *Process Biochem.* **2007**, *42*, 409–414. [CrossRef]
168. Ganesan, D.; Rajendran, A.; Thangavelu, V. An overview on the recent advances in the transesterification of vegetable oils for biodiesel production using chemical and biocatalysts. *Rev. Environ. Sci. Bio/Technol.* **2009**, *8*, 367. [CrossRef]
169. Vicente, G.; Martínez, M.; Aracil, J. Optimisation of integrated biodiesel production. Part I. A study of the biodiesel purity and yield. *Bioresour. Technol.* **2007**, *98*, 1724–1733. [CrossRef]
170. Zhang, J.; Zhang, L.; Jia, L. Variables Affecting Biodiesel Production from Zanthoxylum bungeanum Seed Oil with High Free Fatty Acids. *Ind. Eng. Chem. Res.* **2012**, *51*, 3525–3530. [CrossRef]
171. Chen, L.; Yin, P.; Liu, X.; Yang, L.; Yu, Z.; Guo, X.; Xin, X. Biodiesel production over copper vanadium phosphate. *Energy* **2011**, *36*, 175–180. [CrossRef]
172. Kumari, A.; Mahapatra, P.; Garlapati, V.K.; Banerjee, R. Enzymatic transesterification of Jatropha oil. *Biotechnol. Biofuels* **2009**, *2*, 1. [CrossRef]
173. Diasakou, M.; Louloudi, A.; Papayannakos, N. Kinetics of the non-catalytic transesterification of soybean oil. *Fuel* **1998**, *77*, 1297–1302. [CrossRef]
174. Thangaraj, B.; Solomon, P.R.; Muniyandi, B.; Ranganathan, S.; Lin, L. Catalysis in biodiesel production—A review. *Clean Energy* **2018**, *3*, 2–23. [CrossRef]
175. Kulkarni, M.G.; Gopinath, R.; Meher, L.C.; Dalai, A.K. Solid acid catalyzed biodiesel production by simultaneous esterification and transesterification. *Green Chem.* **2006**, *8*, 1056–1062. [CrossRef]
176. Ranganathan, S.V.; Narasimhan, S.L.; Muthukumar, K. An overview of enzymatic production of biodiesel. *Bioresour. Technol.* **2008**, *99*, 3975–3981. [CrossRef]
177. Aarthy, M.; Saravanan, P.; Gowthaman, M.K.; Rose, C.; Kamini, N.R. Enzymatic transesterification for production of biodiesel using yeast lipases: An overview. *Chem. Eng. Res. Des.* **2014**, *92*, 1591–1601. [CrossRef]
178. Hama, S.; Noda, H.; Kondo, A. How lipase technology contributes to evolution of biodiesel production using multiple feedstocks. *Curr. Opin. Biotechnol.* **2018**, *50*, 57–64. [CrossRef] [PubMed]
179. Lam, M.K.; Lee, K.T.; Mohamed, A.R. Homogeneous, heterogeneous and enzymatic catalysis for transesterification of high free fatty acid oil (waste cooking oil) to biodiesel: A review. *Biotechnol. Adv.* **2010**, *28*, 500–518. [CrossRef]
180. Refaat, A.A. Different techniques for the production of biodiesel from waste vegetable oil. *Int. J. Environ. Sci. Technol.* **2010**, *7*, 183–213. [CrossRef]
181. Schuchardt, U.; Sercheli, R.; Vargas, R.M. Transesterification of vegetable oils: A review. *J. Braz. Chem. Soc.* **1998**, *9*, 199–210. [CrossRef]
182. Endalew, A.K.; Kiros, Y.; Zanzi, R. Inorganic heterogeneous catalysts for biodiesel production from vegetable oils. *Biomass Bioenergy* **2011**, *35*, 3787–3809. [CrossRef]
183. Tang, Y.; Gu, X.; Chen, G. 99% yield biodiesel production from rapeseed oil using benzyl bromide–CaO catalyst. *Environ. Chem. Lett.* **2013**, *11*, 203–208. [CrossRef]
184. Qiu, F.; Li, Y.; Yang, D.; Li, X.; Sun, P. Heterogeneous solid base nanocatalyst: Preparation, characterization and application in biodiesel production. *Bioresour. Technol.* **2011**, *102*, 4150–4156. [CrossRef]
185. Ruhul, A.; Kalam, M.; Masjuki, H.; Fattah, I.R.; Reham, S.; Rashed, M. State of the art of biodiesel production processes: A review of the heterogeneous catalyst. *RSC Adv.* **2015**, *5*, 101023–101044. [CrossRef]
186. Liu, X.; Piao, X.; Wang, Y.; Zhu, S.; He, H. Calcium methoxide as a solid base catalyst for the transesterification of soybean oil to biodiesel with methanol. *Fuel* **2008**, *87*, 1076–1082. [CrossRef]
187. Madhuvilakku, R.; Piraman, S. Biodiesel synthesis by TiO_2–ZnO mixed oxide nanocatalyst catalyzed palm oil transesterification process. *Bioresour. Technol.* **2013**, *150*, 55–59. [CrossRef]
188. Xue, J.; Grift, T.E.; Hansen, A.C. Effect of biodiesel on engine performances and emissions. *Renew. Sustain. Energy Rev.* **2011**, *15*, 1098–1116. [CrossRef]

189. Qi, D.; Geng, L.; Chen, H.; Bian, Y.Z.; Liu, J.; Ren, X.C. Combustion and performance evaluation of a diesel engine fueled with biodiesel produced from soybean crude oil. *Renew. Energy* **2009**, *34*, 2706–2713. [CrossRef]
190. Ganapathy, T.; Gakkhar, R.; Murugesan, K. Influence of injection timing on performance, combustion and emission characteristics of Jatropha biodiesel engine. *Appl. Energy* **2011**, *88*, 4376–4386. [CrossRef]
191. Chauhan, B.S.; Kumar, N.; Cho, H.M. A study on the performance and emission of a diesel engine fueled with Jatropha biodiesel oil and its blends. *Energy* **2012**, *37*, 616–622. [CrossRef]
192. Karthikeyan, A.; Jayaprabakar, J.; Williams, R.D. Experimental Investigations on Diesel engine using Methyl esters of Jatropha oil and fish oil. *IOP Conf. Ser. Mater. Sci. Eng.* **2017**, *197*, 012020. [CrossRef]
193. Canakci, M.; Van Gerpen, J.H. Comparison of engine performance and emissions for petroleum diesel fuel, yellow grease biodiesel, and soybean oil biodiesel. *Trans. ASAE* **2003**, *46*, 937. [CrossRef]
194. Wahlen, B.D.; Morgan, M.R.; McCurdy, A.T.; Willis, R.M.; Morgan, M.D.; Dye, D.J.; Bugbee, B.; Wood, B.D.; Seefeldt, L.C. Biodiesel from microalgae, yeast, and bacteria: Engine performance and exhaust emissions. *Energy Fuels* **2013**, *27*, 220–228. [CrossRef]
195. Tüccar, G.; Aydın, K. Evaluation of methyl ester of microalgae oil as fuel in a diesel engine. *Fuel* **2013**, *112*, 203–207. [CrossRef]
196. Fisher, B.C.; Marchese, A.J.; Volckens, J.; Lee, T.; Collett, J.L. Measurement of gaseous and particulate emissions from algae-based fatty acid methyl esters. *SAE Int. J. Fuels Lubr.* **2010**, *3*, 292–321. [CrossRef]
197. Alptekin, E.; Canakci, M.; Ozsezen, A.N.; Turkcan, A.; Sanli, H. Using waste animal fat based biodiesels–bioethanol–diesel fuel blends in a DI diesel engine. *Fuel* **2015**, *157*, 245–254. [CrossRef]
198. McCormick, R.L.; Graboski, M.S.; Alleman, T.L.; Herring, A.M.; Tyson, K.S. Impact of biodiesel source material and chemical structure on emissions of criteria pollutants from a heavy-duty engine. *Environ. Sci. Technol.* **2001**, *35*, 1742–1747. [CrossRef] [PubMed]
199. Lira, T.A.M.; Santos, A.P.; Moreti, T.C.F.; Lopes, A.; de Oliveira, M.C.J.; Neves, M.C.T.; Iamaguti, P.S.; de Lima, L.P.; Koike, G.H.A.; de Abreu Silva, R. Performance of agricultural tractor consuming diesel and biodiesel derived from babassu ('Orbinya martiana'). *Aust. J. Crop Sci.* **2019**, *13*, 1037–1044. [CrossRef]

Disclaimer/Publisher's Note: The statements, opinions and data contained in all publications are solely those of the individual author(s) and contributor(s) and not of MDPI and/or the editor(s). MDPI and/or the editor(s) disclaim responsibility for any injury to people or property resulting from any ideas, methods, instructions or products referred to in the content.

Article

Deep Eutectic Solvents for Biodiesel Purification in a Microextractor: Solvent Preparation, Selection and Process Optimization

Sara Anđelović [1,†], Marko Božinović [1,†], Željka Ćurić [1,†], Anita Šalić [1,*], Ana Jurinjak Tušek [2], Kristina Zagajski Kučan [1], Marko Rogošić [1], Mia Radović [2], Marina Cvjetko Bubalo [2] and Bruno Zelić [1,3]

[1] Faculty of Chemical Engineering and Technology, University of Zagreb, Marulićev trg 19, HR-10000 Zagreb, Croatia
[2] Faculty of Food Technology and Biotechnology, University of Zagreb, Pierottijeva ul. 6, HR-10000 Zagreb, Croatia
[3] Department of Packaging, Recycling and Environmental Protection, University North, Trg dr. Žarka Dolinara 1, HR-48000 Koprivnica, Croatia
* Correspondence: asalic@fkit.hr; Tel.: +385-1-4597-101
† These authors contributed equally to this work.

Citation: Anđelović, S.; Božinović, M.; Ćurić, Ž.; Šalić, A.; Jurinjak Tušek, A.; Kučan, K.Z.; Rogošić, M.; Radović, M.; Cvjetko Bubalo, M.; Zelić, B. Deep Eutectic Solvents for Biodiesel Purification in a Microextractor: Solvent Preparation, Selection and Process Optimization. *Bioengineering* 2022, 9, 665. https://doi.org/10.3390/bioengineering9110665

Academic Editors: Indra Neel Pulidindi, Aharon Gedanken and Ali Zarrabi

Received: 27 September 2022
Accepted: 4 November 2022
Published: 8 November 2022

Publisher's Note: MDPI stays neutral with regard to jurisdictional claims in published maps and institutional affiliations.

Copyright: © 2022 by the authors. Licensee MDPI, Basel, Switzerland. This article is an open access article distributed under the terms and conditions of the Creative Commons Attribution (CC BY) license (https://creativecommons.org/licenses/by/4.0/).

Abstract: The most important and commonly used process for biodiesel synthesis is transesterification. The main by-product of biodiesel synthesis by transesterification is glycerol, which must be removed from the final product. Recently, deep eutectic solvent (DES) assisted extraction has been shown to be an effective and sustainable method for biodiesel purification. In this study, biodiesel was produced by lipase-catalysed transesterification from sunflower oil and methanol. A total of 12 different eutectic solvents were prepared and their physical properties were determined. Mathematical models were used to define which physical and chemical properties of DES and to what extent affect the efficiency of extraction of glycerol from the biodiesel. After initial screening, cholinium-based DES with ethylene glycol as hydrogen bond donor was selected and used for optimization of extraction process conditions performed in a microsystem. To determine the optimal process conditions (temperature, biodiesel:DES volume ratio, residence time), the experimental three-level-three-factor Box-Behnken experimental design was used. In the end, a combination of a mathematical model and experimental results was used to estimate how many micro-extractors are necessary for the complete removal of glycerol.

Keywords: biodiesel; DES; microextractor; purification; glycerol extraction

1. Introduction

Biodiesel, as one of the representatives of biofuels, still has one of the largest shares of renewable energy sources. The most common method of producing biodiesel is transesterification. When an enzyme is used as a catalyst, the production can be said to be green and environmentally friendly. Since biodiesel produced by transesterification is not suitable for direct use in internal combustion engines, as it may contain traces of soap, catalyst, methanol (and other alcohols), metals, water, oil, and glycerides, it must be purified to meet the appropriate quality standards (ASTM D6751, EN 14214) [1,2]. Moreover, the downstream process of biodiesel production largely determines the final price of biodiesel [3]. This is due to the industrial purification method for biodiesel, wet washing, which has the major drawback of consuming about 10 L of water per 1 L of purified biodiesel [2]. In addition to the fact that wet washing of biodiesel consumes a large amount of fresh water, the wastewater after biodiesel purification must be cleaned before it is discharged into the environment, which requires a large amount of energy. Therefore, it is necessary to find more environmentally friendly methods for the purification of biodiesel and to make the whole production/purification process green.

Because of all the disadvantages of the wet washing, various other methods have been developed, such as dry washing, membrane separation, liquid-liquid extraction, precipitation, reactive distillation, and complexation [4–7]. Liquid-liquid extraction with deep eutectic solvents (DESs) is one of the alternative methods to remove glycerol from biodiesel after the transesterification [8]. DESs have unique properties that include ease of preparation, low cost, environmental friendliness compared to conventional solvents, non-flammability, low volatility, high dissolving power, and good biodegradability. Biodegradability and environmental impact are largely related to the character of the individual pure components that make up DESs [9]. Another important property of DESs is their chemical adaptability, which means that eutectic solvents can be developed for specific applications [10].

Shahbaz et al. [11] synthesized various DESs and removed all the free glycerol from palm oil-based biodiesel with an optimal molar ratio of DES to biodiesel of 1:1. In the work presented by Naiwanti and Zullaikah [12], the authors used cholinium-based DES with ethylene glycol as hydrogen bond donor to remove the total glycerol, but the amount of total glycerol still remained above the specification standard. The same effect was found in the work of Petračić et al. [13] where the same DES (molar ratio 1:2.5) was used as an extraction medium in batch experiments and in the continuous Karr column. It was found that after extraction, the free glycerol content was below the limit for all samples, but the total glycerol and glycerides content was too high to fully meet biodiesel quality standards.

In addition to using DES instead of water to purify biodiesel, switching from batch to continuous processes could also improve the purification efficiency. Microextractors, miniaturized systems, are an example of how smaller is better. Due to their internal properties—large surface-to-volume ratio, intensification of mass transfer, and small dimensions—the interphase diffusion can be neglected [14,15]. In the work of Šalić et al. [8] the authors tested an extraction efficiency of seven different DESs based on mixtures of choline chloride with ethylene glycol or glycerol for glycerol extraction from raw biodiesel produced by transesterification using three microextractors of different sizes. With a residence time of only 13.61 s, glycerol was almost completely removed from biodiesel using a cholinium-based DES with glycerol as a hydrogen bond donor. The results obtained clearly indicate that the microextractors can be used for glycerol removal from raw biodiesel.

As can be seen from the literature, the most commonly used solvents for the removal of glycerol from biodiesel were DESs based on choline chloride and ethylene glycol or glycerol as hydrogen bond donors in various molar ratios. However, DESs can also be formed by mixing other components [9]. The chemical structure of hydrogen donor and acceptor has a significant effect on the formation, properties and stability of the deep eutectic solvent [9]. In the preparation of eutectic solvents, it is necessary to know certain factors such as the purity of the components and the water content of each component, as well as the storage and drying of the prepared DESs, so that their physicochemical properties do not change. Small changes in the water content of DES can lead to significant differences in physicochemical properties such as viscosity, density or polarity. Water can disrupt the hydrogen bond acceptor/donor network because it can act as both a hydrogen bond acceptor and donor. For this reason, and to ensure consistency of properties, different approaches are used for the preparation of DESs [16–20].

In this work, raw biodiesel was produced by transesterification. Glycerol was then removed from the product using 12 different DESs in microextractors, and the efficiency of the extraction was assessed. For the DES exhibiting the highest extraction efficiency, extraction process was optimized using the Box-Behnken model at three levels and with three factors. After process optimization, extraction was performed in a microextractor under optimal process conditions. Finally, a mathematical model was used in order to estimate how many microextractors are required for the complete removal of glycerol from biodiesel.

2. Materials and Methods

2.1. Materials

Chemicals

Edible sunflower oil (Zvijezda, Zagreb, Croatia) was purchased at a local supermarket. Commercial lipase from *Thermomyces lanuginosus* (Lipolase 100L), ethylene glycol, glycerol, betaine and propylene glycol were purchased from Sigma-Aldrich Handels GmbH (Saint Louis, MO, USA). Methanol was purchased from BDH Prolabo (Lutterworth, United Kingdom). Choline chloride, zinc chloride and propylene glycol were purchased from Acros Organics (Geel, Belgium). Ethanol was purchased from Gram-mol (Zagreb, Croatia). The chemicals were of analytical grade and were used without further purification, except for drying.

2.2. Methods

2.2.1. Production of Biodiesel in a Batch Reactor

Biodiesel production from edible sunflower oil using commercial free lipase was performed in a glass jacketed batch reactor (V = 500 mL). The total mass of the mixture was 275.48 g, and the mixture consisted of 225 g sunflower oil, 27.98 g methanol, and 22.5 g lipase enzyme solution [21]. The enzyme solution was prepared by mixing commercial lipase with 0.01 mol/L phosphate buffer pH 7.4 at a volume ratio of 1:10. The transesterification reaction was started by adding the enzyme to the reaction mixture and carried out at a temperature of 40 °C (optimal temperature for the lipase-catalyzed transesterification reaction) which has kept constant by a water bath (Thermomix 1420, Braun, Germany). The process lasted 48 h. Mixing was carried out by a magnetic stirrer (MS-H-S, DLAB, Rowland St, City of Industry, CA, USA) at 600 rpm. At the end of the reaction, the mixture was transferred to a separating funnel to separate glycerol at the bottom. In the further course of the research, the upper phase was used, i.e., partially purified biodiesel [8].

2.2.2. Preparation of Deep Eutectic Solvents

Choline chloride, glycerol, ethylene-glycol, betaine, propylene-glycol and zinc chloride were used in waterless DESs preparation. Prior to DESs preparation, the DESs components were dried in a vacuum oven at 60 °C for 8 h. DESs were prepared by mixing dried chemicals in different molar ratios (Table 1) in a glass vial equipped with a stopper. The components were mixed on a magnetic laboratory stirrer (Rotamix S-10, Tehtnica, Železniki, Slovenia) at 700 rpm and 70–90 °C until a homogeneous, colorless liquid was obtained [22–25].

Table 1. Prepared DESs.

Deep Eutectic Solvent	Abbreviation	Hydrogen Bond Acceptor	Hydrogen Bond Donor	Acceptor:Donor Molar Ratio
Choline chloride:glycerol	ChCl:Gly	[choline chloride structure]	glycerol	1:2 1:3
Choline chloride:ethylene glycol	ChCl:EG	[choline chloride structure]	ethylene glycol	1:2 1:3
Betaine:glycerol	B:Gly	[betaine structure]	glycerol	1:3 1:4

Table 1. Cont.

Deep Eutectic Solvent	Abbreviation	Hydrogen Bond Acceptor	Hydrogen Bond Donor	Acceptor:Donor Molar Ratio
Betaine:ethylene glycol	B:EG			1:3
Betaine:propylene glycol	B:PG			1:3.5
Choline chloride:propylene glycol	ChCl:PG			1:3 1:4
Choline chloride:propylene glycol:zinc chloride	ChCl:PG:ZnCl$_2$			1:4:0.02
Betaine:glycerol:zinc chloride	B:Gly:ZnCl$_2$			1:4:0.02

2.2.3. Measurements/Determination of Physico-Chemical Properties of Prepared Deep Eutectic Solvents

The physico-chemical properties of the prepared DESs were determined at 25 °C. Specific conductivity was determined with a conductivity meter (Schott Instruments Lab 960, Mainz, Germany, resolution 0.01 for a range of 0.00–19.99 mS cm^{-1}, 0.001 for a range of 0.000–1.999 µS cm^{-1}), refractive index was measured with a refractometer (Abbe RL-3, Kern, Myszków, Poland, accuracy ±0.0002), and a density meter (Anton Paar DMA 4500 M, Graz, Austria) was used for density measurements. A programmable rheometer (Brookfield DV-III Ultra, East Lyme, CT, USA, accuracy ±1%) was used to determine dynamic viscosity. As for thermal properties, thermal conductivity, thermal diffusivity and heat capacity were measured with a thermal conductivity meter (Linseis Transient Hot Bridge 1, Selb, Germany, measurement uncertainties according to ISO standards) [22–25].

2.2.4. Calculation of Deep Eutectic Solvents Descriptors

DESs chosen for glycerol extraction were mathematically described using σ-profiles defined with the COSMOtherm software and the generated DES descriptors were further used to mathematically model the glycerol extraction efficiency.

DESs constituents were initially optimized both from an energy and geometry point of view in TURBOMOLE software by adopting DFT (density functional theory) with BP86 functional level of theory and def-TZVP basis set. Molecules consisting of two or more ions (e.g., choline chloride) were treated as ion pairs and their structures were optimized according to Abranches et al. [26]. These quantum chemical calculations resulted in the software-generated .cosmo file for each optimized molecule that was further used in BIOVIA COSMOtherm software. The files contained σ-profile curves that provided a quantitative representation of the molecules' polar surface screen charge on the polarity scale, and therefore included all information necessary for the calculation of the σ-profile function and σ-profile descriptors. As the calculation output, σ-profile for each molecule was created. The σ-profile curve for each molecule was divided into ten regions, making the region width of 0.005 e / Å^2 and covering the total range from –0.025 to +0.025 e/Å^2.

The areas under the curve were integrated separately for each defined region. This was done by simple summation of tabulated σ-profile data point ordinate values as presented by the COSMOtherm software. The ordinate values lying on the boundaries of the regions were split into halves, and each half was attributed to one of the neighboring regions. Thus, 10 S-descriptors (S^1–S^{10}) of σ-profiles were calculated to represent these 10 areas numerically.

For the preparation of the DESs descriptor set, the DESs were modelled as a molar mixture of hydrogen bond acceptor and donor according to Table 1. Any change in DES composition can be described by a change in its σ-profile and the associated numerical value of its descriptors. To obtain a unique descriptor set for each particular DES, the σ-profiles of its constituents were processed as follows. The descriptors of the studied DES (S^i_{mix}) were calculated from the hydrogen bond acceptor and donor descriptors according to Equation (1) proposed by Benguerba et al. [27]:

$$S^i_{mix} = \sum_{j=1}^{NC} X_j S^i_{\sigma-profile,j} \qquad (1)$$

where i denotes the descriptor number (1–10), j stands for the DES constituent number, X_j is the molar fraction of each constituent, $S^i_{\sigma-profile,j}$, is the i-th descriptor of j-th constituent, and NC is the total number of constituents from which DES is prepared.

2.2.5. Two-Phase Liquid-Liquid Extraction in a Microextractor

The extractions (Figure 1) were performed in a PTFE (polytetrafluoroethylene) microextractor with two inlets and a T-junction (L = 30 cm, d = 1000 µm). Liquids were fed to the microextractor using syringes mounted on two piston pumps (Harvard PHD 4400 Programmable, Harvard Apparatus, Inc, Holliston, MA, USA) connected with PTFE tubing. One syringe was filled with crude biodiesel, and the second syringe contained DES. The flows of biodiesel and DES were adjusted to achieve the 1:1 volume ratios of components in a microextractor. Extractions were performed for different residence times (0.05–30 min) depending on the experiment. Samples were collected at the exit of the microextractor in Eppendorf test tubes. The samples collected in this way were centrifuged (Universal 320 R, Hettich, Buford, GA, USA; 14,000 rpm, T = 25 °C, t = 15 min) to completely separate the biodiesel (upper layer) from the DES (lower layer). The samples were collected from upper phases using a needle to avoid phase contamination. Samples were diluted with ethanol and the influence of solvents on glycerol extraction was monitored according to the method described in Section 2.2.6.

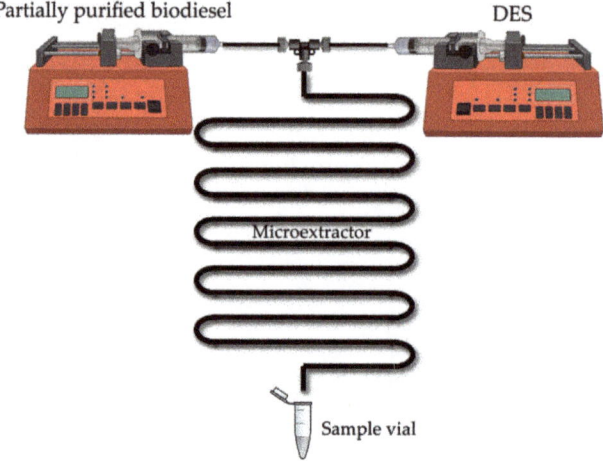

Figure 1. Scheme of the experimental set-up used for the biodiesel purification process in a microextractor.

From the values obtained, biodiesel yield (Y), glycerol extraction efficiency (E), distribution coefficient (K_P) and mass fraction of glycerol in biodiesel (w) were calculated according to the following equations (Equations (2)–(5)):

$$Y = \frac{\gamma_{B, P.B.}}{\gamma_{B, C.B.}} \quad (2)$$

$$E = 1 - \frac{\gamma_{G, P.B.}}{\gamma_{G, C.B.}} \quad (3)$$

$$K = \frac{\gamma_{G, C.B.} - \gamma_{G, P.B.}}{\gamma_{G, C.B.}} \quad (4)$$

$$w = \frac{\gamma_{G, P.B.} \cdot V_B}{m_B} \quad (5)$$

where G denotes glycerol, B denotes biodiesel, P.B. denotes purified biodiesel, γ denotes mass concentration (mg/mL), V denotes volume (mL) and C.B. denotes crude biodiesel.

2.2.6. Determination of Glycerol and FAME Concentration in the Samples

The concentrations of free fatty acids esters (FAME) and glycerol in the samples before and after extraction were determined using a GC (Shimadzu GC-2014, Kyoto, Japan) gas chromatograph equipped with FID detector and Zebron ZB-wax GC capillary column (length 30 m, i.d. 0.53 mm and film thickness 1.00 µm, Phenomenex, Aschaffenburg, Germany) by the method described elsewhere [28]. To confirm repeatability, every sample was analyzed in triplicate. On a 95% confidence interval, the results showed no significant difference.

2.2.7. Data Analysis and Mathematical Modeling
Modeling of Extraction Efficiency Based on DESs Descriptors and Physical Properties

Glycerol extraction efficiency (E) was modeled as the function of σ profile of the DES expressed by a set of S^i_{mix} descriptors according to Equation (6):

$$E = f\left(S^i_{mix}\right) \quad (6)$$

S^i_{mix} descriptors for the description of glycerol extraction efficiency were selected based on the Person correlation matrix. Based on significant correlations, S^1_{mix}, S^3_{mix}, S^4_{mix}, S^5_{mix} and S^6_{mix} were used as the models input variables. The relationship between glycerol extraction efficiency and selected S^i_{mix} descriptors was modeled using multiple linear regression (MLR) according to Equation (7), nonlinear regression (NLR) according to Equation (8) and piecewise linear regression (PLR) according to Equation (9):

$$E = b_0 + b_1 \cdot S^1_{mix} + b_2 \cdot S^3_{mix} + b_3 \cdot S^4_{mix} + b_4 \cdot S^5_{mix} + b_5 \cdot S^6_{mix} \quad (7)$$

$$E = b_0 \cdot \left(S^1_{mix}\right)^{b_1} \cdot \left(S^3_{mix}\right)^{b_2} \cdot \left(S^4_{mix}\right)^{b_3} \cdot \left(S^5_{mix}\right)^{b_4} \cdot \left(S^6_{mix}\right)^{b_5} \quad (8)$$

$$E = \begin{cases} (b_{01} + b_{11} \cdot S^1_{mix} + b_{21} \cdot S^3_{mix} + b_{31} \cdot S^4_{mix} + b_{41} \cdot S^5_{mix} + b_{51} \cdot S^6_{mix}) & \forall \ (E \leq b_n) \\ (b_{02} + b_{12} \cdot S^1_{mix} + b_{22} \cdot S^3_{mix} + b_{32} \cdot S^4_{mix} + b_{42} \cdot S^5_{mix} + b_{52} \cdot S^6_{mix}) & \forall \ (E > b_n) \end{cases} \quad (9)$$

Glycerol extraction efficiency (E) was also modeled as the function of DESs physical properties: density (ρ), dynamic viscosity (η), electrical conductivity (σ), refractive index (n_D), thermal diffusivity (a), thermal conductivity (λ), and specific heat capacity (c_p) according to Equation (10):

$$E = f(\rho, \eta, \sigma, n_D, a, \lambda, c_p) \quad (10)$$

The relationship between glycerol extraction efficiency and analyzed physical properties of DESs was modeled using multiple linear regression (MLR) according to Equation (11), nonlinear regression (NLR) according to Equation (12), and piecewise linear regression (PLR) according to Equation (13):

$$E = b_0 + b_1 \cdot \rho + b_2 \cdot \eta + b_3 \cdot \sigma + b_4 \cdot n_D + b_5 \cdot a + b_6 \cdot \lambda + b_7 \cdot c_P \tag{11}$$

$$E = b_0 \cdot \rho^{b_1} \cdot \eta^{b_2} \cdot \sigma^{b_3} \cdot n_D^{b_4} \cdot a^{b_5} \cdot \lambda^{b_6} \cdot c_P^{b_7} \tag{12}$$

$$E = \begin{cases} (b_{01} + b_{11} \cdot \rho + b_{21} \cdot \eta + b_{31} \cdot \sigma + b_{41} \cdot n_D + b_{51} \cdot a + b_{61} \cdot \lambda + b_{71} \cdot c_P) & \forall \ (E \leq b_n) \\ (b_{02} + b_{12} \cdot \rho + b_{22} \cdot \eta + b_{32} \cdot \sigma + b_{42} \cdot n_D + b_{52} \cdot a + b_{62} \cdot \lambda + b_{72} \cdot c_P) & \forall \ (E > b_n) \end{cases} \tag{13}$$

The parameters of the MLR models (Equations (7) and (11)), NLR models (Equations (8) and (12)), and PLR models (Equations (9) and (13)) were estimated using the Levenberg-Marquardt algorithm implemented in Statistica 13.0 (Tibco Software Inc, Palo Alto CA, USA). The algorithm searches for numerical solutions in the function parameter space using the least squares method. Calculations were performed in 50 iterations with the convergence parameter of 10^{-6} and 95% confidence interval [29,30].

Optimization of Biodiesel Purification in a Microextractor

To optimize the biodiesel purification, the experiments were carried out in the microextractor as described in Section 2.2.5, according to the Box-Behnken experimental design at three levels (−1, 0, 1). The effects of the (i) extraction temperature X_1 (T = 25, 40, 55 °C), (ii) residence time X_2 (τ = 0.05, 0.5, 0.95 min) and (iii) volume ratio of biodiesel:DES X_3 (9:1, 1:1, 1:9 v/v) were analyzed. Experimental data were fitted to the second order polynomial equation (Equation (14)):

$$Z = \beta_0 + \sum_{i=1}^{3} \beta_i \cdot X_i + \sum_{i=1}^{3} \beta_{ii} \cdot X_i^2 + \sum_{i=1}^{2} \sum_{j=i+1}^{3} \beta_{ij} \cdot X_i \cdot X_j \tag{14}$$

Therein, Z is the predicted response, β_0, β_i, β_{ii} and β_{ij} are the regression coefficients for intercept, linear, quadratic and interaction terms, and X_i and X_j are the independent variables. The response surface analysis was performed using Statistica 13.0 (Tibco Software Inc, Palo Alto, CA, USA)

Mathematical Modeling of Glycerol Extraction in a Microextractor

The glycerol separation in a microextractor was described with a 2D model including convection in the flow direction (x) and diffusion in two directions (x and y). The mathematical model for steady-state conditions in a microextractor was composed of dimensionless partial differential equations for glycerol concentrations in biodiesel and DES phase and corresponding boundary and initial conditions. The proposed model was described elsewhere [8].

3. Results and Discussion
3.1. Biodiesel Production in the Batch Reactor

To obtain larger amounts of biodiesel to be used in all purification experiments in this work, biodiesel was synthesized in the batch reactor (V = 500 mL) using edible sunflower oil, methanol, and the enzyme lipase from *Thermomyces lanuginosus*. At the end of the reaction, the yield of the obtained biodiesel in the form of FAME was 97.96 ± 2.25% and the glycerol concentration was 117.19 ± 2.70 mg/mL. The obtained results were consistent with those of previous studies [7,21]. Although the yield of FAME in the sample was within the range prescribed by the standards for the quality of biodiesel, the content of glycerol was much higher than allowed (w < 0.02%), so biodiesel thus obtained had to be purified. The first and simplest method of glycerol separation was by gravitational settling since glycerol was insoluble in biodiesel. In order to remove most of glycerol, a separation funnel was used. After a settling time of 24 h, samples of biodiesel were taken and the glycerol concentration decreased by 96.96 ± 0.59%, which means that the glycerol concentration in partially purified biodiesel was 3.57 ± 0.78 mg/mL. The same percentage of removed glycerol was obtained in the work of Šalić et al. [8]. The authors reported that several additional steps could be used for biodiesel purification to further increase this percentage, such as second gravitational settling, centrifugation, and filtration, but the total amount of biodiesel removed by these processes would not justify the cost of their application. For this reason, no additional purification steps were performed in this study.

3.2. Deep Eutectic Solvents Preparation and Physico-Chemical Properties

As mentioned in the introduction, the most commonly used DESs for the removal of glycerol from biodiesel were DESs based on choline chloride and ethylene glycol or glycerol as hydrogen bond donors in various molar ratios [8,11–13]. However, by combining different hydrogen donors and acceptors, DESs with different properties can be obtained. In this research 12 water-free DESs were prepared. They were selected based on their ability to denitrify diesel fuels [22–25]. Water was not used for DESs preparation to prevent water entering biodiesel because this would require an additional step in the purification, since water must also be removed from biodiesel before use. After preparation, DESs properties: density (ρ, g/mL), dynamic viscosity (η, Pa s), specific conductance (σ, mS/cm), refractive index (n_D, -), thermal diffusivity (a, mm^2/s), thermal conductivity (λ, W/(m K)), and specific heat capacity (c_p, J/(g K)) were determined. The selected properties were chosen according to Rogošić and Zagajski Kučan [25], as being among the most important properties relevant to extraction processes. Viscosity is very important when it comes to microfluidics since it is directly connected to the fluid behavior in flow. Also, if the fluids (DESs) are too viscous, they cannot be used in microextractors [31]. As for the density, it determines the settling of the layers of primary and secondary solvent after extraction in macroextractors. Without the differences in the density, two liquids will not separate in the gravitational field. This property is important for macroextractors; however, in a microextractor even the two liquids of the same density flow side by side with a clearly defined boundary when two liquids are introduced into a microchannel by two inlets [32]. The refractive index is important in order to observe the phase separation more easily. With this, and by regulating the flow, it is possible to position the phase boundary between the two liquids at a specific place in the microchannel, which enables complete phase separation at the exit from the microchannel. Knowledge of the electrical conductivity of DES is important for any DES uses that involve electric current, such as the separation by electrocoagulation. Thermal properties are important when extraction is performed at a temperature that is higher or lower than 25 °C [25]. Furthermore, increasing the temperature decreases the viscosity and density of all DESs. In addition, the increase of specific conductance was noted with increasing temperature of the DESs. In the research performed by Abbott et al. [18], it was found that the lower was the viscosity, the higher was the observed specific conductance. Besides heating, a possible strategy to reduce DES viscosity is the simple dilution of DES with water [33–36] which was not an option in this research.

The obtained results for selected DESs are presented in Table 2.

Table 2. Properties of selected DESs at 25 °C.

DES	ρ, g/mL	η, Pa s	σ, mS/cm	n_D	a, mm^2/s	λ, W/(m K)	c_p, J/(g K)	Ref.
ChCl:Gly (1:2)	1.188 ± 0.002	0.369 ± 0.053	1.130 ± 0.010	1.448 ± 0.000	0.097 ± 0.011	0.232 ± 0.006	2.010 ± 0.193	[22]
ChCl:Gly (1:3)	1.204 ± 0.001	0.316 ± 0.011	1.122 ± 0.008	1.448 ± 0.000	0.097 ± 0.009	0.241 ± 0.005	2.057 ± 0.140	[22]
ChCl:EG (1:2)	1.115 ± 0.002	0.042 ± 0.000	8.610 ± 0.005	1.448 ± 0.000	0.167 ± 0.002	0.227 ± 0.002	1.205 ± 0.003	[23]
ChCl:EG (1:3)	1.113 ± 0.000	0.028 ± 0.000	9.410 ± 0.010	1.448 ± 0.000	0.195 ± 0.020	0.231 ± 0.008	1.055 ± 0.100	[23]
B:Gly (1:3)	1.223 ± 0.000	1.103 ± 0.012	0.001 ± 0.000	1.478 ± 0.000	0.151 ± 0.011	0.270 ± 0.004	1.455 ± 0.091	[24]
B:EG (1:3)	1.131 ± 0.000	0.062 ± 0.002	0.006 ± 0.001	1.456 ± 0.000	0.189 ± 0.005	0.231 ± 0.002	1.071 ± 0.018	[24]

Table 2. Cont.

DES	ρ, g/mL	η, Pa s	σ, mS/cm	n_D	a, mm^2/s	λ, W/(m K)	c_p, J/(g K)	Ref.
B:PG (1:3.5)	1.074 ± 0.000	0.139± 0.003	0.000 ± 0.000	1.452 ± 0.000	0.116 ± 0.003	0.206 ± 0.093	1.642 ± 0.048	[24]
ChCl:PG (1:3)	1.078 ± 0.003	0.066 ± 0.004	3.380 ± 0.009	1.458 ± 0.000	0.145 ± 0.023	0.208 ± 0.013	1.347 ± 0.147	[23]
ChCl:PG (1:4)	1.075 ± 0.134	0.049 ± 0.000	3.093 ± 0.033	1.455 ± 0.000	0.217 ± 0.021	0.213 ± 0.007	0.915 ± 0.063	
ChCl:PG:ZnCl$_2$ (1:4:0.02)	1.079 ± 0.000	0.054 ± 0.002	2.290 ± 0.073	1.455± 0.000	0.219 ± 0.001	0.211 ± 0.003	0.896 ± 0.034	
B:Gly (1:4)	1.232 ± 0.006	2.431 ± 0.006	2.920 ± 0.021	1.456 ± 0.000	0.138 ± 0.017	0.278 ± 0.006	1.653 ± 0.173	
B:Gly:ZnCl$_2$ (1:4:0.02)	1.233 ± 0.000	1.406 ± 0.396	2.770 ± 0.029	1.456 ± 0.000	0.094 ± 0.016	0.257 ± 0.100	2.257 ± 0.355	

As can be seen, as the ethylene glycol or glycerol mole fraction increases, the viscosity of the DESs gradually decreases, which is related to the lower viscosity of ethylene glycol and glycerol in comparison to choline chloride. Also, according to literature [37–44], the viscosity of DES is usually higher than 0.1 Pa s which is related to the large network of hydrogen bonds in DESs [45]. Among synthetized DESs, few of them, based on propylene glycol and ethylene glycol, had viscosity below that value which was attributed to the small molecular size of glycol components [46]. As for DESs density, the molar ratio of DES components has a significant effect on their density, which is related to molecular arrangements in DES structure [47].

The comparison of the obtained results with the literature ones was not possible. Namely, the changes in the DES composition (i.e., component ratio, amount of water, purity of components) significantly affect the properties of DES and no literature data were available for DES compositions studied.

3.3. Glycerol Extraction in a Microextractor with Different Deep Eutectic Solvents

In order to select the DES which would enable the highest glycerol extraction efficiency, the extraction process was carried out at a temperature of 25 °C, a biodiesel:DES volume ratio of 1:1 and a residence time of τ = 0.05 min. The selected volume ratio of biodiesel and DES was based on the literature data [11] where all free glycerol from palm oil-based biodiesel was removed exactly with that ratio. The temperature was selected in order to minimize process costs by avoiding heating of the system. The short residence time was chosen to ensure fast screening of selected DESs and based on the previous research [8] where glycerol was removed from biodiesel in 0.22 min. In addition, a longer residence time was not chosen at that point simply to avoid complete glycerol extraction because in that way it would not be possible to choose the best DES. The obtained results are presented in Figure 2A,B.

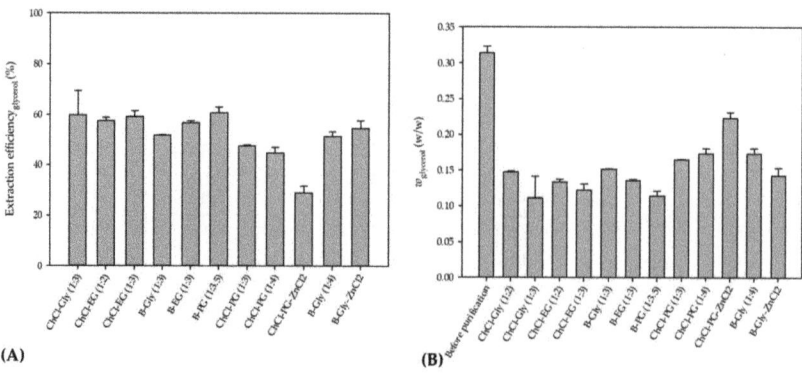

Figure 2. (A) Glycerol extraction efficiencies using different DESs and (B) mass fraction of glycerol before and after purification for tested DESs.

As can be seen from these preliminary results, all of the selected eutectic solvents removed glycerol from biodiesel, with the lowest extraction efficiency obtained with ChCl:PG:ZnCl$_2$, while the highest extraction efficiency was observed when ChCl:Gly (1:3), ChlCl:EG (1:3) and B:PG (1:3.5) were used. The presented results are in correspondence with the results of our previous research [8,48] where ChCl:Gly and ChlCl:EG were successfully used for glycerol removal.

In addition, all prepared DESs have a very low freezing point [11]. Due to the polarity of DESs, the existence of hydroxyl groups in both DES (solvent) and glycerol (solute), and the solvation power for glycerol in biodiesel, DESs have a high affinity for attracting glycerol through hydrogen bonding and dipole-dipole attraction. All this leads to improved extractability of the solvent [11].

When the obtained results (Figure 2B) for mass fractions of glycerol are observed, it can be seen that the glycerol amounts in biodiesel samples do not correspond to the values prescribed by the ASTM D6751 and EN 14214 standards. Due to that, further optimization of the extraction process was necessary. Among several DES candidates that proved effective in the removal of glycerol, ChCl:EG was chosen for further extraction optimization experiments due to its lowest viscosity, which assured the simplest possible operation of a microextractor. As the ethylene glycol mole fraction increases, the viscosity of the DESs gradually decreases, which is related to the intrinsically low viscosity of ethylene glycol.

3.4. Influence of Deep Eutectic Solvents Properties on Glycerol Extraction

The extraction efficiency is related to the properties of the DES used and it was defined which physicochemical properties of DES have the most influence on the extraction efficiency. To get insight into relationship between DESs properties and extraction efficiency, the regression analysis was used. Firstly, the relationship between DESs molecular descriptors and extraction efficiency was evaluated. Molecular descriptors can be defined as mathematical representations of molecules properties that are generated by algorithms [49] and there is still a big challenge in development and selecting of the specific one for each particular application [50]. According to Abranches et al. [26], σ-profile was presented as a potentially useful choice for a universal molecular descriptor. In this work, firstly, it was assumed that the extraction efficiency can be expressed as a function of the σ-profile of the mixture, expressed by a set of S^i_{mix} descriptors. To select the appropriate S^i_{mix} descriptors, mostly contributing the extraction efficiency, Spearman correlations between S^i_{mix} descriptors and extraction efficiency were analyzed. As presented in Table 3, significant correlations were noticed between S^1_{mix}, S^3_{mix}, S^4_{mix}, S^5_{mix}, S^6_{mix} and extraction efficiency and therefore those S^i_{mix} descriptors were used for further modeling. The similar approach to model input variables selection was previously described by Benguerba et al. [27] and Panić et al. [30]. By reducing the number of model input variables by excluding the non-significant ones, a low dispersion between observed DES viscosity data and multiple linear region model data was achieved.

Table 3. Correlation matrix for S^i_{mix} descriptors and extraction efficiencies (correlations significant for $p < 0.05$ are marked bold).

	S^1_{mix}	S^2_{mix}	S^3_{mix}	S^4_{mix}	S^5_{mix}	S^6_{mix}	S^7_{mix}	S^8_{mix}	S^9_{mix}	S^{10}_{mix}	E
S^1_{mix}	1.000	0.260	0.038	−0.439	0.287	**0.357**	**0.341**	0.095	−0.299	0.085	**−0.489**
S^2_{mix}	0.260	1.000	**0.848**	−0.050	−0.114	−0.012	**0.852**	**0.843**	0.131	0.170	0.247
S^3_{mix}	0.038	**0.848**	1.000	**0.454**	−0.342	−0.258	**0.841**	**0.984**	**0.636**	−0.219	**0.312**
S^4_{mix}	−0.439	−0.050	**0.454**	1.000	−0.647	−0.689	0.011	**0.373**	**0.903**	−0.418	**0.348**
S^5_{mix}	0.287	−0.114	−0.342	−0.647	1.000	**0.984**	0.077	−0.242	−0.419	0.005	**−0.438**
S^6_{mix}	**0.357**	−0.012	−0.258	−0.689	**0.984**	1.000	0.207	−0.137	−0.409	−0.062	**−0.470**
S^7_{mix}	**0.341**	**0.852**	**0.841**	0.011	0.077	0.207	1.000	**0.913**	**0.356**	−0.323	−0.001
S^8_{mix}	0.095	**0.843**	**0.984**	**0.373**	−0.242	−0.137	**0.913**	1.000	**0.620**	−0.330	0.218
S^9_{mix}	−0.299	0.131	**0.636**	**0.903**	−0.419	−0.409	**0.356**	**0.620**	1.000	−0.681	0.201
S^{10}_{mix}	0.085	0.170	−0.219	−0.418	0.005	−0.062	−0.323	−0.330	−0.681	1.000	0.248
E	**−0.489**	0.247	**0.312**	**0.348**	**−0.438**	**−0.470**	−0.001	0.218	0.201	0.248	1.000

The success of MLR, NLR and PLR models to describe the extraction efficiency of DESs was evaluated using R^2, R^2_{adj} and RMSE. As stated by Sanquetta et al., [51] criteria for model selection must incorporate goodness-of-fit allowing that several models examined can be simultaneously compared. Estimated model coefficients are given in Table 4 and comparisons between observed and model predicted data are shown in Figure 3. It can be noticed that the best agreement between observed and model predicted data was obtained with the PLR model, while the biggest dispersion between observed and model predicted data was found with the MLR model. It is also important to emphasize that, for all the three analyzed models, all estimated coefficients were significant, indicating that the most important descriptors were used as the model's input variables. It can also be noticed that estimated coefficients of the MLR, NLR and PLR models have the same trend. Coefficients b_1, b_3 and b_5 related with S^1_{mix}, S^4_{mix} and S^6_{mix} had negative values, while coefficients related with S^3_{mix}, and S^5_{mix} had positive values. As described by Zhang and Li [52], in the piecewise-regression analysis (also known as segmented regression) a dataset is split at a defined break point, and regression parameters (intercept and slopes) are calculated separately for data before and after the breakpoint. Piecewise linear regression is applicable if the data exhibit different linear trends over different domains [53]; thus, the regression can be made more accurate, as it is the case in this work (Figure 3C). By analyzing Figure 3A,B, it can be noticed that data behave differently below and above the 50% efficiency threshold, which was confirmed by the estimated break point at 52.9 ± 5.1 PLR models were also previously shown to be more efficient that MLR models for the prediction of the DESs pH values based on the σ-descriptors [30].

Table 4. MLR, NLR and PLR coefficients for prediction of extraction efficiencies based on DESs S^i_{mix} descriptors (significant coefficients for $p < 0.05$ are marked bold).

	MLR	NLR	PLR
Break point			**52.9 ± 5.1**
b_0	−6.1 ± 4.1	**3.9 ± 1.7**	**−9.0 ± 1.1** **40.7 ± 7.9**
b_1 (S^1_{mix})	**−184.1 ± 59.2**	**−0.08 ± 0.03**	**−201.2 ± 10.3** **−103.1 ± 11.8**
b_2 (S^3_{mix})	**3.6 ± 1**	**1 ± 0.4**	**4.6 ± 0.6** **0.3 ± 1 × 10^{-2}**
b_3 (S^4_{mix})	**−1.8 ± 0.6**	**−0.5 ± 0.1**	**−1.1 ± 0.1** **−0.6 ± 1 × 10^{-2}**
b_4 (S^5_{mix})	**5.5 ± 2.2**	**0.4 ± 0.1**	**2.3 ± 0.4** **0.8 ± 2 × 10^{-2}**
b_5 (S^6_{mix})	**−9.2 ± 3.2**	**−0.03 ± 0.01**	**−3.0 ± 0.2** **−0.4 ± 4 × 10^{-3}**
R^2	0.52	0.67	0.73
R^2_{adj}	0.46	0.52	0.68
RMSE	6.39	5.36	4.47
F−value	F (5,36) = 7.93	F (5,36) = 7.93	F (5,36) = 7.93
p−value	$p < 0.001$	$p < 0.001$	$p < 0.001$

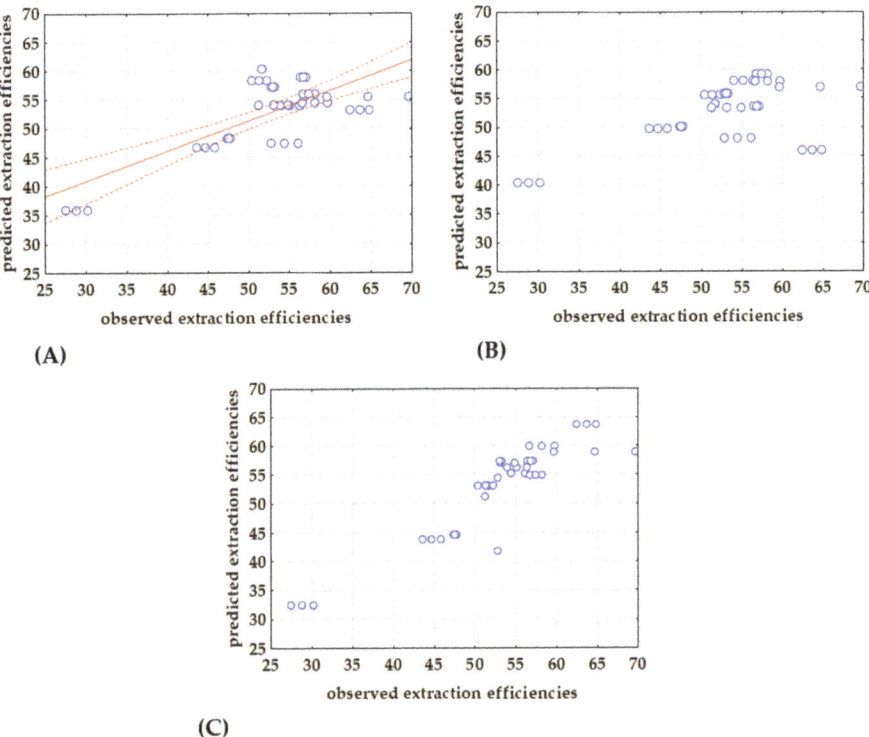

Figure 3. Comparison between observed and (**A**) MLR, (**B**) NLR and (**C**) PLR predicted extraction efficiencies based on DESs S^i_{mix} descriptors (—regression line, ···· 95 % confidence interval).

Taking into account the statement of Le Man et al. [54] that a regression model can be considered applicable if R^2 is greater than 0.75, the models based on σ-descriptors should be further improved. Therefore, a second approach was evaluated, where glycerol extraction efficiency (E) was modeled as the function of DESs physical properties including: density (ρ), dynamic viscosity (η), specific conductance (σ), refractive index (n_D), thermal diffusivity (a), dynamic viscosity (λ), and specific heat capacity (c_p). Estimated model coefficients are given in Table 5 and comparisons between observed and model predicted data are shown in Figure 4. As for the models based on σ-descriptors, the best agreement between experimental data and model predicted data (the highest $R^2 > 0.9$ and the lowest RMSE) was obtained for the PLR model (Figure 4C). Furthermore, estimated regression coefficients showed that density, electrical conductivity, refractive index, thermal conductivity and specific heat capacity had negative effects on the extraction efficiency, while dynamic viscosity and thermal diffusivity exhibited positive effects. Also, in the cases of MLR and PLR, all coefficients were significant, while in the case of NLR model, coefficients b_3 and b_4 describing the effects of specific conductance and refractive index, respectively, were non-significant. Moreover, ANOVA analysis showed that all developed models, including those based on σ-descriptors and those based on DESs physical properties, were significant with $p < 0.05$ and F-values higher that F-critical = 2.29 (Table 5).

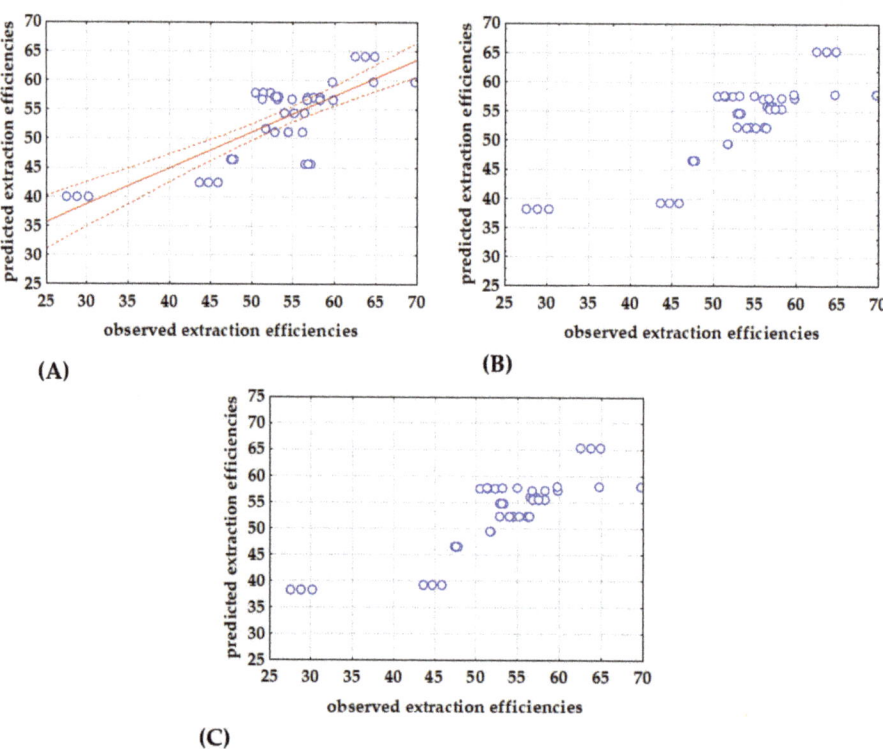

Figure 4. Comparison between observed and (**A**) MLR, (**B**) NLR and (**C**) PLR predicted extraction efficiencies based on DESs physical properties (—regression line, ···· 95 % confidence interval).

Table 5. MLR, NLR and PLR coefficients for prediction of extraction efficiencies based on DESs physical properties (significant coefficients for $p < 0.05$ are marked bold).

	MLR	NLR	PLR
Break point			52.9 ± 5.1
b_0	2489 ± 561.4	1.4 ± 0.9	1310.6 ± 157.1 -4315.7 ± 237.2
$b_1\ (\rho)$	-292.3 ± 83.8	-17.9 ± 4.3	-2604.4 ± 188.9 -367.9 ± 22.5
$b_2\ (\eta)$	$2 \times 10^{-2} \pm 1 \times 10^{-3}$	$-0.1 \pm 3 \times 10^{-2}$	$2 \times 10^{-2} \pm 1 \times 10^{-3}$ -46.9 ± 21.1
$b_3\ (\sigma)$	-0.7 ± 0.5	$-1 \times 10^{-2} \pm 3 \times 10^{-3}$	5.2 ± 2.1 1.5 ± 0.6
$b_4\ (n_D)$	-1573.7 ± 371.3	20.7 ± 1.7	-7991.6 ± 255.6 3084.7 ± 154.1
$b_5\ (a)$	-220.5 ± 97.3	-13.9 ± 4.1	694.9 ± 115.7 -317.5 ± 59.7
$b_6\ (\lambda)$	738.1 ± 18.4	16.6 ± 4.22	4789.4 ± 321.5 1455.2 ± 118.7

Table 5. Cont.

	MLR	NLR	PLR
b_7 (c_p)	-3.2 ± 1.2	-13.5 ± 3.9	217.2 ± 15.8 24.2 ± 3.8
R^2	0.62	0.73	0.97
R^2_{adj}	0.51	0.68	0.97
RMSE	5.89	4.47	1.45
F-value	F (7,34) = 7.9	F (7,34) = 7.9	F (7,34) = 7.9
p-value	$p < 0.001$	$p < 0.001$	$p < 0.001$

3.5. Extraction Optimization

After selecting the best DES for glycerol extraction, the next step was the optimization of the extraction conditions since at this point mass fraction of glycerol was still above the values prescribed by the ASTM D6751 and EN 14214 standards. The extraction optimization was performed in a continuously operated PTFE microextractor with two inlets and a T-junction. The effects of the (i) extraction temperature X_1 (T = 25, 40, 55 °C), (ii) residence time X_2 (τ = 0.05, 0.5, 0.95 min) and (iii) volume ratio of biodiesel:DES X_3 (9:1, 1:1, 1:9 v/v) were analyzed. The results obtained according to Box-Benkhen experimental design are presented in Table 6. Extraction efficiencies achieved were in the range from $E = 26.3 \pm 3.9$ % to $E = 57.2 \pm 0.5$%. The lowest extraction efficiency was obtained for T = 25 °C, τ = 0.50 min and biodiesel:DES ratio of 90:10, while the highest extraction efficiency was obtained for T = 40 °C, τ = 0.50 min and biodiesel:DES volume ratio of 1:9.

Table 6. Optimization of extraction by Box-Behnken experimental design (observed response factor levels are shown in parentheses).

Exp.	T/°C	τ/min	Volume Ratio of Biodiesel:DES	E/%
1	25.00 (−1)	0.50 (0)	9:1 (1)	26.9 ± 3.9
2	55.00 (1)	0.50 (0)	9:1 (1)	52.3 ± 0.2
3	40.00 (0)	0.50 (0)	1:1 (0)	53.1 ± 0.2
4	40.00 (0)	0.50 (0)	1:1 (0)	51.2 ± 0.6
5	55.00 (1)	0.95 (1)	1:1 (0)	50.6 ± 0.4
6	40.00 (0)	0.05 (−1)	9:1 (1)	44.4 ± 0.1
7	40.00 (0)	0.95 (1)	1:9 (−1)	46.2 ± 0.5
8	40.00 (0)	0.50 (0)	1:1 (0)	53.2 ± 0.2
9	25.00 (−1)	0.05 (−1)	1:1 (0)	53.5 ± 0.1
10	55.00 (1)	0.05 (−1)	1:1 (0)	51.7 ± 0.4
11	25.00 (−1)	0.50 (0)	1:9 (−1)	55.3 ± 0.5
12	40.00 (0)	0.50 (0)	1:1 (0)	53.6 ± 0.1
13	40.00 (0)	0.05 (−1)	1:9 (−1)	57.2 ± 0.5
14	40.00 (0)	0.50 (0)	1:1 (0)	54.6 ± 0.7
15	55.00 (1)	0.50 (0)	1:9 (−1)	40.1 ± 1.6
16	25.00 (−1)	0.95 (1)	1:1 (0)	52.9 ± 0.6
17	40.00 (0)	0.95 (1)	9:1 (1)	55.7 ± 0.3

The obtained results were as expected. Several factors influence the extraction processes while the extraction temperature, extraction time, characteristics of solid particles (size, shape, and condition), and type of solvent are the most important ones [8].

Moreover, the DES/biodiesel molar ratio has a positive effect on the overall glycerol removal efficiency. This means that if more DES is available, more glycerol will be removed.

To identify variables with a significant effect on the response variable, a second-order polynomial model was used to describe the experimental data. The regression coefficients of the developed models are given in Table 7. The results showed the significant influence of residence time and biodiesel:DES volume ratio in linear and quadratic coefficients and the significant effect of temperature in the quadratic coefficient of RSM model for extraction efficiency. By analyzing results presented in Table 7, it can be noticed that the residence time and biodiesel:DES volume ratio have a negative effect on the extraction efficiency, while temperature has a positive effect. Figure 5 depicts 3D response surfaces, which illustrate the interaction effects of the independent factors on the extraction efficiency. The graphs were created by correlating the response of two independent variables (the third was kept constant). It can be noticed that the extraction efficiency increases with temperature until optimum temperature is reached. Moreover, according to the R^2 value and RMSE of the proposed RSM model, it could be assumed that the RSM model accurately describes the experimental data. ANOVA revealed that the proposed model was significant ($p < 0.05$) and that the F-value for the model was higher than the F-critical (2.8).

Table 7. RSM model for description of glycerol extraction efficiency.

Coefficient	Regression Coefficient ± St. Error	p-Value
β_0	61.9 ± 5.2	<0.001
β_1 (T)	0.7 ± 0.2	0.005
β_2 (τ)	$-2 \times 10^{-2} \pm 3 \times 10^{-3}$	<0.001
β_3 (v/v)	-30.9 ± 5.2	<0.001
β_4 (T^2)	15.9 ± 3.0	<0.001
β_5 (τ)	$-0.6 \pm 6 \times 10^{-2}$	<0.001
β_6 (v/v)	$-3 \times 10^{-3} \pm 1 \times 10^{-3}$	<0.001
β_7 ($T \times \tau$)	$-2 \times 10^{-2} \pm 1 \times 10^{-2}$	0.823
β_8 ($T \times v/v$)	$2 \times 10^{-2} \pm 1 \times 10^{-3}$	<0.001
β_9 ($\tau \times v/v$)	$0.3 \pm 4 \times 10^{-2}$	<0.001
RSM model	R^2	0.93
	R^2_{adj}	0.91
	RMSE	1.96
	F-value	10.12
	p-value	<0.001

Given that a high R^2 value does not ensure that the model would fit the data well, residual analysis was also carried out. The results of the residual analysis are shown in Figure 6. Residuals were distributed approximately around the line (Figure 6A), and histograms depicting residual classification (Figure 6C) exhibited a distinctive bell shape, confirming the assumption of normality. Furthermore, by examining the plots of residuals vs. estimated values (Figure 6B), it is clear that the residuals were randomly distributed, indicating high agreement between the model and the experimental data. The residual analysis further revealed that the sequence of the experimental runs had no effect on the results, since the residuals distributed themselves around zero (Figure 6D). The acquired results suggest that the proposed response surface model was reliable for the examined range of input variables.

The optimal conditions were those that resulted in the highest extraction efficiency. According to RSM model, the optimal extraction conditions were T = 55 °C, τ = 0.95 min and biodiesel:DES volume ratio of 1:9. The optimal extraction conditions were estimated based on the desirability profiles obtained from the RSM predictions. The desirability scale ranged from 0 (undesirable) to 1 (very desirable). The estimated optimal conditions represent the local maximum for the selected range of input variables.

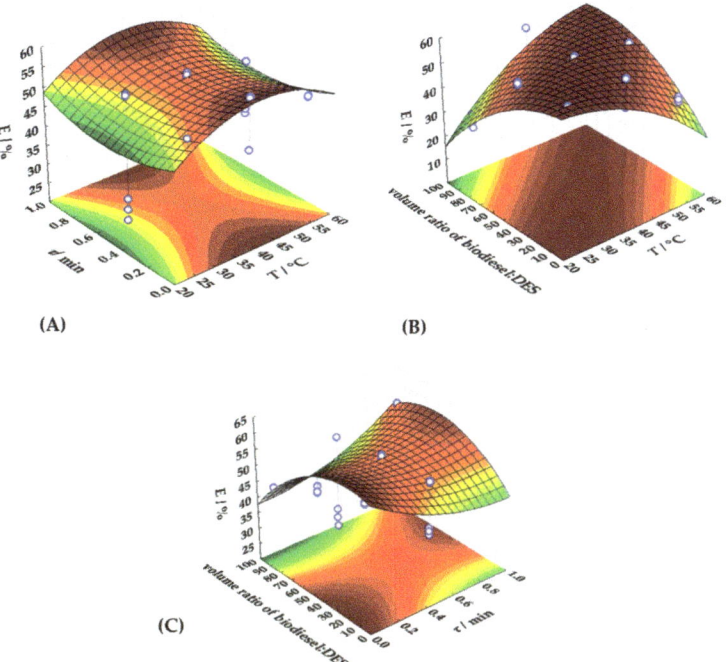

Figure 5. Three dimensional response surface plot for the interaction between (**A**) extraction temperature and residence time, (**B**) extraction temperate and biodiesel:DES volume ratio and (**C**) residence time and biodiesel:DES volume ratio.

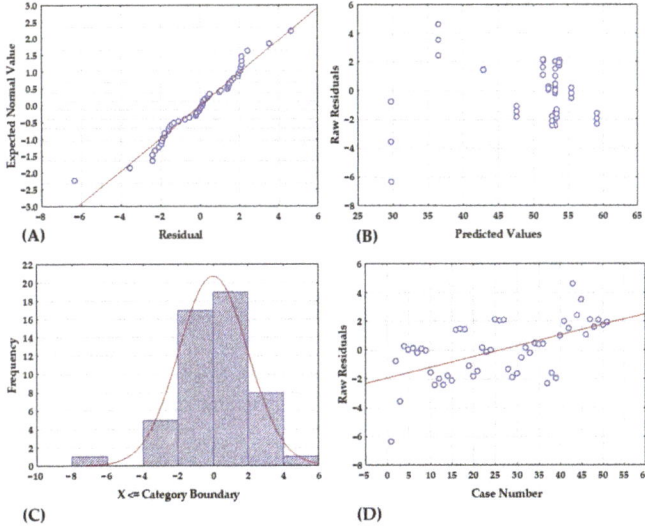

Figure 6. RSM model residual analysis (**A**) normality plot, (**B**) dependence between model predicted values and residuals, (**C**) histogram of residuals and (**D**) dependence between experiment number and residual value.

After optimizing the process conditions, the biodiesel purification process was carried out in a microextractor with the selected DES under optimal conditions to validate the mathematical model of the process. The obtained experimental results were compared with the simulation of the mathematical model. According to the mathematical model, it was predicted that 60.6% of glycerol would be removed from the biodiesel under optimal conditions. The experimental results showed that $53 \pm 5.2\%$ ($w = 0.2 \pm 0.1\ w/w$) of glycerol was removed during extraction under optimal conditions. Although there was a good agreement between the model and the experimental results, the results were not satisfactory since the mass fraction of glycerol was still higher than the values prescribed by ASTM D6751 and EN 14214 standards. Considering that the results under optimal conditions did not meet the standard requirements, the process of purification of biodiesel with eutectic solvent was carried out in a microextractor with a smaller channel diameter to increase the rate of glycerol transfer as a result of a shorter diffusion path, which should lead to higher extraction efficiency. This assumption was based on our previous work [8], where glycerol extraction was performed in three different microextractor channel sizes (250, 350 and 500 μm). Therein it was demonstrated that the channel diameter had a significant effect on the extraction efficiency. When the extraction was performed in the narrowest channel, the maximum extraction efficiency was obtained in 13.61 s and in the widest channel it took 55.03 s to obtain the same efficiency. In this study, up to this point, only the microextractor with the channel diameter of 1000 μm was used; however, new experiments were performed with the 500 μm diameter microchannel. When extraction was performed in a narrower channel, only a 4 percent point increase in extraction efficiency was observed. In addition to analyzing the influence of the diameter of the microextractor channel on the extraction efficiency, the influence of the residence time on the success of glycerol extraction was also investigated. It was found that the residence time in the studied range (0.5–30 min) had no significant influence on the extraction process in the microextractor, i.e., the achieved extraction efficiencies did not differ significantly among the residence times. At this point, it became clear that despite the change in extraction conditions, a higher efficiency of the process would not be achieved. However, comparing the obtained results with the previous studies [8,26,46], in which successful glycerol purification was achieved, it was observed that initial glycerol concentration in this study was significantly higher than in the previous ones. In this study, the glycerol concentration at the inlet of microextractor was 3.57 ± 0.78 mg/mL, while in the previous studies it was 0.8 mg/mL or 1.5 mg/mL. One has to understand that the compositions of the biodiesel and DES phase will approach equilibrium in the cases of sufficiently long microchannels or sufficiently low flowrates. Thus, molar fractions cannot exceed the equilibrium ones and, the higher is the initial molar fraction of glycerol on biodiesel, the higher are the equilibrium molar fractions of glycerol in both phases. But such high molar fractions of glycerol in biodiesel might exceed the standard requirements. It might be necessary to perform the extraction in more than a single stage. To estimate how many microextractors are required for the sufficient removal of glycerol, a mathematical model was used that has been described and validated elsewhere [8]. According to the mathematical model, two microextractors should be connected in series to remove the glycerol so that biodiesel can meet ASTM D6751 and EN 14214 standards (Figure 7A,B). To validate this, the outlet stream from the 1st microextractor was collected and biodiesel was separated from DES. Collected biodiesel, now containing 1.75 ± 0.21 mg/mL glycerol was introduced into the 2nd microextractor together with the fresh DES. As it can be seen from Figure 7B, predictions of mathematical model were validated with the experimental results and biodiesel of sufficient purity was obtained at the exit of microextractor.

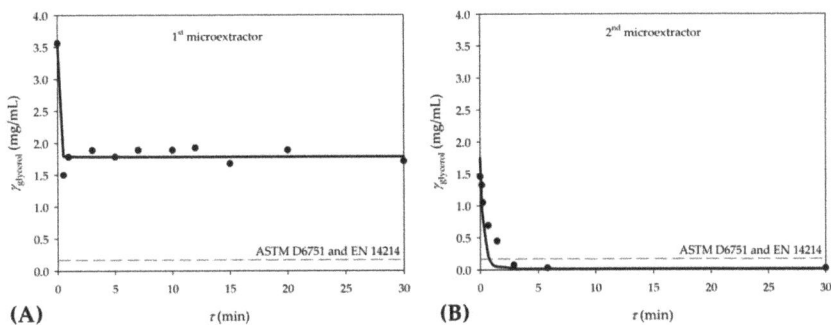

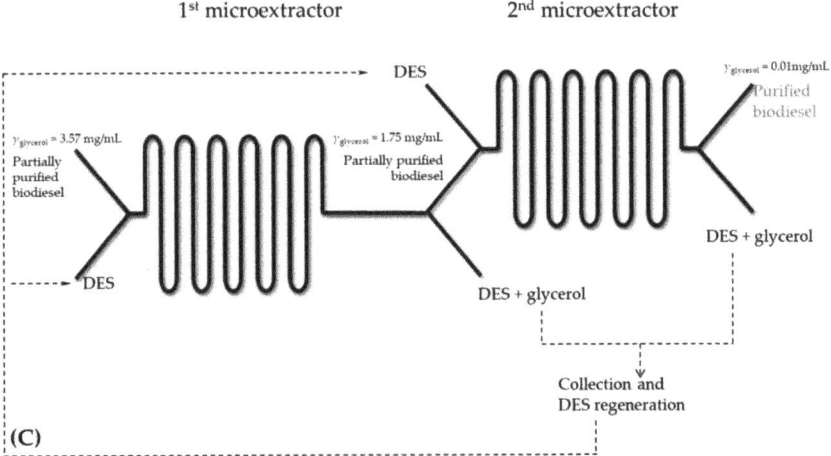

Figure 7. Influence of the residence time on the extraction of glycerol from biodiesel using DES as a solvent (**A**) glycerol concentration in 1st microextractor, (**B**) glycerol concentration in a 2nd microextractor and (**C**) proposed system composed of two microextractor connected into series for extraction of glycerol from biodiesel [— mathematical model, • experimental results (DES phase)].

Based on the obtained results, an integrated system (Figure 7C) was proposed as a possible solution for continuous glycerol extraction. In that system, partially purified biodiesel would enter the microextractor as one process stream and DES as the other one. As reported in our previous work [8], a stable and parallel flow is formed in this system, with the interface located exactly in the middle of the microextractor channel. This allows for complete phase separation at the outlet of the microextractor. Biodiesel that now contains 50% less glycerol than partially purified biodiesel at the inlet of the 1st microseparator is fed into the 2nd microextractor along with fresh DES, resulting with purified biodiesel at the exit. All DES waste streams could be collected, regenerated and returned to the process, making the proposed system more sustainable.

It is also noted that the extraction of glycerol is very fast in both microextractors. As it can be seen, the desirable extraction efficiency was reached after only 0.5 min in the 1st microextractor, as well as in the 2nd microextractor. This indicates that if the flow rate is chosen correctly and biodiesel can be completely purified in about one minute with microextractors connected in series, making this process significantly faster than other biodiesel purification processes.

4. Conclusions

In this work, optimization of the biodiesel purification process using eutectic solvents in a microextractor was carried out to obtain biodiesel suitable for use in internal combustion engines. Mathematical models were used to determine which physical and chemical properties of DES and to what extent affect the efficiency of extraction of glycerol from biodiesel. It was found that the physical and chemical properties of DES have an impact on glycerol separation. It was found that increasing density, electrical conductivity, refractive index, thermal conductivity and specific heat capacity negatively impact the extraction efficiency, while increasing dynamic viscosity and thermal diffusivity affects it positively. Among the 12 DES candidates, which allowed for the effective removal of glycerol, the highest extraction efficiency was observed when ChCl:Gly (1:3), ChlCl:EG (1:3) and B:PG (1:3.5) DESs were used. DES ChCl:EG (1:3) was chosen for further extraction optimization experiments due to its high extraction efficiency and the lowest viscosity compared to three mentioned DESs, which makes the microextractor operation rather simple. To determine the optimal process conditions (temperature, biodiesel:DES volume ratio, residence time), the three-level-three-factor Box-Behnken experimental design was used. The highest extraction efficiency was obtained at T = 40 °C, τ = 0.50 min, and biodiesel:DES volume ratio of 1:9. After optimizing the conditions, the biodiesel purification process was carried out in a microextractor with the selected DES under optimal conditions and only 52.99 ± 5.23% (w = 0.19 ± 0.12 w/w) of glycerol was removed. In addition, different microchannel diameters and different residence times were tested, which unfortunately also led to unsatisfactory results. As a possible solution, an integrated system, consisting of two microextractors connected in a series, was proposed for glycerol removal. The results obtained indicated that this new approach could be a good solution for glycerol removal, since the biodiesel was almost completely purified in about one minute, making this process much faster than other biodiesel purification methods.

Author Contributions: Conceptualization, A.Š. and A.J.T.; methodology, A.Š. and A.J.T.; software, A.J.T. and M.R. (Mia Radović); formal analysis, A.Š., M.C.B., A.J.T. and K.Z.K.; investigation, S.A., M.B., Ž.Ć., A.Š., A.J.T. and K.Z.K.; resources, B.Z.; data curation, S.A., M.B., Ž.Ć., A.Š., A.J.T. and K.Z.K.; writing—original draft preparation, A.Š. and A.J.T.; writing—review and editing, A.Š., A.J.T., K.Z.K., M.R. (Marko Rogošić), M.C.B. and B.Z.; visualization, A.Š. and A.J.T.; supervision, B.Z. All authors have read and agreed to the published version of the manuscript.

Funding: This research received no external funding.

Institutional Review Board Statement: Not applicable.

Informed Consent Statement: Not applicable.

Data Availability Statement: Not applicable.

Conflicts of Interest: The authors declare no conflict of interest.

References

1. Franjo, M.; Šalić, A.; Zelić, B. Microstructured devices for biodiesel production by transesterification. *Biomass. Convers. Biorefin.* **2018**, *8*, 1005–1020. [CrossRef]
2. Karaosmanoğlu, F.; Cığızoğlu, K.B.; Tüter, M.; Ertekin, S. Investigation of the refining step of biodiesel production. *Energy Fuels* **1996**, *10*, 890–895. [CrossRef]
3. Atadashi, I.M.; Aroua, M.K.; Aziz, A.A. Biodiesel separation and purification: A review. *Renew. Energy* **2011**, *36*, 437–443. [CrossRef]
4. Bateni, H.; Saraeian, A.; Able, C. A comprehensive review on biodiesel purification and upgrading. *Biofuel Res. J.* **2017**, *15*, 668–690. [CrossRef]
5. Jaber, R.; Shirazi, M.M.A.; Toufaily, J.; Hamieh, A.T.; Noureddin, A.; Ghanavati, H.; Ghaffari, A.; Zenouzi, A.; Karout, A.; Ismail, A.F.; et al. Biodiesel wash-water reuse using microfiltration: Toward zero-discharge strategy for cleaner and economized biodiesel production. *Biofuel Res. J.* **2015**, *2*, 148–151. [CrossRef]
6. Catarino, M.; Ferreira, E.; Soares Dias, A.P.; Gomes, J. Dry washing biodiesel purification using fumed silica sorbent. *Chem. Eng. J.* **2014**, *256*, 372–379. [CrossRef]

7. Sokač, T.; Gojun, M.; Jurinjak Tušek, A.; Šalić, A.; Zelić, B. Purification of biodiesel produced by lipase catalyzed transesterification by ultrafiltration: Selection of membranes and analysis of membrane blocking mechanisms. *Renew. Energy* **2020**, *159*, 642–651. [CrossRef]
8. Šalić, A.; Jurinjak Tušek, A.; Gojun, M.; Zelić, B. Biodiesel purification in microextractors: Choline chloride based deep eutectic solvents vs water. *Sep. Purif. Technol.* **2020**, *242*, 116783. [CrossRef]
9. Socas-Rodríguez, B.; Santana-Mayor, Á.; Herrera-Herrera, A.V.; Rodríguez-Delgado, M.Á. Deep eutectic solvents. *Green Sustan. Process Chem. Environ. Eng. Sci.* **2020**, *2020*, 123–177. [CrossRef]
10. Moura, L.; Kollau, L.; Gomes, M.C. Solubility of Gases in Deep Eutectic Solvents. In *Deep Eutectic Solvents for Medicine, Gas Solubilization and Extraction of Natural Substances*; Fourmentin, S., Costa Gomes, M., Lichtfouse, E., Eds.; Springer: Cham, Switzerland, 2021; pp. 131–155.
11. Shahbaz, K.; Mjalli, F.S.; Hashim, M.A.; AlNashef, I.M. Using deep eutectic solvents for the removal of glycerol from palm oil-based biodiesel. *Energy Fuels* **2011**, *25*, 2671–2678. [CrossRef]
12. Niawanti, H.; Zullaikah, S. Effect of extraction time on unreacted oil removal in biodiesel purification using deep eutectic solvent. *Reaktor* **2018**, *18*, 122–127. [CrossRef]
13. Petračić, A.; Gavran, M.; Škunca, B.; Štajduhar, L.; Sander, A. Deep eutectic solvents for purification of waste cooking oil and crude biodiesel. *Sci. J. Energy Eng.* **2017**, *5*, 87–94. [CrossRef]
14. Šalić, A.; Tušek, A.; Fabek, D.; Rukavina, I.; Zelić, B. Aqueous two-phase extraction of polyphenols using a microchannel system—Process optimization and intensification. *Food Technol. Biotechnol.* **2011**, *49*, 495–501.
15. Šalić, A.; Zelić, B. Synergy of microtechnology and biotechnology: Microreactors as an effective tool for biotransformation processes. *Food Technol. Biotechnol.* **2018**, *56*, 464–479. [CrossRef] [PubMed]
16. Farooq, M.Q.; Abbasi, N.M.; Anderson, J.L. Deep eutectic solvents in separations: Methods of preparation, polarity, and applications in extractions and capillary electrochromatography. *J. Chromatogr. A* **2020**, *1633*, 461613. [CrossRef]
17. Abbott, A.P.; Capper, G.; Davies, D.L.; Rasheed, R.K.; Tambyrajah, V. Novel solvent properties of choline chloride/urea mixtures. *Chem. Commun.* **2003**, *2003*, 70–71. [CrossRef]
18. Abbott, A.P.; Boothby, D.; Capper, G.; Davies, D.L.; Rasheed, R.K. Deep Eutectic Solvents formed between choline chloride and carboxylic acids: Versatile alternatives to ionic liquids. *J. Am. Chem. Soc.* **2004**, *126*, 9142–9147. [CrossRef] [PubMed]
19. Ruggeri, S.; Poletti, F.; Zanardi, C.; Pigani, L.; Zanfrognini, B.; Corsi, E.; Dossi, N.; Salomäki, M.; Kivelä, H.; Lukkari, H.J.; et al. Chemical and electrochemical properties of a hydrophobic deep eutectic solvent. *Electrochim. Acta* **2019**, *295*, 124–129. [CrossRef]
20. Plastiras, O.E.; Andreasidou, E.; Samanidou, V. Microextraction techniques with deep eutectic solvents. *Molecules* **2020**, *25*, 6026. [CrossRef]
21. Budžaki, S.; Šalić, A.; Zelić, B.; Tišma, M. Enzyme-catalyzed biodiesel production from edible and waste cooking oils. *Chem. Biochem. Eng. Q.* **2015**, *29*, 329–333. [CrossRef]
22. Zagajski Kučan, K.; Rogošić, M. Purification of motor fuels by means of extraction using deep eutectic solvent based on choline chloride and glycerol. *J. Chem. Technol. Biotechnol.* **2018**, *94*, 1282–1293. [CrossRef]
23. Rogošić, M.; Kučan, K.Z. Deep eutectic solvents based on choline chloride and ethylene glycol as media for extractive denitrification/desulfurization/dearomatization of motor fuels. *J. Ind. Eng. Chem.* **2018**, *72*, 87–99. [CrossRef]
24. Kučan, K.Z.; Perković, M.; Cmrk, K.; Načinović, D.; Rogošić, M. Betaine + (glycerol or ethylene glycol or propylene glycol) deep eutectic solvents for extractive purification of gasoline. *ChemistrySelect* **2018**, *3*, 12582–12590. [CrossRef]
25. Rogošić, M.; Zagajski Kučan, K. Deep eutectic solvent based on choline chloride and propylene glycol as a potential medium for extraction denitrification of hydrocarbon fuels. *Chem. Eng. Res. Des.* **2020**, *161*, 45–57. [CrossRef]
26. Abranches, D.O.; Zhang, Y.; Maginn, E.J.; Colón, Y.J. Sigma profiles in deep learning: Towards a universal molecular descriptor. *Chem. Commun.* **2022**, *58*, 5630–5633. [CrossRef]
27. Benguerba, Y.; Alnashef, I.; Erto, A.; Balsamo, M.; Ernst, B. A quantitative prediction of the viscosity of amine based DESs using Sσ-profile molecular descriptors. *J. Mol. Struct.* **2019**, *1184*, 357–363. [CrossRef]
28. Gojun, M.; Šalić, A.; Zelić, B. Integrated microsystems for lipase-catalysed biodiesel production and glycerol removal by extraction or ultrafiltration. *Renew. Energy* **2021**, *180*, 213–221. [CrossRef]
29. Jurinjak Tušek, A.; Jurina, T.; Benković, M.; Valinger, D.; Belščak-Cvitanović, A.; Gajdoš Kljusurić, J. Application of multivariate regression and artificial neural network modelling for prediction of physical and chemical properties of medicinal plants aqueous extracts. *J. Appl. Res. Med. Aromat. Plants* **2020**, *16*, 100229. [CrossRef]
30. Panić, M.; Radović, M.; Cvjetko Bubalo, M.; Radošević, K.; Rogošić, M.; Coutinho, J.A.P.; Radojčić Redovniković, I.; Jurinjak Tušek, A. Prediction of pH value of aqueous acidic and basic deep eutectic solvent using COSMO-RS σ Profiles' molecular descriptors. *Molecules* **2022**, *27*, 4489. [CrossRef]
31. Perry, S.L.; Higdon, J.; Kenis, P.J.A. Microfluidic Strategies to Mix Highly Viscous and/or Non-Newtonian Fluids. In Proceedings of the 2009 AIChE Annual Meeting, Nashville, TN, USA, 8–13 November 2009.
32. Tišma, M.; Zelić, B.; Vasić-Rački, Đ.; Žnidaršič-Plazl, P.; Plazl, I. Modelling of laccase-catalyzed L-DOPA oxidation in a microreactor. *Chem. Eng. J.* **2009**, *149*, 383–388. [CrossRef]
33. Aroso, I.M.; Paiva, A.; Reis, R.L.; Rita, A.; Duarte, C. Natural deep eutectic solvents from choline chloride and betaine—Physicochemical properties. *J. Mol. Liq.* **2017**, *241*, 654–661. [CrossRef]

34. Cardellini, F.; Tiecco, M.; Germani, R.; Cardinali, G.; Corte, L.; Roscini, L.; Spreti, N. Novel zwitterionic deep eutectic solvents from trimethylglycine and carboxylic acids: Characterization of their properties and their toxicity. *RSC Adv.* **2014**, *4*, 55990–56002. [CrossRef]
35. Craveiro, R.; Aroso, I.; Flammia, V.; Carvalho, T.; Viciosa, M.T.; Dionısio, M.; Barreiros, S.; Reis, R.L.; Duarte, A.R.C.; Paiva, A. Properties and thermal behavior of natural deep eutectic solvents. *J. Mol. Liq.* **2016**, *215*, 534–540. [CrossRef]
36. Dai, Y.; Witkamp, G.-J.; Verpoorte, R.; Choi., Y.H. Tailoring properties of natural deep eutectic solvents with water to facilitate their applications. *Food Chem.* **2015**, *187*, 14–19. [CrossRef] [PubMed]
37. Lapeña, D.; Lomba, L.; Artal, M.; Lafuente, C.; Giner, B. Thermophysical characterization of the deep eutectic solvent choline chloride: Ethylene glycol and one of its mixtures with water. *Fluid Phase Equilib.* **2019**, *492*, 1–9. [CrossRef]
38. Su, H.Z.; Yin, J.M.; Liu, Q.S.; Li, C.P. Properties of four deep eutectic solvents: Density, electrical conductivity, dynamic viscosity and refractive index. *Acta Phys. Chim. Sin.* **2015**, *31*, 1468–1473. [CrossRef]
39. Zhu, J.; Yu, K.; Zhu, Y.; Zhu, R.; Ye, F.; Song, N.; Xu, Y. Physicochemical properties of deep eutectic solvents formed by choline chloride and phenolic compounds at T = (293.15 to 333.15) K: The influence of electronic effect of substitution group. *J. Mol. Liq.* **2017**, *232*, 182–187. [CrossRef]
40. Lapeña, D.; Bergua, F.; Lomba, L.; Giner, B.; Lafuente, C. A comprehensive study of the thermophysical properties of reline and hydrated reline. *J. Mol. Liq.* **2020**, *303*, 112679. [CrossRef]
41. Škulcová, A.; Majová, V.; Dubaj, T.; Jablonský, M. Physical properties and thermal behavior of novel ternary green solvents. *J. Mol. Liq.* **2019**, *287*, 110991. [CrossRef]
42. Sedghamiz, M.A.; Raeissi, S. Physical properties of deep eutectic solvents formed by the sodium halide salts and ethylene glycol, and their mixtures with water. *J. Mol. Liq.* **2018**, *269*, 694–702. [CrossRef]
43. Siongco, K.R.; Leron, R.B.; Li, M. Densities, refractive indices, and viscosities of N,N-diethylethanol ammonium chloride-glycerol or ethylene glycol deep eutectic solvents and their aqueous solutions. *J. Chem. Thermodyn.* **2013**, *65*, 65–72. [CrossRef]
44. Sánchez, P.B.; González, B.; Salgado, J. José; J. Domínguez, Á. Physical properties of seven deep eutectic solvents based on L-proline or betaine. *J. Chem. Thermodyn.* **2019**, *131*, 517–523. [CrossRef]
45. Jafari, K.; Fatemi, M.H.; Patrice Estellé, P. Deep eutectic solvents (DESs): A short overview of the thermophysical properties and current use as base fluid for heat transfer nanofluids. *J. Mol. Liq.* **2021**, *321*, 114752. [CrossRef]
46. Abbott, A.P.; Capper, G.; Gray, S. Design of improved deep eutectic solvents using hole theory. *Chemphyschem.* **2006**, *7*, 803–806. [CrossRef]
47. Zhang, Q.; De Oliveira Vigier, K.; Royer, S.; Jérôme, F. Deep eutectic solvents: Syntheses, properties and applications. *Chem. Soc. Rev.* **2012**, *41*, 7108–7146. [CrossRef] [PubMed]
48. Šalić, A.; Jurinjak Tušek, A.; Sander, A.; Zelić, B. Lipase catalysed biodiesel synthesis with integrated glycerol separation in continuously operated microchips connected in series. *New Biotechnol.* **2018**, *47*, 80–88. [CrossRef] [PubMed]
49. Chandrasekaran, B.; Sara Nidal Abed, S.N.; Al-Attraqchi, O.; Kuche, K.; Tekade, R.T. Computer-Aided Prediction of Pharmacokinetic (ADMET) Properties. In *Advances in Pharmaceutical Product Development and Research, Dosage Form Design Parameters*; Tekade, R.K., Ed.; Academic Press: London, UK, 2018; pp. 731–755. [CrossRef]
50. Mullins, E.; Oldland, R.; Liu, Y.A.; Wang, S.; Sandler, S.I.; Chen, C.-C.; Zwolak, M.; Seavey, K.C. Sigma-profile database for using COSMO-Based thermodynamic methods. *Ind. Eng. Chem. Res.* **2006**, *45*, 4389–4415. [CrossRef]
51. Sanquetta, C.R.; Dalla Corte, A.P.; Behling, A.; de Oliveria Piva, L.R.; Netto, S.P.; Rodrigues, A.L.; Sanquetta, M.N.I. Selection criteria for linear regression models to estimate individual tree biomasses in the Atlantic Rain Forest, Brazil. *Carbon Balance Manag.* **2018**, *13*, 25. [CrossRef]
52. Zhang, F.; Li, Q. Robust bent line regression. *J. Stat. Plan. Inference* **2017**, *185*, 41–55. [CrossRef]
53. Ryan, S.E.; Porth, L.S. *A Tutorial on the Piecewise Regression Approach Applied to Bedload Transport Data*; U.S. Department of Agriculture, Forest Service, Rocky Mountain Research Station: Fort Collins, CO, USA, 2007; pp. 41–51.
54. Le Man, H.; Behera, S.K.; Park, H.S. Optimization of operational parameters for ethanol production from Korean food waste leachate. *Int. J. Environ. Sci. Technol.* **2010**, *7*, 157–164. [CrossRef]

Article

Upgrading of Biobased Glycerol to Glycerol Carbonate as a Tool to Reduce the CO_2 Emissions of the Biodiesel Fuel Life Cycle

Biagio Anderlini [1], Alberto Ughetti [1], Emma Cristoni [1], Luca Forti [2], Luca Rigamonti [1,3,4] and Fabrizio Roncaglia [1,3,4,*]

1. Department of Chemical and Geological Sciences, University of Modena and Reggio Emilia, Via G. Campi 103, 41125 Modena, Italy
2. Department of Life Sciences, University of Modena and Reggio Emilia, Via G. Campi 103, 41125 Modena, Italy
3. Interdepartmental Centre H2-MORE, University of Modena and Reggio Emilia, Via Università 4, 41121 Modena, Italy
4. INSTM Research Unit of Modena, Via G. Campi 103, 41125 Modena, Italy
* Correspondence: fabrizio.roncaglia@unimore.it

Abstract: With regards to oil-based diesel fuel, the adoption of bio-derived diesel fuel was estimated to reduce CO_2 emissions by approximately 75%, considering the whole life cycle. In this paper, we present a novel continuous-flow process able to transfer an equimolar amount of CO_2 (through urea) to glycerol, producing glycerol carbonate. This represents a convenient tool, able to both improve the efficiency of the biodiesel production through the conversion of waste streams into added-value chemicals and to beneficially contribute to the whole carbon cycle. By means of a Design of Experiments approach, the influence of key operating variables on the product yield was studied and statistically modeled.

Keywords: biodiesel; glycerol; flow chemistry; glycerol carbonate; carbon cycle; CO_2 capture

Citation: Anderlini, B.; Ughetti, A.; Cristoni, E.; Forti, L.; Rigamonti, L.; Roncaglia, F. Upgrading of Biobased Glycerol to Glycerol Carbonate as a Tool to Reduce the CO_2 Emissions of the Biodiesel Fuel Life Cycle. *Bioengineering* **2022**, *9*, 778. https://doi.org/10.3390/bioengineering9120778

Academic Editors: Indra Neel Pulidindi, Aharon Gedanken and Giorgos Markou

Received: 7 November 2022
Accepted: 25 November 2022
Published: 6 December 2022

Publisher's Note: MDPI stays neutral with regard to jurisdictional claims in published maps and institutional affiliations.

Copyright: © 2022 by the authors. Licensee MDPI, Basel, Switzerland. This article is an open access article distributed under the terms and conditions of the Creative Commons Attribution (CC BY) license (https://creativecommons.org/licenses/by/4.0/).

1. Introduction

In addition to different natural dynamics outside of our control, such as solar activity, developments over the last hundred years have been accompanied by a general and progressive increase in the mean global temperature, collectively known as 'global warming'. This variation is strictly related to several climate alterations such as melting glaciers and rising sea levels, but also to extreme weather events, a change in wildlife habitats, and an array of other impacts [1]. Beyond the heating due to direct irradiation from the sun, the atmosphere plays a determinant role in retaining part of the thermal energy close to the Earth's surface [1]. This heat-trapping phenomenon, known as the "greenhouse effect", comes from some naturally occurring gases, such as CO_2, methane, and nitrous oxides, and helps our planet to maintain stable conditions for life.

The continuous and ever-increasing exploitation of fossil carbon resources to sustain our development brought a symmetrical increase in greenhouse gas (GHG) emissions, especially CO_2, which have been linked to an increased greenhouse effect.

More sustainable strategies to drive our development without hampering the progress of future generations are urgently required and, in this context, a great deal of attention has recently been devoted to the study and implementation of biorefineries. Biorefineries are production facilities able to exploit renewable biomass instead of a fossil carbon resource and convert it into carbon-based products, such materials, fuels, or energy [2]. CO_2 fixation during plant growth can considerably reduce the impact on carbon balance when a vegetable source is exploited. This provides biorefineries with a greatly improved GHG emissions profile compared to traditional refineries. Among the different biorefinery types, biodiesel biorefineries focus on the production of carboxylic acid esters that can be directly blended with regular diesel fuel, starting from vegetable or animal fats. The latter are subjected to acid- or base-catalyzed transesterification in the presence of an excess of a short chain alcohol, to give a mixture of fatty acid methyl (or ethyl) esters, which is biodiesel. Depending on the triglyceride source, it has been estimated that the adoption of bio-derived diesel fuel brings between a 63% (energy crops) and 86% (waste fat) reduction in GHG emission compared to oil-based diesel fuel [3]. Moreover, a common factor in all transesterification processes is the co-production of glycerol (Gly) that is usually considered a waste stream, especially in small-scale productions where further refining costs are not justified. The present crude Gly price is as low as between USD 0.04 and 0.09/lb and it is expected to decrease further according to the growing industrial availability. Current global production is approximately 42 billion liters per year [4].

Gly shows some direct potential uses in the pharma, food, cosmetic, and polymer industries; however, its features (limited miscibility with organics, high boiling point, chemoselectivity of primary vs. secondary hydroxyl), as well as the presence of contaminants in the raw material, pose several barriers to its free applicability. As a result, the identification of effective processes able to economically convert Gly into value-added chemicals represents a key development needed to improve the sustainability of the whole biodiesel industry [5].

Various Gly-derived upgraded structures with enhanced properties and extended application profiles are well-described. Solketal [6] and other acetals, 1,3-propanediol [7,8] and dihydroxy acetone [9], represent known examples, and among these, a leading position is occupied by glycerol carbonate (GC), which represents the main focus of this work. As a multiple-site electrophile, GC can efficiently and selectively interact with diverse nucleophiles, such as amines [10–12], amino acids [13,14], phenols [15], carboxylates [16], and others [17]. The nucleophilic property of the free hydroxyl function can be directly exploited [18,19] or can be used to obtain a new electrophilic site capable of additional reactivity [20,21]. Valuable GC-derived oligomers or polymers, especially polyglycerols [22,23], polycarbonates, and polyesters [24], find application in biomedical tissue engineering [25] and in the controlled release of active pharmaceutical ingredients [26]. Polyglycerols are also applied in water-dispersible polymers, adhesives [27,28], and bio-based surfactants [16]. As a building block, GC has been included in diverse classes of polymeric materials, such as acrylates [29,30], polyesters [31–33], isocyanate-free polyurethanes [33–37], hybrid structures [38,39], and others [40,41]. Finally, GC can also be used as a carbonate carrier able to transfer the same function to other polyols, including sugars [42,43], through transcarbonylation processes. The carbonate function can also be considered as a precursor to highly electrophilic (and valuable) epoxy function [44,45].

Thanks to the excellent properties of GC, a number of methods able to convert Gly into GC were devised [46,47]. Based on the nature of the main interacting reagent involved, they can be divided into three groups, as follows:

(i). processes using activated phosgene-sourced reagents, such as phosgene itself, chloroformates, or carbonyldiimidazole;
(ii). processes using activated reagents not sourced from phosgene, such as dialkyl carbonates, diaryl carbonates, or $CO + O_2$;
(iii). processes using non-activated CO_2-sourced reagents, such as CO_2 itself or urea.

Class (i) processes are characterized by high Gly conversion and product selectivity, but also by health and environmental issues due to the toxicity of the involved reagent (or precursor). The reactivity of class (ii) reagents allows high efficiency as well and, thanks to the better sustainability profile, the related processes are the most investigated [48–50]. Different implementations of these processes are described, extending through various catalysts and plant engineering, including continuous-flow techniques [51,52]. Organic carbonates can, in principle, be prepared from CO_2, however, aside from promising research advancements [53,54], no industrial process is currently available, possibly due to known issues such as process reversibility and hydrolytic instability of the product [55].

Dealing with class (iii) reagents, CO_2 certainly represents the more atom economical choice, but the same hurdles described to form alkyl carbonates also hamper the direct carbonylation of glycerol to GC [56,57]. The option to use highly toxic glycidol as an activated Gly-sourced substrate [58,59] able to overcome the chemical inertness of CO_2 presents critical toxicity features similar to phosgene or gaseous CO [60]. To pursue chemical upgrading routes characterized by a non-toxic and stable reactant as well as featuring high atom economy, our attention was devoted to urea. Urea is a white crystalline solid containing 46 wt% of nitrogen and, being non-toxic, is largely used as an animal feed additive and fertilizer [61]. It is even more resonance-stabilized [62] than CO_2, thus, its reactivity with polyols is expected to not be very manifest; in fact, thermally stable mixtures between urea and polyols are well-described [63] and are used as deep eutectic solvents. What makes urea an attractive carbonylation agent is its multi-basic nature [64], which makes the coordination of metal cations possible [65]. These urea–cation complexes are the key to converting urea into a highly electrophilic "masked" isocyanate species [66,67], where the resonance is strongly reduced by the interaction with the metal center. This allowed the easy nucleophilic attack by Gly, as shown in Figure 1. The process is reversible [68] and proceeds stepwise, with the formation of a carbamic acid intermediate (**I**) (Figure 1) [69], which is converted in the final product through the elimination of a second ammonia molecule [70].

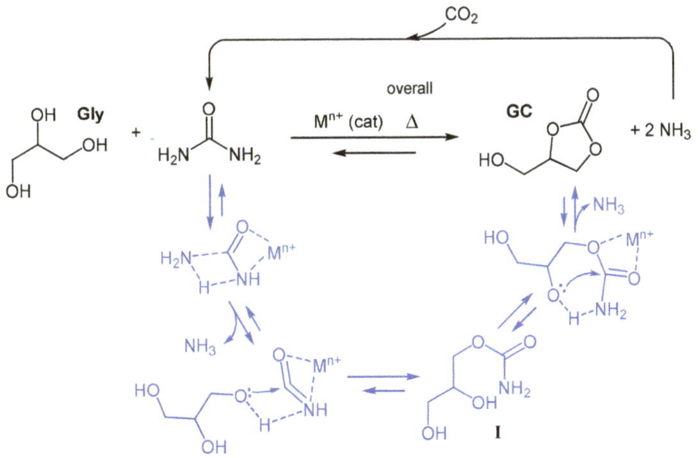

Figure 1. Metal-catalyzed carbonylation of Gly with urea.

Urea is industrially prepared from ammonia and CO_2 [71], therefore, it can be considered a CO_2 carrier [72] and promises additional advantages regarding the overall carbon balance, especially when the proper recycling of ammonia is implemented. In other words, the upgrading of Gly to GC involves the fixation of one mole of CO_2 per mole of vegetable oil (i.e., ~5 wt%, when 880 g/mol is taken as a mean molecular mass of a vegetable oil), a fact that, together with refined farming practices [3], can give additional support to reducing GHG emissions within the whole biodiesel industry.

In this paper, we present a novel continuous-flow process able to convert Gly into GC. The influence of key operating variables, such as time, temperature, and urea/Gly molar ratio (MR), on the product yield and selectivity are studied and statistically modeled by means of a Design of Experiments (DoE) approach.

2. Materials and Methods

2.1. General Information

Solvents and reagents were commercial grade and used as received. Glycerol (from vegetable source) was purchased from Merck (Milano, Italy). $ZnSO_4 \cdot H_2O$ was obtained by heating $ZnSO_4 \cdot 7H_2O$ overnight in an oven at 130 °C, which was then stored in a desiccator. Gly was vacuum dried for 4 h in the rotavapor (water bath 40 °C) and stored in a capped bottle. Urea was dried at 65 °C overnight and stored in a desiccator. ^{1}H NMR spectra were acquired with a Bruker Avance 400 spectrometer (Billerica, MA, USA).

2.2. Procedure for Preliminary Batch Reactions

In a 10 mL Schlenk tube with a screw cap and equipped with a stirring bar, Gly (1.0 g, 10.9 mmol), solid catalyst (0.05 mol/mol Gly, see later), and urea (0.82 g, 13.65 mmol, 1.25 mol/mol Gly) were inserted. A membrane vacuum pump was connected through the side arm and the vacuum was set at 400 mmHg. The capped tube was inserted in an oil bath (pre-heated at 150 °C) and stirred for 4 h. The crude reaction mixture, once cooled to room temperature (r.t.), was extracted with EtOAc:Et$_2$O (4:1 mixture, 3 × 5 mL), and the crude colorless product, obtained from the dried organic phase, was analyzed with ^{1}H NMR using CDCl$_3$ or D$_2$O as the solvent.

2.3. Procedure for Continuous Flow Reactions

The plant assembly is depicted in Figure 2. The reacting mixture was continuously recirculated through a heated tubular reactor (R) by means of a Bellco (model: BL 758, Mirandola, Italy) peristaltic pump (P). A mixing chamber (M), composed of a 25 mL two-necked round bottom flask, was heated and stirred through a standard stirring plate and an oil bath. A three-arm distillation connector acted as an expanding chamber (E) and connected the reactor output to an air-cooled condenser (A, also connected to M) and a vacuum line, set at 400 mm Hg. The said reactor was composed of a metallic AISI 316 stainless steel tube (1/16 in od × 1.2 mm id, 3 m long, ~4 mL internal volume, sourced from Restek, Milano, Italy) coiled around a 4 cm diameter cylindrical aluminum block, featuring a slot for a heating resistor and temperature sensor. Both of these heating and measuring elements were connected to a PID controller (Rex-C100, sourced from RobotDigg, Shangai, China), which allowed for setting and maintaining the desired temperature. Some thermal insulation (not shown in figure) was wrapped around the external side of the reactor.

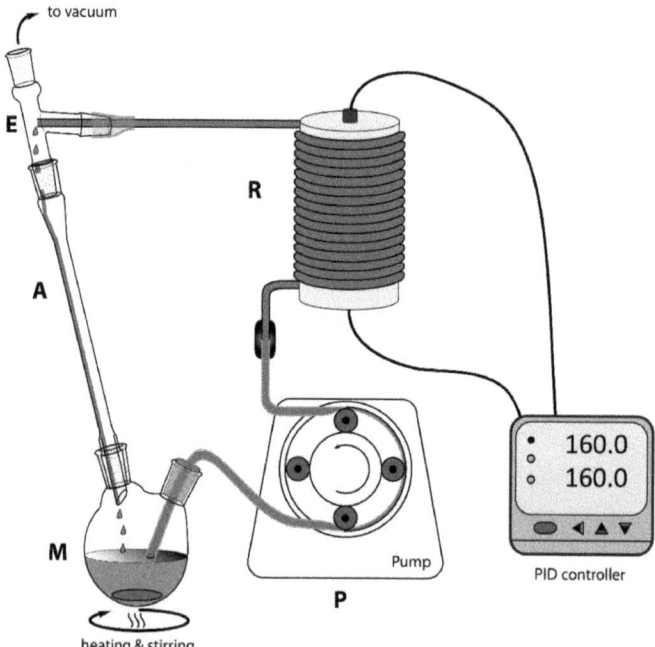

Figure 2. Layout of the continuous flow reactor. M = mixing chamber; P = peristaltic pump; R = coiled tubular reactor; E = expansion chamber; A = air-cooled condenser.

Gly (11.0 g, 0.119 mol), ZnSO$_4$·H$_2$O (0.643 g; 3.58 mmol, 0.03 mol/mol Gly), and urea (e.g., 8.97 g, 0.149 mol for 1.25:1.00 urea:Gly molar ratio) were stirred at 65 °C within the mixing chamber (M, Figure 2) until a homogeneous solution was obtained. By that time, the temperature of the coiled reactor (R) was set at the desired value. Then, the peristaltic pump was started at constant flow (3.0 mL/min), and the mixture was recirculated for the desired time. To assess yield and conversion of each experiment, a 2.0 g portion of the crude reaction mixture was extracted with EtOAc:Et$_2$O (4:1 mixture, 3 × 5 mL), and the raw colorless product, obtained from the organic phase, was carefully dried at the rotavapor. Purity was directly assessed with ^{1}H NMR spectroscopy (see Supplementary Materials). GC gave the following signals (Figure S1): ^{1}H NMR (400 MHz, CDCl$_3$, 298 K): δ_H (ppm) = 4.81 (1H, CH, m), 4.49 (2H, CH$_2$, m), 3.99 (1H, CH$_2$, dd, J = 12.84 Hz; 3.1 Hz), 3.72 (1H, CH$_2$, dd, J = 12.84; 3.5 Hz), 2.21 (1H, OH, br). The metal catalyst and some carbamic acid intermediates (currently not characterized) constituted the denser, polar phase insoluble in the extracting mixture. No residual Gly was detected.

2.4. Design of Experiments (DoE)

Nineteen experiments were suggested, and the relative output data were statistically analyzed by means of Design Expert® v.12 software (Stat-Ease Inc., 1300 Godward Street NE, Suite 6400, Minneapolis, MN 55413 USA). A three-level face-centered cubic design with five replicates of the central point was implemented. The investigated independent parameters were time (90–210 min), temperature (175–195 °C), and urea/Gly ratio (1.2–1.8), while output parameters were GC yield (%) and GC selectivity (%). Reagent flow was demonstrated to be not significant during the preliminary experiments, therefore, it was kept constant at 3.0 mL/min. Pressure was set to 400 mmHg because of hardware technical limits (squeezing of peristaltic pump tube) as well as urea losses through sublimation.

3. Results

Gly carbonylation with urea commonly involves solventless operation, temperatures ranging from 120 °C to 160 °C, reaction times from 3 to 24 h, and the presence of a catalyst (metal salt or metal oxide). Some degree of vacuum is also commonly applied in order to favor the removal of ammonia, obtaining a desired equilibrium shift [73,74] (Figure 1). A number of catalysts were described for this transformation [75–78] and, among these, zinc-containing species are by far the most active and popular. Irrespective of its initial form, soluble zinc-based catalysts are expected to give the same performance thanks to the formation of the metal glycerolate [79–81], while insoluble forms [67,82–84] are involved in heterogeneous reactions.

It is interesting to note that most of the recent papers have focused on an accurate description of the catalyst but have not given enough attention to the separation and isolation of GC, which represent key steps for considering the industrial feasibility of the chemical process [74].

Our investigations started with some preliminary experiments in which mixtures composed of Gly:urea:catalyst in a 1.00:1.25:0.05 molar ratio (MR) were heated in batch at 150 °C for 4 h, keeping the pressure at 400 mmHg (see materials and methods). After extraction with organic solvents, the amount of GC was evaluated with ^{1}H NMR spectroscopy [85]. This screening, of which the results are collected in Table 1, let us draw some general observations, such as (i) yields are limited to 30%; (ii) longer reaction times do not improve the result (entry 3 vs. 2, Table 1); (iii) temperatures higher than 150 °C induce degradation (reaction mixture became brown and lower yields of GC were obtained); (iv) pressures lower than 400 mmHg induce urea losses through sublimation [86]; (v) catalyst anhydrification is beneficial (entry 2 vs. 1).

Table 1. Screening of alternative catalysts in the batch conversion of Gly to GC [a].

Entry	Catalyst	Isolated Yield (%)
1	$ZnSO_4 \cdot 7H_2O$	27
2	$ZnSO_4 \cdot H_2O$	30
3 [b]	$ZnSO_4 \cdot H_2O$	27
4	$ZnCl_2$	30
5	$FeCl_3$	20
6	MgO	22

[a] Gly (1.00 g, 10.9 mmol), urea (0.815 g, 13.6 mmol), catalyst (5 mol% resp. to Gly), 150 °C, 4 h, 400 mmHg.
[b] reaction time extended to 6 h.

Cheap and easily available $ZnSO_4 \cdot H_2O$ was chosen as the reference catalyst, especially because of its ability to give homogeneous mixtures that, being simpler reacting systems, let us focus on engineering approaches towards process improvement.

Some attempts to set up a reactive distillation [87] were carried out, however, the high boiling point of GC (~140 °C at 0.5 mmHg) was a considerable hurdle to its isolation. Better success comes from the known implementation of microwave-heated batch reactors, giving improved isolated yields and reduced reaction times [88,89]. As this suggests, a main problem of the batch process featuring traditional heating is likely to be the supply of thermal energy to the reacting mixture that, being subjected to an endothermic event [73], is affected by self-cooling, especially in the core (far from the heating bath). Therefore, to improve such heat transfer limitations, our focus was directed to the implementation of a high area-to-volume ratio reactor, such as a small diameter tubular reactor, together with a continuous flow of reactants inside it.

Solventless operation is a valuable feature of this process, but the high viscosity of Gly/urea mixtures poses serious troubles concerning their pumpability. Different HPLC pumps were not able to give a reliable flow, possibly due to the inability of check valves to promptly block the backflow. The addition of a high boiling point solvent such as DMSO was proposed as a solution [90], but we judge this practice undesirable as it nullifies the "solventless advantage" and makes the isolation of reaction products troublesome. Fortunately, the use of a peristaltic pump in addition to a heated mixing chamber (M, Figure 2) allowed us to obtain a steady flow of reactants. In fact, by keeping the Gly/urea mixture at 65 °C, a useful reduction in viscosity was observed and, more importantly, the precipitation of urea within the tubes was avoided.

Using the flow chemistry technique in small volume reactors, thanks to strongly increased heat and mass transfer, typically benefits enhanced reaction control with respect to batch processes, especially in fast reactions [91]. Nevertheless, the process under study (Figure 1) features slow kinetics as evidenced by the long reaction times required for complete conversion, such as 2 h at 150 °C with microwave irradiation [89]; the meagre conversion obtained with reaction times of a few minutes, even with the implementation of a capillary reactor [90], gives evidence of that. Using a coiled 3 m long tubular reactor, we first tried to get the best conversion through a single pass, keeping flow at a minimum (0.2 mL/min) and evaluating increased temperatures. Unluckily, unsatisfactory conversion was obtained even at 195 °C, while greater temperatures resulted in brown (degraded) mixtures, containing low amounts of GC. A possible solution was to consider multiple passes through the reactor, thus, a recirculating layout was set up, as shown in Figure 2.

Gly, urea, and the catalyst were stirred at 65 °C in the mixing chamber (M, Figure 2) to obtain a homogeneous mixture. This was continuously pumped through the heated coiled tubular reactor (R, Figure 2) and the outflow was recirculated through an air-cooled condenser (A) into the mixing chamber. The total volume of the reacting mixture was set to ~1.2 times of the total piping volume (reactor and interconnections) to maximize the number of passes through R. The entire system was maintained at reduced pressure by a membrane pump connected to E. At time intervals, a sample of reacting mixture was withdrawn from the mixing chamber to assess conversions and yields. Several trials let us delineate the general features of the system, as follows:

- temperatures greater than 175 °C resulted in better GC yields;
- a minimum time of 90 min was required to obtain complete conversion of Gly;
- pressures lower than 400 mmHg resulted in unreliable flow due to peristaltic pump malfunction (tube squeezing) and minor urea losses due to sublimation;
- yields were unaffected by changes in flow rate, within the range from 0.5–5.0 mL/min;
- yields were unaffected by changes in catalyst amount, within the range from 0.03–0.05 $mol_{ZnSO_4 \cdot H_2O}/mol_{Gly}$;
- diglycerol tricarbonate (DGTC) was identified as the main by-product [23].

In order to get a better understanding of the influence of multiple process parameters, we decided to implement a multivariate statistical evaluation based on a DoE approach. In particular, a three-level face-centered central composite design (CCD) was chosen and was used to define an appropriate number of experiments within the variable domains arising from the above observations. The included independent variables were the reactor temperature (ranging from 175 to 195 °C), the recirculation time (from 90 to 210 min), and the urea:Gly MR (from 1.2 to 1.8); while the other process parameters such as pressure, flow, amount of catalyst, and mixing chamber temperature were fixed at 400 mmHg, 3.0 mL/min, 0.03 $mol_{ZnSO_4 \cdot H_2O}/mol_{Gly}$, and 65 °C, respectively. As suggested by the DoE CCD model, we planned nineteen experiments, with five replicates of the central point, as shown in Table 2. All the experiments were conducted by the same operator to minimize systematic errors, while the order of experiments was randomized. The two monitored responses were GC yield and GC purity. A quantitative evaluation of the reaction selectivity (100·mol GC/(mol GC + mol DGTC)) was obtained by the integration of isolated ^{1}H NMR signals of the product and isolated ^{1}H NMR signals of the main by-product (DGTC), as described in the Supplementary Materials (Figure S2). This was supported by the fact that the ^{1}H NMR signals due to other substances always presented at a very low intensity.

Table 2. Experiments suggested by the DoE model [a].

Exp. n.	Temperature (°C)	Urea:Gly MR	Time (min)	GC Yield (%)	GC Selectivity (%)
1	195	1.8	210	25.8	57
2	175	1.2	90	33.6	93
3	175	1.8	210	37.9	68
4	185	1.5	210	35.2	85
5	185	1.2	150	40.3	91
6	175	1.2	210	37.9	90
7	195	1.2	90	41.9	86
8	185	1.5	150	42.9	88
9	185	1.8	150	42.3	80
10	175	1.8	90	33.4	82
11	195	1.2	210	27.0	84
12	185	1.5	150	45.1	88
13	185	1.5	150	45.7	86
14	185	1.5	150	42.8	89
15	185	1.5	150	41.3	89
16	195	1.8	90	38.7	64
17	175	1.5	150	39.0	87
18	185	1.5	90	41.0	82
19	195	1.5	150	40.3	83

[a.] fixed parameters: P = 400 mmHg; flow = 3.0 mL/min; catalyst amount 0.03 $mol_{ZnSO_4 \cdot H_2O}/mol_{Gly}$.

4. Discussion

The ability to statistically evaluate the interrelation between process variables is one of the main advantages offered by the DoE approach. For instance, the influence of the three independent variables on the output responses are shown in Figure 3. GC-isolated yields (Figure 3a–c) strongly depends upon recirculation time and reactor temperature, while less marked is the influence of the urea:Gly MR. Best GC yields are obtained at temperatures between 180 and 190 °C, times between 110 and 160 min, and for urea:Gly MR in the range from 1.3–1.7. GC selectivity (Figure 3d–f) strongly depends upon urea:Gly MR and reactor temperature, while recirculation time is found to be the less influent parameter. Best selectivities are obtained at lower temperatures, lower urea:Gly MR, and times between 110 and 170 min.

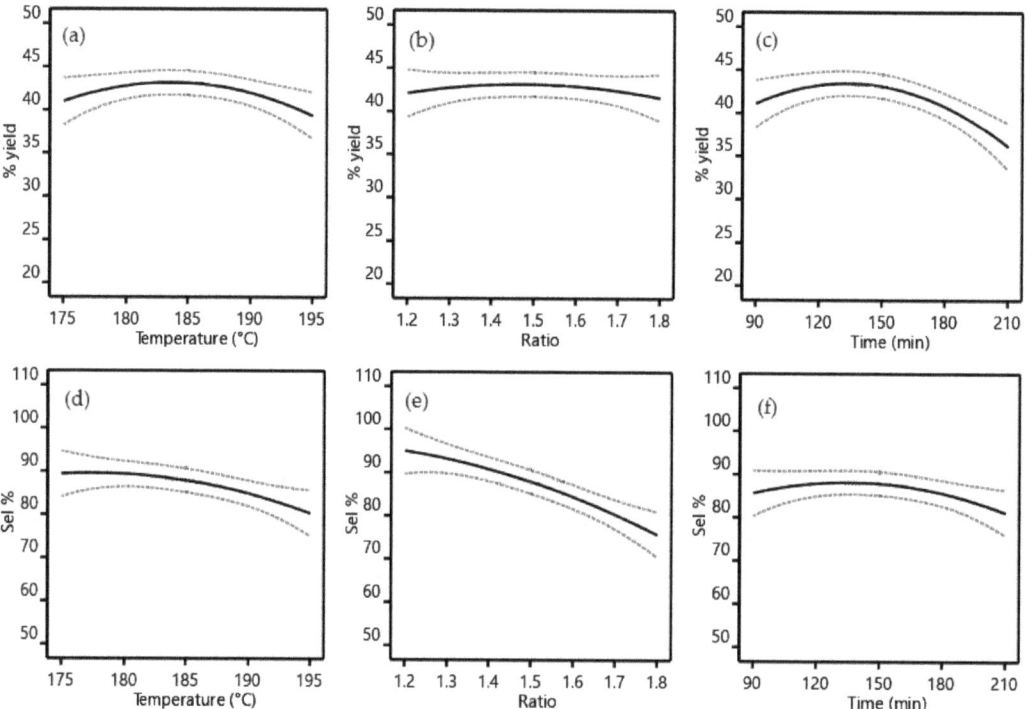

Figure 3. All factors graphs. (**a**) GC yield% vs. temperature, (**b**) GC yield% vs. urea/Gly MR, (**c**) GC yield% vs. recirculation time, (**d**) GC selectivity% vs. temperature, (**e**) GC selectivity% vs. urea/Gly ratio and (**f**) GC selectivity% vs. recirculation time. Dashed blue lines refers to minimum and maximum values of the data set. Continuous black lines refers to the mean values.

The simultaneous influence of two independent variables on each physical property is of particular value. Figure 4 shows some "two factors response surfaces", which put into evidence the dependency of GC yield% (left of Figure 4) from time–temperature or urea:Gly MR–time couples. On the right of the same figure, the influence of time–temperature or ratio–temperature couples on GC selectivity% are shown.

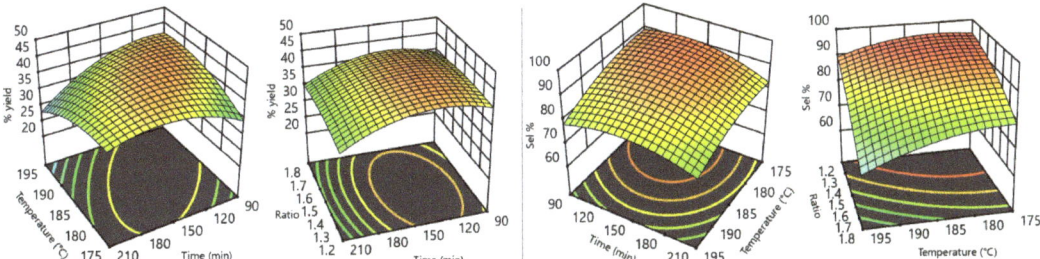

Figure 4. Two factors response surfaces.

Finally, both GC yield% and GC selectivity% response parameters can be conveniently composed in a single factor, called "desirability", which is able to describe the best process conditions in a simple and effective way. In the present case, desirability is obtained by the product of the said response parameters, normalized to unity. A desirability of 1.0 means that both parameters are maximized, while the lowest desirability (0.0) describes conditions where both responses are at minimum levels. In Figure 5, some contour heatmap plots of desirability as a function of independent variable couples are shown.

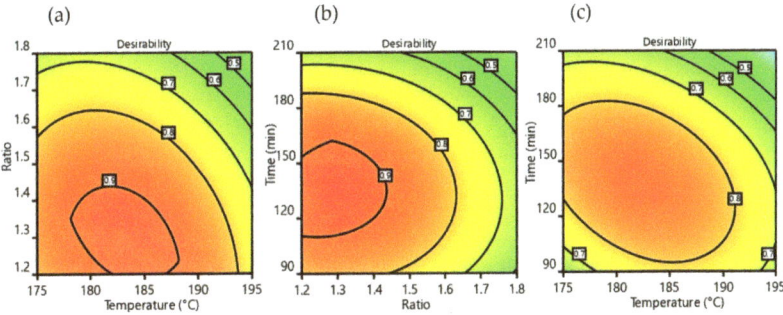

Figure 5. Desirability contour maps, (**a**) vs. urea/Gly MR-temperature, (**b**) vs. time-urea/Gly MR and (**c**) vs. time-temperature.

The strong reduction in process performance at high urea:Gly MR (Figure 5a,b) comes from a loss of selectivity (see also Figure 3b,e), meaning an increased formation of DGTC. This behavior is not surprising, as DGTC is described as an overreaction product of GC when the carbonylation reagent is in high molar excess [23,92,93]. This also suggests good potential for DGTC production as a possible future implementation of the process. The significant yield decrement at high times and temperatures (Figures 3 and 5c) found good correlation with the experimental observation of the brownish color acquired by the reacting mixture in these circumstances, a sign of partial degradation. Overall, best operating conditions of the process were found to be a reactor temperature from 180 to 185 °C, a recirculation time from 120 to 150 min, and a urea:Gly MR of 1.25.

Regarding the potential risk of isocyanic acid emission coming from carbamate decomposition [94,95], the process conditions (low operating temperatures, low urea:Gly MR, presence of glycerol) make this event highly unlikely. As an additional safety measure, it might be considerable, for further developments, to add a "water trap" on the output of the vacuum pump.

5. Conclusions

The most challenging issue we faced during the development of the process was the pumping of the viscous reacting mixture. This was solved by means of a peristaltic pump (P, Figure 2) and by a heated mixing chamber (M) able to lower the viscosity and increase urea solubility. The technical limits of this implementation were evidenced when we attempted to operate at a pressure lower than 400 mmHg. This pressure constraint is thought to be one of the factors limiting the GC-isolated yield to ~42%; therefore, the employment of different pumping hardware could significatively improve performance. Moreover, to the best of our knowledge, this is the first example of a multiple pass tubular reactor applied to the conversion of Gly into GC. This plant layout (Figure 2) allowed for the management of a kinetically slow process within a confined, highly thermally controlled reactor (R). The same layout also features a closed loop with a single evacuation point (E), which has the particular advantage of collecting the co-produced ammonia and recycling it to urea (Figure 1). This feature, coupled with the modest molar excess of urea required for the transformation, demonstrate the opportunity to sustainably transfer an equimolar amount of CO_2 to Gly, beneficially contributing to the entire carbon cycle of the biodiesel industry.

Supplementary Materials: The following supporting information can be downloaded at https://www.mdpi.com/article/10.3390/bioengineering9120778/s1, Figure S1: Typical ^{1}H NMR spectra ($CDCl_3$, 400 MHz) of a GC-rich raw mixture; Figure S2: Typical ^{1}H NMR spectra ($CDCl_3$, 400 MHz) of a DGTC-rich raw mixture.

Author Contributions: Conceptualization, F.R. and B.A.; validation, A.U. and L.F.; investigation, B.A. and E.C.; data curation, B.A.; writing—original draft preparation, F.R.; writing—review and editing, L.R., B.A. and L.F.; visualization, A.U.; supervision, F.R. All authors have read and agreed to the published version of the manuscript.

Funding: This research received no external funding.

Data Availability Statement: The data presented in this study are available in the article and supplementary material.

Conflicts of Interest: The authors declare no conflict of interest.

References

1. Pörtner, H.-O.; Roberts, D.C.; Tignor, M.; Poloczanska, E.S.; Mintenbeck, K.; Alegría, A.; Craig, M.; Langsdorf, S.; Löschke, S.; Möller, V.; et al. (Eds.) Contribution of Working Group II to the Sixth Assessment Report of the Intergovernmental Panel on Climate Change. In *IPCC 2022: Climate Change 2022: Impacts, Adaptation and Vulnerability*; Cambridge University Press: Cambridge, UK; New York, NY, USA, 2022; p. 3056. [CrossRef]
2. Ubando, A.T.; Felix, C.B.; Chen, W.-H. Biorefineries in circular bioeconomy: A comprehensive review. *Bioresour. Technol.* **2020**, *299*, 122585. [CrossRef]
3. Xu, H.; Ou, L.; Li, Y.; Hawkins, T.R.; Wang, M. Life Cycle Greenhouse Gas Emissions of Biodiesel and Renewable Diesel Production in the United States. *Environ. Sci. Technol.* **2022**, *56*, 7512–7521. [CrossRef] [PubMed]
4. Nanda, M.R.; Zhang, Y.; Yuan, Z.; Qin, W.; Ghaziaskar, H.S.; Xu, C. Catalytic conversion of glycerol for sustainable production of solketal as a fuel additive: A review. *Renew. Sustain. Energy Rev.* **2015**, *56*, 1022–1031. [CrossRef]
5. Zandalinas, S.I.; Fritschi, F.B.; Mittler, R. Global Warming, Climate Change, and Environmental Pollution: Recipe for a Multifactorial Stress Combination Disaster. *Trends Plant Sci.* **2021**, *26*, 588–599. [CrossRef] [PubMed]
6. Roncaglia, F.; Forti, L.; D'Anna, S.; Maletti, L. An Expedient Catalytic Process to Obtain Solketal from Biobased Glycerol. *Processes* **2021**, *9*, 141. [CrossRef]
7. Wu, F.; Jiang, H.; Zhu, X.; Lu, R.; Shi, L.; Lu, F. Effect of Tungsten Species on Selective Hydrogenolysis of Glycerol to 1,3-Propanediol. *ChemSusChem* **2020**, *14*, 569–581. [CrossRef]
8. Sittijunda, S.; Reungsang, A. Valorization of crude glycerol into hydrogen, 1,3-propanediol, and ethanol in an up-flow anaerobic sludge blanket (UASB) reactor under thermophilic conditions. *Renew. Energy* **2020**, *161*, 361–372. [CrossRef]
9. Zheng, Z.; Luo, M.; Yu, J.; Wang, J.; Ji, J. Novel Process for 1,3-Dihydroxyacetone Production from Glycerol. 1. Technological Feasibility Study and Process Design. *Ind. Eng. Chem. Res.* **2012**, *51*, 3715–3721. [CrossRef]
10. Selva, M.; Fabris, M. The reaction of glycerol carbonate with primary aromatic amines in the presence of Y- and X-faujasites: The synthesis of N-(2,3-dihydroxy)propyl anilines and the reaction mechanism. *Green Chem.* **2009**, *11*, 1161–1172. [CrossRef]
11. Nohra, B.; Candy, L.; Blanco, J.; Raoul, Y.; Mouloungui, Z. Synthesis of five and six-membered cyclic glycerilic carbonates bearing exocyclic urethane functions. *Eur. J. Lipid Sci. Technol.* **2012**, *115*, 111–122. [CrossRef]
12. Quienne, B.; Poli, R.; Pinaud, J.; Caillol, S. Enhanced aminolysis of cyclic carbonates by β-hydroxylamines for the production of fully biobased polyhydroxyurethanes. *Green Chem.* **2021**, *23*, 1678–1690. [CrossRef]
13. Fricke, N.; Keul, H.; Möller, M. Carbonate Couplers and Functional Cyclic Carbonates from Amino Acids and Glucosamine. *Macromol. Chem. Phys.* **2009**, *210*, 242–255. [CrossRef]

14. Bao, Y.-M.; Shen, G.-R.; He, J.; Li, Y.-S. Water-soluble hyperbranched poly(ester urethane)s based on d,l-alanine: Isocyanate-free synthesis, post-functionalization and application. *Green Chem.* 2012, *14*, 2243–2250. [CrossRef]
15. Galletti, G.; Prete, P.; Vanzini, S.; Cucciniello, R.; Fasolini, A.; De Maron, J.; Cavani, F.; Tabanelli, T. Glycerol Carbonate as a Versatile Alkylating Agent for the Synthesis of β-Aryloxy Alcohols. *ACS Sustain. Chem. Eng.* 2022, *10*, 10922–10933. [CrossRef]
16. Ghandi, M.; Mostashari, A.; Karegar, M.; Barzegar, M. Efficient Synthesis of α-Monoglycerides via Solventless Condensation of Fatty Acids with Glycerol Carbonate. *J. Am. Oil Chem. Soc.* 2007, *84*, 681–685. [CrossRef]
17. Simao, A.-C.; Lynikaite-Pukleviciene, B.; Rousseau, C.; Tatibouet, A.; Cassel, S.; Sackus, A.; Rauter, A.; Rollin, P. 1,2-Glycerol Carbonate: A Versatile Renewable Synthon. *Lett. Org. Chem.* 2006, *3*, 744–748. [CrossRef]
18. da Costa, P.L.F.; Melo, V.N.; Guimarães, B.M.; Schuler, M.; Pimenta, V.; Rollin, P.; Tatibouët, A.; de Oliveira, R.N. Glycerol carbonate in Ferrier reaction: Access to new enantiopure building blocks to develop glycoglycerolipid analogues. *Carbohydr. Res.* 2016, *436*, 1–10. [CrossRef]
19. Carré, C.; Zoccheddu, H.; Delalande, S.; Pichon, P.; Avérous, L. Synthesis and characterization of advanced biobased thermoplastic nonisocyanate polyurethanes, with controlled aromatic-aliphatic architectures. *Eur. Polym. J.* 2016, *84*, 759–769. [CrossRef]
20. Rousseau, J.; Rousseau, C.; Lynikaite, B.; Sackus, A.; de Leon, C.; Rollin, P.; Tatibouet, A. Tosylated glycerol carbonate, a versatile bis-electrophile to access new functionalized glycidol derivatives. *Tetrahedron* 2009, *65*, 8571–8581. [CrossRef]
21. Legros, V.; Taing, G.; Buisson, P.; Schuler, M.; Bostyn, S.; Rousseau, J.; Sinturel, C.; Tatibouët, A. Activated Glycerol Carbonates, Versatile Reagents with Aliphatic Amines: Formation and Reactivity of Glycidyl Carbamates and Trialkylamines. *Eur. J. Org. Chem.* 2017, *2017*, 5032–5043. [CrossRef]
22. Parzuchowski, P.G.; Świderska, A.; Roguszewska, M.; Frączkowski, T.; Tryznowski, M. Amine functionalized polyglycerols obtained by copolymerization of cyclic carbonate monomers. *Polymer* 2018, *151*, 250–260. [CrossRef]
23. Rokicki, G.; Rakoczy, P.; Parzuchowski, P.; Sobiecki, M. Hyperbranched aliphatic polyethers obtained from environmentally benign monomer: Glycerol carbonate. *Green Chem.* 2005, *7*, 529–539. [CrossRef]
24. Vogt, L.; Ruther, F.; Salehi, S.; Boccaccini, A.R. Poly(Glycerol Sebacate) in Biomedical Applications—A Review of the Recent Literature. *Adv. Heal. Mater.* 2021, *10*, e2002026. [CrossRef] [PubMed]
25. Zhang, H.; Grinstaff, M.W. Recent Advances in Glycerol Polymers: Chemistry and Biomedical Applications. *Macromol. Rapid Commun.* 2014, *35*, 1906–1924. [CrossRef]
26. Ekladious, I.; Liu, R.; Zhang, H.; Foil, D.H.; Todd, D.A.; Graf, T.N.; Padera, R.F.; Oberlies, N.H.; Colson, Y.L.; Grinstaff, M.W. Synthesis of poly(1,2-glycerol carbonate)–paclitaxel conjugates and their utility as a single high-dose replacement for multi-dose treatment regimens in peritoneal cancer. *Chem. Sci.* 2017, *8*, 8443–8450. [CrossRef]
27. Kundys, A.; Plichta, A.; Florjańczyk, Z.; Zychewicz, A.; Lisowska, P.; Parzuchowski, P.; Wawrzyńska, E. Multi-arm star polymers of lactide obtained in melt in the presence of hyperbranched oligoglycerols. *Polym. Int.* 2016, *65*, 927–937. [CrossRef]
28. Mamiński, M.; Czarzasta, M.; Parzuchowski, P. Wood adhesives derived from hyperbranched polyglycerol cross-linked with hexamethoxymethyl melamines. *Int. J. Adhes. Adhes.* 2011, *31*, 704–707. [CrossRef]
29. Jansen, J.F.G.A.; Dias, A.A.; Dorschu, M.; Coussens, B. Fast Monomers: Factors Affecting the Inherent Reactivity of Acrylate Monomers in Photoinitiated Acrylate Polymerization. *Macromolecules* 2003, *36*, 3861–3873. [CrossRef]
30. Mhanna, A.; Sadaka, F.; Boni, G.; Brachais, C.-H.; Brachais, L.; Couvercelle, J.-P.; Plasseraud, L.; Lecamp, L. Photopolymerizable Synthons from Glycerol Derivatives. *J. Am. Oil Chem. Soc.* 2013, *91*, 337–348. [CrossRef]
31. Sacripante, G.G.; Zhou, K.; Farooque, M. Sustainable Polyester Resins Derived from Rosins. *Macromolecules* 2015, *48*, 6876–6881. [CrossRef]
32. Ibn El Alami, M.S.; Suisse, I.; Fadlallah, S.; Sauthier, M.; Visseaux, M. Telomerization of 1,3-butadiene with glycerol carbonate and subsequent ring-opening lactone co-polymerization. *Comptes Rendus. Chim.* 2016, *19*, 299–305. [CrossRef]
33. Wu, Z.; Tang, L.; Dai, J.; Qu, J. Synthesis and properties of aqueous cyclic carbonate dispersion and non-isocyanate polyurethanes under atmospheric pressure. *Prog. Org. Coatings* 2019, *136*, 5209. [CrossRef]
34. Ke, J.; Li, X.; Jiang, S.; Liang, C.; Wang, J.; Kang, M.; Li, Q.; Zhao, Y. Promising approaches to improve the performances of hybrid non-isocyanate polyurethane. *Polym. Int.* 2018, *68*, 651–660. [CrossRef]
35. Annunziata, L.; Diallo, A.K.; Fouquay, S.; Michaud, G.; Simon, F.; Brusson, J.-M.; Carpentier, J.-F.; Guillaume, S.M. α,ω-Di(glycerol carbonate) telechelic polyesters and polyolefins as precursors to polyhydroxyurethanes: An isocyanate-free approach. *Green Chem.* 2013, *16*, 1947–1956. [CrossRef]
36. Duval, C.; Kébir, N.; Jauseau, R.; Burel, F. Organocatalytic synthesis of novel renewable non-isocyanate polyhydroxyurethanes. *J. Polym. Sci. Part A: Polym. Chem.* 2015, *54*, 758–764. [CrossRef]
37. Quienne, B.; Kasmi, N.; Dieden, R.; Caillol, S.; Habibi, Y. Isocyanate-Free Fully Biobased Star Polyester-Urethanes: Synthesis and Thermal Properties. *Biomacromolecules* 2020, *21*, 1943–1951. [CrossRef] [PubMed]
38. Ekin, A.; Webster, D.C. Synthesis and Characterization of Novel Hydroxyalkyl Carbamate and Dihydroxyalkyl Carbamate Terminated Poly(dimethylsiloxane) Oligomers and Their Block Copolymers with Poly(ε-caprolactone). *Macromolecules* 2006, *39*, 8659–8668. [CrossRef]
39. Tachibana, Y.; Shi, X.; Graiver, D.; Narayan, R. The Use of Glycerol Carbonate in the Preparation of Highly Branched Siloxy Polymers. *Silicon* 2014, *7*, 5–13. [CrossRef]
40. Nomanbhay, S.; Ong, M.Y.; Chew, K.W.; Show, P.-L.; Lam, M.K.; Chen, W.-H. Organic Carbonate Production Utilizing Crude Glycerol Derived as By-Product of Biodiesel Production: A Review. *Energies* 2020, *13*, 1483. [CrossRef]

41. Magniont, C.; Escadeillas, G.; Oms-Multon, C.; De Caro, P. The benefits of incorporating glycerol carbonate into an innovative pozzolanic matrix. *Cem. Concr. Res.* **2010**, *40*, 1072–1080. [CrossRef]
42. Hough, L.; Priddle, J.E.; Theobald, R.S. 363. Carbohydrate carbonates. Part II. Their preparation by ester-exchange methods. *J. Chem. Soc.* **1962**, *9*, 1934–1938. [CrossRef]
43. Shen, Y.; Yang, X.; Song, Y.; Tran, D.K.; Wang, H.; Wilson, J.; Dong, M.; Vazquez, M.; Sun, G.; Wooley, K.L. Complexities of Regioselective Ring-Opening vs. Transcarbonylation-Driven Structural Metamorphosis during Organocatalytic Polymerizations of Five-Membered Cyclic Carbonate Glucose Monomers. *JACS Au* **2022**, *2*, 515–521. [CrossRef] [PubMed]
44. Li, Y.; Wang, L.; Cao, Y.; Xu, S.; He, P.; Li, H.; Liu, H. Tris-imidazolinium-based porous poly(ionic liquid)s as an efficient catalyst for decarboxylation of cyclic carbonate to epoxide. *RSC Adv.* **2021**, *11*, 14193–14202. [CrossRef] [PubMed]
45. Ochoa Gómez, J.R.; Gómez de Miranda Jiménez de Aberastul, O.; Blanco Pérez, N.; Maestro Madurga, B.; Prieto Fernández, S. Glycidol Synthesis Method. Patent WO2017017307A1, 30 July 2015.
46. Teng, W.K.; Ngoh, G.C.; Yusoff, R.; Aroua, M.K. A review on the performance of glycerol carbonate production via catalytic transesterification: Effects of influencing parameters. *Energy Convers. Manag.* **2014**, *88*, 484–497. [CrossRef]
47. Van Mileghem, S.; De Borggraeve, W.M.; Baxendale, I.R. A Robust and Scalable Continuous Flow Process for Glycerol Carbonate. *Chem. Eng. Technol.* **2018**, *41*, 12. [CrossRef]
48. Singh, D.; Reddy, B.; Ganesh, A.; Mahajani, S. Zinc/Lanthanum Mixed-Oxide Catalyst for the Synthesis of Glycerol Carbonate by Transesterification of Glycerol. *Ind. Eng. Chem. Res.* **2014**, *53*, 18786–18795. [CrossRef]
49. Okoye, P.; Abdullah, A.; Hameed, B. Glycerol carbonate synthesis from glycerol and dimethyl carbonate using trisodium phosphate. *J. Taiwan Inst. Chem. Eng.* **2016**, *68*, 51–58. [CrossRef]
50. Wang, X.; Zhang, P.; Cui, P.; Cheng, W.; Zhang, S. Glycerol carbonate synthesis from glycerol and dimethyl carbonate using guanidine ionic liquids. *Chin. J. Chem. Eng.* **2017**, *25*, 1182–1186. [CrossRef]
51. Nascimento, M.A.D.; Gotardo, L.E.; Leão, R.A.C.; de Castro, A.M.; de Souza, R.O.M.A.; Itabaiana, J.I. Enhanced Productivity in Glycerol Carbonate Synthesis under Continuous Flow Conditions: Combination of Immobilized Lipases from Porcine Pancreas and *Candida antarctica* (CALB) on Epoxy Resins. *ACS Omega* **2019**, *4*, 860–869. [CrossRef]
52. Gérardy, R.; Estager, J.; Luis, P.; Debecker, D.P.; Monbaliu, J.-C.M. Versatile and scalable synthesis of cyclic organic carbonates under organocatalytic continuous flow conditions. *Catal. Sci. Technol.* **2019**, *9*, 6841–6851. [CrossRef]
53. Zheng, Q.; Nishimura, R.; Sato, Y.; Inomata, H.; Ota, M.; Watanabe, M.; Camy, S. Dimethyl carbonate (DMC) synthesis from methanol and carbon dioxide in the presence of ZrO_2 solid solutions and yield improvement by applying a natural convection circulation system. *Chem. Eng. J.* **2021**, *429*, 132378. [CrossRef]
54. Liu, H.; Zhu, D.; Jia, B.; Huang, Y.; Cheng, Y.; Luo, X.; Liang, Z. Study on catalytic performance and kinetics of high efficiency CeO_2 catalyst prepared by freeze drying for the synthesis of dimethyl carbonate from CO_2 and methanol. *Chem. Eng. Sci.* **2022**, *254*, 117614. [CrossRef]
55. Huang, S.; Yan, B.; Wang, S.; Ma, X. Recent advances in dialkyl carbonates synthesis and applications. *Chem. Soc. Rev.* **2015**, *44*, 3079–3116. [CrossRef] [PubMed]
56. Liu, J.; He, D. Transformation of CO_2 with glycerol to glycerol carbonate by a novel $ZnWO_4$-ZnO catalyst. *J. CO_2 Util.* **2018**, *26*, 370–379. [CrossRef]
57. Park, C.-Y.; Nguyen-Phu, H.; Shin, E.W. Glycerol carbonation with CO_2 and $La_2O_2CO_3$/ZnO catalysts prepared by two different methods: Preferred reaction route depending on crystalline structure. *Mol. Catal.* **2017**, *435*, 99–109. [CrossRef]
58. Chaugule, A.A.; Tamboli, A.H.; Kim, H. Ionic liquid as a catalyst for utilization of carbon dioxide to production of linear and cyclic carbonate. *Fuel* **2017**, *200*, 316–332. [CrossRef]
59. He, Y.; Lu, H.; Li, X.; Wu, J.; Pu, T.; Du, W.; Li, H.; Ding, J.; Wan, H.; Guan, G. Insight into the reversible behavior of Lewis–Brønsted basic poly(ionic liquid)s in one-pot two-step chemical fixation of CO_2 to linear carbonates. *Green Chem.* **2021**, *23*, 8571–8580. [CrossRef]
60. Giannoccaro, P.; Casiello, M.; Milella, A.; Monopoli, A.; Cotugno, P.; Nacci, A. Synthesis of 5-membered cyclic carbonates by oxidative carbonylation of 1,2-diols promoted by copper halides. *J. Mol. Catal. A Chem.* **2012**, *365*, 162–171. [CrossRef]
61. Glibert, P.M.; Harrison, J.; Heil, C.; Seitzinger, S. Escalating Worldwide use of Urea—A Global Change Contributing to Coastal Eutrophication. *Biogeochemistry* **2006**, *77*, 441–463. [CrossRef]
62. Estiu, G.; Merz, J.K.M. The Hydrolysis of Urea and the Proficiency of Urease. *J. Am. Chem. Soc.* **2004**, *126*, 6932–6944. [CrossRef] [PubMed]
63. Zdanowicz, M. Deep eutectic solvents based on urea, polyols and sugars for starch treatment. *Int. J. Biol. Macromol.* **2021**, *176*, 387–393. [CrossRef] [PubMed]
64. Newman, M.S.; Lala, L.K. Urea as a base in organic reactions. *Tetrahedron Lett.* **1967**, *8*, 3267–3269. [CrossRef]
65. Rodríguez-Santiago, L.; Noguera, M.; Sodupe, M.; Salpin, J.Y.; Tortajada, J. Gas Phase Reactivity of Ni+ with Urea. Mass Spectrometry and Theoretical Studies. *J. Phys. Chem. A* **2003**, *107*, 9865–9874. [CrossRef]
66. Li, Q.; Zhao, N.; Wei, W.; Sun, Y. Catalytic performance of metal oxides for the synthesis of propylene carbonate from urea and 1,2-propanediol. *J. Mol. Catal. A: Chem.* **2007**, *270*, 44–49. [CrossRef]
67. Nguyen-Phu, H.; Shin, E.W. Disordered structure of $ZnAl_2O_4$ phase and the formation of a Zn NCO complex in ZnAl mixed oxide catalysts for glycerol carbonylation with urea. *J. Catal.* **2019**, *373*, 147–160. [CrossRef]

68. Elman, A.R.; Davydov, I.E.; Stepanov, A.A. Synthesis of Urea by Ammonolysis of Propylene Carbonate. *J. Chem. Chem. Eng.* **2018**, *12*, 26–30. [CrossRef]
69. Calvino-Casilda, V.; Mul, G.; Fernández, J.; Rubio-Marcos, F.; Bañares, M. Monitoring the catalytic synthesis of glycerol carbonate by real-time attenuated total reflection FTIR spectroscopy. *Appl. Catal. A Gen.* **2011**, *409*, 106–112. [CrossRef]
70. Park, J.-H.; Choi, J.S.; Woo, S.K.; Lee, S.D.; Cheong, M.; Kim, H.S.; Lee, H. Isolation and characterization of intermediate catalytic species in the Zn-catalyzed glycerolysis of urea. *Appl. Catal. A Gen.* **2012**, *433*, 35–40. [CrossRef]
71. Wang, H.; Xin, Z.; Li, Y. Synthesis of Ureas from CO_2. *Top. Curr. Chem.* **2017**, *375*, 49. [CrossRef]
72. Ishaq, H.; Siddiqui, O.; Chehade, G.; Dincer, I. A solar and wind driven energy system for hydrogen and urea production with CO_2 capturing. *Int. J. Hydrogen Energy* **2020**, *46*, 4749–4760. [CrossRef]
73. Li, J.; Wang, T. Chemical equilibrium of glycerol carbonate synthesis from glycerol. *J. Chem. Thermodyn.* **2011**, *43*, 731–736. [CrossRef]
74. Ochoa-Gómez, J.R.; Gómez-Jiménez-Aberasturi, O.; Ramírez-López, C.; Belsué, M. A Brief Review on Industrial Alternatives for the Manufacturing of Glycerol Carbonate, a Green Chemical. *Org. Process. Res. Dev.* **2012**, *16*, 389–399. [CrossRef]
75. Indran, V.P.; Saud, A.S.H.; Maniam, G.P.; Yusoff, M.M.; Taufiq-Yap, Y.H.; Rahim, M.H.A. Versatile boiler ash containing potassium silicate for the synthesis of organic carbonates. *RSC Adv.* **2016**, *6*, 34877–34884. [CrossRef]
76. Fernandes, G.P.; Yadav, G.D. Selective glycerolysis of urea to glycerol carbonate using combustion synthesized magnesium oxide as catalyst. *Catal. Today* **2018**, *309*, 153–160. [CrossRef]
77. Chaves, D.M.; Da Silva, M.J. A selective synthesis of glycerol carbonate from glycerol and urea over $Sn(OH)_2$: A solid and recyclable *in situ* generated catalyst. *New J. Chem.* **2019**, *43*, 3698–3706. [CrossRef]
78. Mallesham, B.; Rangaswamy, A.; Rao, B.G.; Rao, T.V.; Reddy, B.M. Solvent-Free Production of Glycerol Carbonate from Bioglycerol with Urea Over Nanostructured Promoted SnO_2 Catalysts. *Catal. Lett.* **2020**, *150*, 3626–3641. [CrossRef]
79. Turney, T.W.; Patti, A.; Gates, W.; Shaheen, U.; Kulasegaram, S. Formation of glycerol carbonate from glycerol and urea catalysed by metal monoglycerolates. *Green Chem.* **2013**, *15*, 1925–1931. [CrossRef]
80. Fujita, S.-I.; Yamanishi, Y.; Arai, M. Synthesis of glycerol carbonate from glycerol and urea using zinc-containing solid catalysts: A homogeneous reaction. *J. Catal.* **2013**, *297*, 137–141. [CrossRef]
81. Kulasegaram, S.; Shaheen, U.; Turney, T.W.; Gates, W.P.; Patti, A.F. Zinc monoglycerolate as a catalyst for the conversion of 1,3- and higher diols to diurethanes. *RSC Adv.* **2015**, *5*, 47809–47812. [CrossRef]
82. Nguyen-Phu, H.; Do, L.T.; Shin, E.W. Investigation of glycerolysis of urea over various ZnMeO (Me = Co, Cr, and Fe) mixed oxide catalysts. *Catal. Today* **2020**, *352*, 80–87. [CrossRef]
83. Nguyen-Phu, H.; Park, C.-Y.; Shin, E.W. Dual catalysis over ZnAl mixed oxides in the glycerolysis of urea: Homogeneous and heterogeneous reaction routes. *Appl. Catal. A Gen.* **2018**, *552*, 1–10. [CrossRef]
84. Endah, Y.K.; Kim, M.S.; Choi, J.; Jae, J.; Lee, S.D.; Lee, H. Consecutive carbonylation and decarboxylation of glycerol with urea for the synthesis of glycidol via glycerol carbonate. *Catal. Today* **2017**, *293*, 136–141. [CrossRef]
85. Kaur, A.; Prakash, R.; Ali, A. 1H NMR assisted quantification of glycerol carbonate in the mixture of glycerol and glycerol carbonate. *Talanta* **2018**, *178*, 1001–1005. [CrossRef] [PubMed]
86. Chen, J.P.; Isa, K. Thermal Decomposition of Urea and Urea Derivatives by Simultaneous TG/(DTA)/MS. *J. Mass Spectrom. Soc. Jpn.* **1998**, *46*, 299–303. [CrossRef]
87. Lertlukkanasuk, N.; Phiyanalinmat, S.; Kiatkittipong, W.; Arpornwichanop, A.; Aiouache, F.; Assabumrungrat, S. Reactive distillation for synthesis of glycerol carbonate via glycerolysis of urea. *Chem. Eng. Process. Process Intensif.* **2013**, *70*, 103–109. [CrossRef]
88. Zhang, L.; Zhang, Z.; Wu, C.; Qian, Q.; Ma, J.; Jiang, L.; Han, B. Microwave assisted synthesis of glycerol carbonate from glycerol and urea. *Pure Appl. Chem.* **2017**, *90*, 1–6. [CrossRef]
89. Romano, G.; Paradisi, E.; Rosa, R.; Leonelli, C.; Roncaglia, F. Synthesis of Glycerol Carbonate from Glycerol and Urea Using a Microwave Reactor. *AMPERE Newsl.* **2022**, *111*, 1–8. Available online: www.ampereeurope.org/issue-111 (accessed on 22 November 2022).
90. She, Q.M.; Liu, J.H.; Aymonier, C.; Zhou, C.H. In situ fabrication of layered double hydroxide film immobilizing gold nanoparticles in capillary microreactor for efficient catalytic carbonylation of glycerol. *Mol. Catal.* **2021**, *513*, 111825. [CrossRef]
91. Plutschack, M.B.; Pieber, B.; Gilmore, K.; Seeberger, P.H. The Hitchhiker's Guide to Flow Chemistry II. *Chem. Rev.* **2017**, *117*, 11796–11893. [CrossRef]
92. Tennant, A.J.; Krause, M.J.; Kujawski, M.P.; Sherren, B.T. Compositions de Polyuréthane. Patent WO2021222192A1, 29 April 2020.
93. Aresta, M.; Dibenedetto, A.; di Bitonto, L. New efficient and recyclable catalysts for the synthesis of di- and tri-glycerol carbonates. *RSC Adv.* **2015**, *5*, 64433–64443. [CrossRef]
94. Atlantic Richfield Company. Patent GB1 458 595 A, 15 December 1976.
95. Arne Them Jensen. Subtilases. Patent EP0 611 243 A1, 21 March 2021.

Review

Sustainable Feedstocks and Challenges in Biodiesel Production: An Advanced Bibliometric Analysis

Misael B. Sales [1], Pedro T. Borges [1], Manoel Nazareno Ribeiro Filho [1], Lizandra Régia Miranda da Silva [1], Alyne P. Castro [2], Ada Amelia Sanders Lopes [1], Rita Karolinny Chaves de Lima [1], Maria Alexsandra de Sousa Rios [2] and José C. S. dos Santos [1,3,*]

[1] Instituto de Engenharias e Desenvolvimento Sustentável, Universidade da Integração Internacional da Lusofonia Afro-Brasileira, Campus das Auroras, Redenção 62790970, CE, Brazil
[2] Departamento de Engenharia Mecânica, Grupo de Inovações Tecnológicas e Especialidades Químicas—GRINTEQUI, Universidade Federal do Ceará, Bloco 715, Campus do Pici, Fortaleza 60440554, CE, Brazil
[3] Departamento de Engenharia Química, Universidade Federal do Ceará, Campus do Pici, Bloco 709, Fortaleza 60455760, CE, Brazil
* Correspondence: jcs@unilab.edu.br

Citation: Sales, M.B.; Borges, P.T.; Ribeiro Filho, M.N.; Miranda da Silva, L.R.; Castro, A.P.; Sanders Lopes, A.A.; Chaves de Lima, R.K.; de Sousa Rios, M.A.; Santos, J.C.S.d. Sustainable Feedstocks and Challenges in Biodiesel Production: An Advanced Bibliometric Analysis. *Bioengineering* **2022**, *9*, 539. https://doi.org/10.3390/bioengineering9100539

Academic Editors: Indra Neel Pulidindi, Aharon Gedanken and Giorgos Markou

Received: 9 September 2022
Accepted: 4 October 2022
Published: 10 October 2022

Publisher's Note: MDPI stays neutral with regard to jurisdictional claims in published maps and institutional affiliations.

Copyright: © 2022 by the authors. Licensee MDPI, Basel, Switzerland. This article is an open access article distributed under the terms and conditions of the Creative Commons Attribution (CC BY) license (https://creativecommons.org/licenses/by/4.0/).

Abstract: Biodiesel can be produced from vegetable oils, animal fats, frying oils, and from microorganism-synthesized oils. These sources render biodiesel an easily biodegradable fuel. The aim of this work was to perform an advanced bibliometric analysis of primary studies relating to biodiesel production worldwide by identifying the key countries and regions that have shown a strong engagement in this area, and by understanding the dynamics of their collaboration and research outputs. Additionally, an assessment of the main primary feedstocks employed in this research was carried out, along with an analysis of the current and future trends that are expected to define new paths and methodologies to be used in the manufacture of biodegradable and renewable fuels. A total of 4586 academic outputs were selected, including peer-reviewed research articles, conference papers, and literature reviews related to biodiesel production, in the time period spanning from 2010 to 2021. Articles that focused on feedstocks for the production of biodiesel were also included, with a search that returned 330 papers. Lastly, 60 articles relating to biodiesel production via sewage were specifically included to allow for an analysis of this source as a promising feedstock in the future of the biofuel market. Via the geocoding and the document analyses performed, we concluded that China, Malaysia, and India are the largest writers of articles in this area, revealing a great interest in biofuels in Asia. Additionally, it was noted that environmental concerns have caused authors to conduct research on feedstocks that can address the sustainability challenges in the production of biodiesel.

Keywords: biodiesel; biofuel; feedstocks; sustainability; research articles; bibliometric analysis

1. Introduction

The process of deescalating the use of conventional fuels faces several hurdles worldwide, and despite the growing scientific evidence of the viability of other biofuel production routes, governmental and regulatory limitations accentuate these difficulties and forestalls further progress [1–7]. Despite these hindrances, biodiesel production shows a rapid growth globally owing to the security offered by this type of fuel and its considerably smaller environmental footprint [8–13]. There is also an increased pressure on governments to adopt and implement more sustainable production processes. Thus, the demand for biofuel consumption is predicted to grow significantly in the coming years due to governmental policies in some countries pushing for a shift towards the use of renewable energy, to the rising prices of petroleum-based fuels, and to the emerging concerns related to pollutants [14].

Bibliometric analyses are a well-known and rigorous method for the analysis of large amounts of scientific information [15]. It allows for the visualization of the development of a specific research topic, while also revealing important information about the areas that are rapidly evolving in that field [16,17]. The data sets encompassing the structure of a bibliometric analysis are usually extremely large, in the scale of hundreds or even thousands, and they can describe specific research numbers, such as the volume of citations and/or publications and the number of occurrences of certain keywords, among others. However, the interpretations gathered and discussed are not only objective in nature, as in performance analyses, but they can also comprise more subjective considerations, such as topical analyses [18].

This article aimed to critically evaluate, from a qualitative and quantitative standpoint, academic papers published on themes related to the specific aforementioned area via the creation of a highly specific database. Articles in this database included those related to biodiesel production and those that were published between 2010 and 2021. Quantitative and qualitative conditions were set and entered within the search tool available on the Web of Science Core Collection website (https://www-webofscience.ez373.periodicos.capes.gov.br/wos/woscc/basic-search, accessed on 8 August 2022). The articles selected were specifically related to feedstocks employed in the manufacture of biodiesel (e.g., animal oils, vegetable oils, oils from microorganisms, etc.).

In addition to the considerations above, the relevance of sewage sludge as a specific feedstock for biofuel production was analyzed, due to its strong future potential in this specific market. Sewage sludge has been recognized worldwide as being a strong lipid feedstock for biofuel production due to its wide availability and good concentration of lipids. In the US alone, around 6.2 million dry metric tons of sludge are produced annually by sewage treatment plants, and this number is expected to increase in the future due to the ever-growing urbanization and industrialization observed in most countries [19]. This raw material became attractive in the biofuel market in mid-2010, mainly owing to its increased production at that time. In the European Union alone, sludge production is forecasted to have reached the 13-million-ton mark in 2020 [20].

The main aim of this article was to showcase the evolution of the literature related to the research on biofuel production worldwide via an analysis process that facilitates the visualization of future trends in related emerging studies. Thus, the raw materials that presented the highest relevance in the database built were assessed, along with the frequency of studies citing these sources. Given the accumulation of a great number of studies across the years, we had to set boundaries concerning search criteria and targeted conditions, as described in Section 2.

As presented in Figure 1, the growth of publications on the manufacture of biodiesel is noticeable. It is also noteworthy that, especially in the year 2021, the number of publications related to feedstocks for biodiesel production was considerably higher than in previous years, with about 9.8% of the total number of publications coming from that year (44 documents). The second most relevant year is 2018, with 8.9% of the total publications (41 documents), followed by 2017, with approximately 7.2% (32 documents). Regarding articles on sewage sludge, 2016 saw the greatest number of publications (with a total of 14).

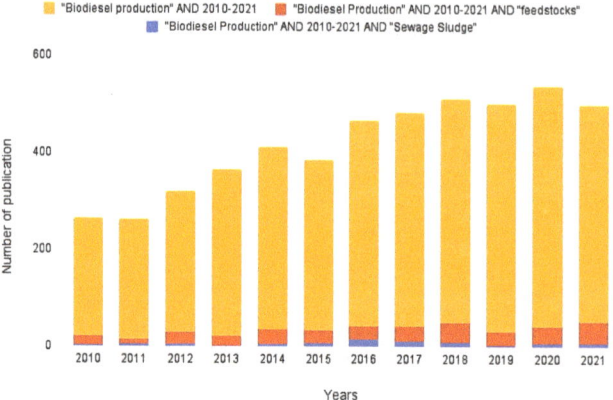

Figure 1. Annual publications on biodiesel production, relevant raw materials, and on sewage sludge as a resource. The Google Sheets tool was used to produce this figure.

2. Materials and Methods

2.1. Data Sources

Some search parameters were previously set on the Web of Science Core Collection website when using programs to collect the data to compose our advanced bibliometric analysis. The database files were accessed and downloaded via the login credentials provided by the CAPES PERIODICOS platform (https://www-periodicos-capes-gov-br.ezl.periodicos.capes.gov.br/index.php, accessed on 9 August 2022).

As shown in Figure 2, first, the term "Biodiesel Production" was entered in the search bar, along with the filter "Title", into their central database, which contained all the subsequent databases. Then, a new row was added and the term "2010–2021" was entered within the filter "Year Published". In total, 4917 articles fitting these search parameters were returned, but to ensure a more refined analysis, we proceeded to discard the publications that showed little relevance to this bibliometric analysis. Refinements were made for "Document Types" (limiting the search to articles, conference papers, and review articles only) and Language (English). The new, targeted search returned 4586 articles.

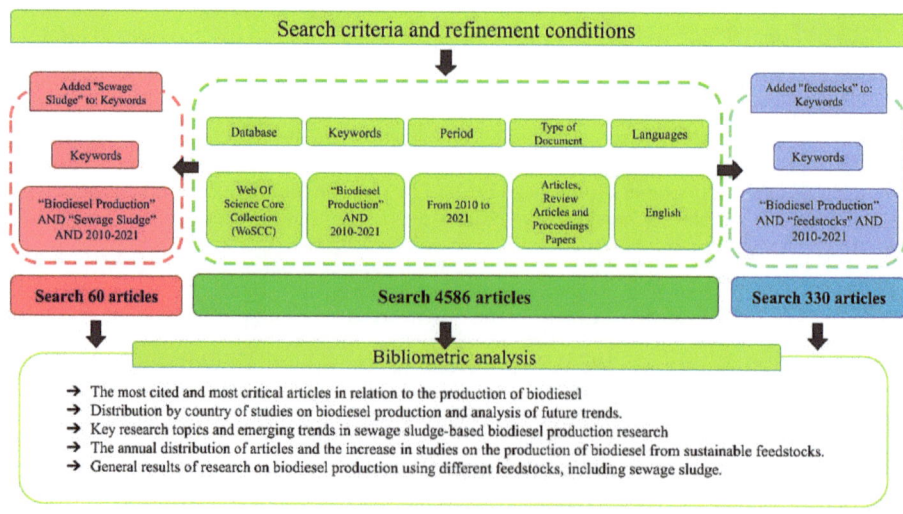

Figure 2. Search and analysis criteria used for the data collection stage.

In addition, two further databases were generated from the main one above in order to allow us to delve deeper into other subtopics of interest. The first database was related to feedstocks for biodiesel production, where a third search line was added with the term "feedstocks" in the "All Fields" section and with the same previous refinements. Overall, 330 articles published between 2010 and 2021 matched these criteria. The second derived database was related to biodiesel production using sewage sludge, with a third row being added with these terms. A total of 60 related articles were then identified.

The methodology described and used in this research was chosen based on the literature evidenced by the article *A bibliometric analysis of sustainable oil and gas production research using VOSviewer* published in 2022 [21]. The authors reinforce that such a strategy facilitates the construction of a database that is consistent with the aim being proposed and that improves the visualization and exploration of results in this area of research. The three databases obtained in our work were all crucial for this analysis, with the largest of them providing the comparison basis that created, with the delimitation and search conditions shown in Figure 2, the second and third databases, which returned 330 and 60 documents, respectively. As the central objective of this work was to explore mainly the second database, which deals with raw materials, the third could then serve as an additional dataset to explore the importance of the topic it covers in the advancement of this research topic. Despite the limitations observed, another differentiating characteristic of this work was the inclusion of articles that provide important information, but that are not necessarily a part of the repertoire usually treated in the programs used to process these data for analysis. The literature on biodiesel is characterized by a wide variety of works scattered through various areas of study, and those were used to enrich this work.

Figure 2 above is a categorical representation of the search process followed for building the databases used in this work. Each block shows one step taken in our research, from entering the key terms and constraints used in the Web of Science search engine, to the analysis methodology to be followed. Such visual representations serve as more simplified demonstrations, aiming to describe how the research was carried out, since they do not contain any specific descriptions that can only be understood through reading the main body of text.

2.2. Data Analysis

For the bibliometric analysis, three programs were used: CiteSpace (v.5.8.R3 Philadelphia, PA, USA), CiteSpace was developed by Dr. Chaomei Chen from Drexel University, USA. VOSviewer (version 1.6.17 Leiden, South Holland, The Netherlands), VOSviewer software, developed at Leiden University's Centre for Science and Technology Studies (CWTS), Leiden University, the Netherlands. ArcGIS (version 10.5 Redlands, California, USA), ArcGIS is a product suite developed by Environmental Systems Research Institute (Esri), Redlands, California, United States. ArcGIS 10.5 was used to analyze the geographical distribution of publications, while VOSviewer was used for data visualization [22]. CiteSpace identified and predicted the future possible research subareas in this area by using keywords and clusters [23].

3. Results and Discussion
3.1. Bibliometric Analysis
3.1.1. Publications: General Results

The initial search on the Web of Science Database Core Collection website resulted in 4586 research outputs from 2010 to 2021, the most relevant of which was published in March 2012. The title of this specific article is *Biodiesel production from microalgae*, in which the use of microalgae in the manufacture of different energy sources, such as oil and biofuels, including biodiesel, were discussed [24]. Furthermore, the authors presented the various factors of relevance in the area of microalgae valorization for biodiesel, microalgae cultivation, lipid extraction, and transesterification reactions, among other aspects. By analyzing all the articles in the database, it was possible to observe that, among the raw materials commonly mentioned, microalgae were the most widely employed, being reported in 40 articles. This is higher than the number of mentions of soybean oil, a raw material with a relatively high rate of employment in the production of biofuels.

Microalgae, however, as well as other feedstocks, still face a cost–benefit challenge. Nevertheless, great advances are expected to be achieved in the areas of cultivation, harvesting, lipid extraction, transesterification, and biomass-processing technologies that could render biodiesel production from microalgae more viable and more easily commercialized in the near future [25]. Optimizing the manufacture of biofuels is a paramount step towards market acceptance and implementation, as this directly influences the economics and yields of the final product. The optimization of the variables and conditions in these processes are also of high relevance for the entire production process [26]. Additionally, the studies carried out on the topics above have been successful in demonstrating how key environmental concerns have been addressed through the steps that have been taken to make the biofuel market more sustainable. This trend is confirmed by the prominence of related articles present in the literature repertoire used in this research, mainly when considering the number of citations and the relevance of the authors of the papers. These factors are described in subsequent sections of this article.

3.1.2. Distribution of Scientific Journals

From the reference database created, we observed that the manuscripts relating to raw materials for biodiesel were published across 154 journals, with an average of approximately 2.1 publications per journal. This indicates that a great interest in this research theme is shared by several areas, but the number of relevant publications can still be considered low. Judging by the vast number of journals analyzed, it is possible to note that distinct groups of researchers have explored this topic via different methodologies of scientific analysis.

Table 1 shows the list of the ten most relevant scientific journals in the biodiesel research area, classified by number of citations. Together, these ten journals are responsible for more than a quarter of the total number of the publications analyzed. The journal *Renewable & Sustainable Energy Reviews*, which ranks first on this list, has an impact factor of 14.982 and a total of 33 articles published in the area. These account for 9.9% of the total number of documents, and the articles published in this periodical have amassed 8250 citations over the years. The second journal on the list is *Bioresource Technology*, with an impact factor of 9.642 and 16 articles published (4.8% of the total documents) that gathered 1785 citations. The ranking list is mostly composed of European journals, with *Renewable & Sustainable Energy Reviews* being the only American journal. Nevertheless, it has the highest impact factor of all the journals listed and the highest number of citations and publications, with an average of 250 citations per article. This shows a high concentration of articles in the area of feedstocks for biodiesel production in a single scientific journal.

Table 1. Top ten scientific journals with publications in the field of feedstocks for biodiesel production.

Rank	Journal	C	IF	NP	NC	AC	P
1	Renewable & sustainable energy reviews	USA	14.982	33	8250	250	9.9%
2	Bioresource technology	NL	9.642	16	1785	111.5	4.8%
3	Energy conversion and management	EN	9.709	13	789	61.4	3.9%
4	Biomass & bioenergy	EN	5.061	6	750	125	1.8%
5	Fuel	EN	6.609	18	729	40.5	5.4%
6	Biofuels, bioproducts & biorefining—Biofpr	EN	4.102	5	151	30.2	1.5%
7	Journal of environmental chemical engineering	EN	5.909	4	63	15.7	1.2%
8	Catalysts	CH	4.146	6	59	9.8	1.8%
9	Biofuels-UK	EN	2.956	4	57	14.2	1.2%
10	Environmental chemistry letters	GER	9.027	4	53	13.2	1.2%

C = Country; IF = Impact Factor in 2020; NP = Number of Publications; NC = Number of Citations; AC = Average Citations; P = Percentage in Relation to the Total Number of Papers. USA = United States of America; NL = Netherlands; EN = England; CH = Switzerland; GER = Germany.

3.1.3. Distribution by Country and Institution

Analyses made from the country information declared by the authors across the articles in our database allowed us to verify that the ten countries that published most articles on biodiesel feedstocks contributed 64.5% of the total amount of publications from the 56 countries identified (Table 2).

Table 2. The 10 most prolific countries in the area of feedstocks for biodiesel production.

Rank	Country	NP	NC	AC	Total Link Strength	AC
1	China	53	2555	48.2	169	250
2	Malaysia	48	3715	77.4	353	111.5
3	India	48	2190	45.6	252	61.4
4	United States of America	33	2030	61.5	116	125
5	Thailand	17	164	9.6	55	40.5
6	Indonesia	16	1309	81.8	134	30.2
7	Brazil	15	531	35.4	51	15.7
8	Italy	13	164	12.6	42	9.8
9	Serbia	12	920	76.6	82	14.2
10	Saudi Arabia	12	492	41.0	100	13.2

NP = Number of Publications; NC = Number of citations; AC = Average citations (NC/NP).

It is noteworthy that, although there is a great demand and interest for this research area, the resulting publications are concentrated within just a few regions of the world. China is responsible for 12.8% of the published articles (53 articles), thus occupying first place on the list. It is followed by Malaysia, with 11.6% (48 articles), and India, with 11.6% (48 articles).

To evaluate the relevance and impact of the published articles, it is necessary to also consider the number of citations. Malaysia has the highest number of citations (3715) across its 48 published papers, followed by Portugal, with 10 published papers that have amassed 3323 citations, and China, with 53 published papers and 2555 citations. Figure 3 shows the distribution of articles by country and specifies the 27 geographical regions that published at least five articles on the topic.

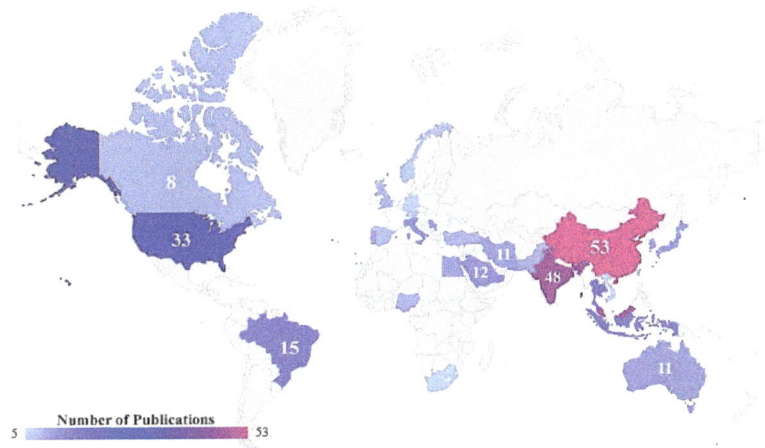

Figure 3. Representation of the distribution of articles by country.

Figure 4 illustrates publication collaborations among countries, within the same constraints as above regarding research area and time period. China and the United States, for example, present a strong collaborative relationship between authors despite being located in different continents. This can be explained by the notable leadership of these two countries concerning world economic power, so there is a natural interest in forming scientific partnerships to address shared environmental concerns in the search for sustainable products. This can be observed in Figure 1, where there is a clear growing interest in studies related to feedstock for biodiesel production in both these countries. The most significant collaborative relationship, however, is between Malaysia and Indonesia. This is most likely due to their geographical proximity in Southeast Asia, which demonstrates the similarity between their economic and socio-environmental interests towards the production of biofuels.

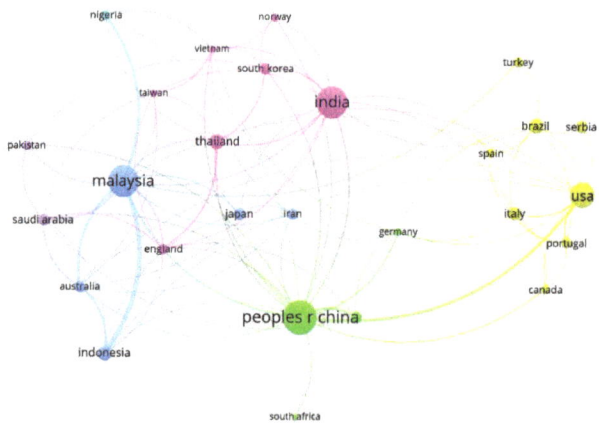

Figure 4. Network visualization map showing the collaboration among countries with at least five published papers. The thickness of the lines connecting two countries indicates the accumulation of co-authorships (thicker lines means more published articles), and the color clusters illustrate the groups of countries with a high level of collaboration.

It can also be noted that the articles published were the result of the work of 481 organizations across 56 different countries, which also confirms the high interest in research in this area. Nevertheless, the number of academic outputs is concentrated primarily in only ten countries, which produced two-thirds of the total number of documents. Among them are China, India, the United States, and Brazil, which indicates that there is a strong interest for this area especially in countries with large economies and a notable involvement in socio-environmental policies.

Another point worthy of note was that more than 50% of the organizations identified had only one publication in the field. In addition, only a small number of organizations were perceived to conduct consistent research and constantly publish in the area. Among the identified papers, there is a peculiarity related to the Polytechnic Institute of Porto (IPP). This organization has published only one article, *Microalgae for biodiesel production and other applications: A review* [27], but this output alone has gathered 3160 citations, the highest number of citations among all documents. This shows that the importance of each article is dictated by the high quality of the scientific research performed, thus some of them receive a more significant number of citations than others. The University of Agriculture in Faisalabad (UAF) can also be highlighted with three publications [28–30] that obtained only seven citations (an average of 2.3 citations per document), also confirming the above trend. Naturally, there may be other factors that cause this discrepancy, such as the point in time when the article was published, among others.

It is clear that many of the collaboration links are geographically related. However, there are some countries that have overcome the regional barrier and sought extracontinental academic cooperation, which is the case of (and between) the United States, China, and Malaysia. Brazil, for example, is a country that cooperates strongly with European countries such as Portugal and Spain. This is likely to happen due to their linguistic proximities and the fact that the country has a great relevance for world agribusiness, apart from showing high interest in using sustainable and renewable raw materials for the manufacture of various chemicals [31].

A network map representation was built to enable a better visualization of the links between organizations, and it is shown in Figure 5. To build the map, a minimum requirement of 250 citations per institution was set for the period analyzed, which identified 27 institutions from the database. The most relevant of them include the University of Porto, the Polytechnic Institute of Porto (IPP), and the University of Malaya.

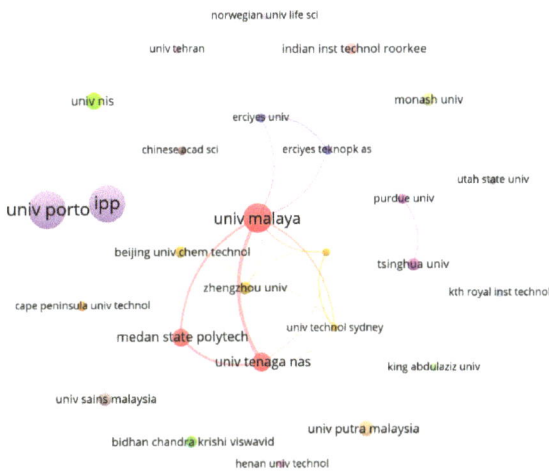

Figure 5. Network visualization map showing the collaboration among organizations with at least 250 accumulated citations. The thickness of the lines connecting two organizations indicates the accumulation of co-authorships (thicker lines means more joint published articles), and the color clusters highlight the groups of institutions with a high level of collaboration.

In addition to the analyses performed by studying the above clusters, geocoding was also used by transforming locations into points on the map. This allowed for a deeper observation of the geographical reference of the institutions identified in the collated database. Figure 6 illustrates the representation of the geocoded locations, which reveals a large concentration of these organizations in North America, Europe, and South Asia. The United States and India, which are countries of continental proportions, clearly stand out when exploring the data from this perspective. Furthermore, Figure 5 shows that, of the most relevant institutions, only a few have a co-authorship relationship among the analyzed articles, showing that there may be a sense of individualism in the scientific groups highlighted in this study. From the 27 institutions that published papers containing at least 250 cumulative citations, only 11 (40.7%) showed collaboration with other organizations.

A total of 1304 different authors were identified across the 332 documents in the database, indicating a wide range of scientific collaboration among the researchers with the most significant influence in this area. Figure 7, which was built by selecting authors with 450+ citations, depicts these collaborative relationships. The article with the highest number of citations is entitled *Microalgae for biodiesel production and other applications: A review* [27], with 3160 citations. However, it can be noted that its author, Mata, had little collaborative input in other works. On the other hand, the author Mahlia collaborated in nine highly cited documents, including *Patent landscape review on biodiesel production: Technology updates* [32], published in 2020 and presenting 152 citations as of the writing of this review. Of the 977 authors with more than five citations, only two had no co-authorship relationship, i.e., were found isolated in two different clusters, which represented 1.07% of the collaboration sets. It can be concluded that most of the research in this specific area is achieved through collaborative clusters, highlighting a high appreciation for cooperative research despite the geographical dispersion of researchers.

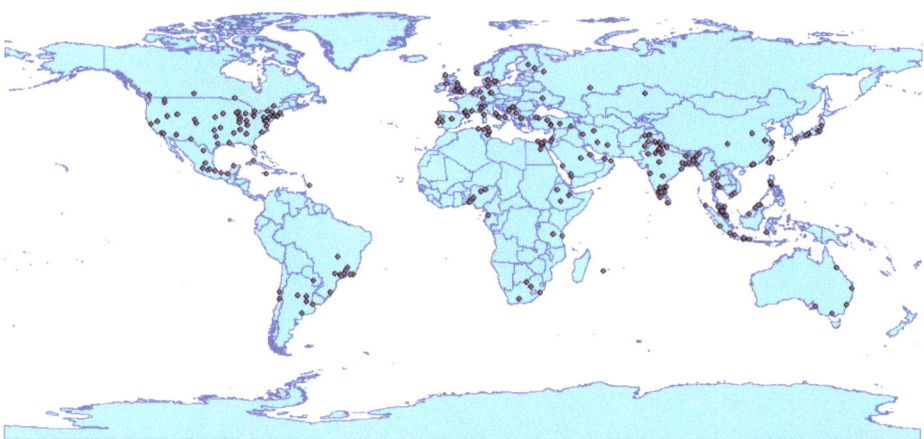

Figure 6. Geocoding of the organizations responsible for the publication of the 332 articles analyzed.

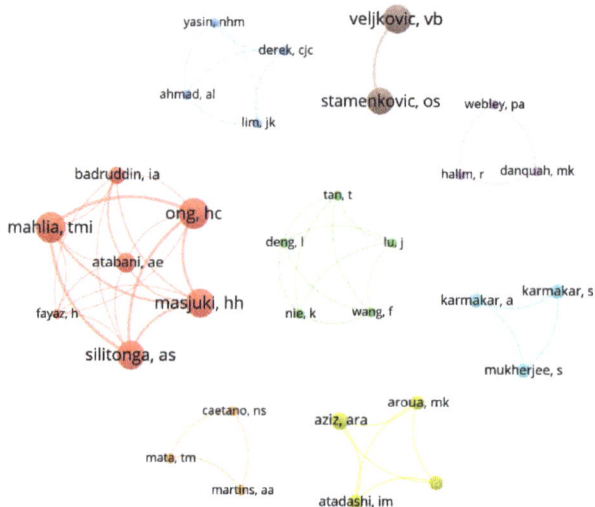

Figure 7. Network map showing the collaboration among authors with at least four publications. The thickness of the lines connecting two authors indicates the accumulation of co-authorships (thicker lines means more published articles), and the color clusters illustrate the groups of authors with a high level of collaboration.

3.1.4. Most Cited Articles

The ten most cited articles on raw materials for biodiesel production have gained a collective number of 7219 citations, with the document occupying first place in this ranking amassing 3160 (43.7%) of these. As mentioned earlier, Mata's work discussed the current status of biodiesel production by microalgae, focusing on the cultivation, harvesting, and processing techniques of the main microalgae species, along with the advantages and disadvantages of their use [27]. This article is of paramount relevance for the literature on this specific feedstock, given its use in other articles that compose the list presented in Table 3. Apart from the first position, microalgae are also mentioned in the titles of the articles in positions three, four, and nine. However, we can also note that outputs using these materials are now considerably old (2010–2012), which may hint at a changing scenario and a newfound appreciation of other raw materials that may currently be more advantageous or relevant.

Table 3. Most cited papers in the field of feedstocks for biodiesel production.

Rank	Article Title	Authors	Year Published	Citations
1	Microalgae for biodiesel production and other applications: A review [27].	Mata, Teresa M.; Martins, Antonio A.; Caetano, Nidia. S.	2010	3162
2	Non-edible vegetable oils: A critical evaluation of oil extraction, fatty acid compositions, biodiesel production, characteristics, engine performance, and emissions production [33].	Atabani, A. E.; Silitonga, A. S.; Ong, H. C.; Mahlia, T. M. I.; Masjuki, H. H.; Badruddin, Irfan Anjum; Fayaz, H.	2013	645
3	Microalgae as a sustainable energy source for biodiesel production: A review [34].	Ahmad, A. L.; Yasin, N. H. Mat; Derek, C. J. C.; Lim, J. K.	2011	590
4	Extraction of oil from microalgae for biodiesel production: A review [35].	Halim, Ronald; Danquah, Michael K.; Webley, Paul A.	2012	586
5	Biodiesel production with immobilized lipase: A review [36].	Tan, Tianwei; Lu, Jike; Nie, Kaili; Deng, Li; Wang, Fang	2010	457
6	Properties of various plants and animal feedstocks for biodiesel production [37].	Karmakar, Aninidita; Karmakar, Subrata; Mukherjee, Souti	2010	421
7	A review of current technology for biodiesel production: State of the art [38].	Aransiola, E. F.; Ojumu, T. V.; Oyekola, O. O.; Madzimbamuto, T. F.; Ikhu-Omoregbe, D. I. O.	2014	368
8	A review of biodiesel production from jatropha curcas L. oil [39].	Koh, May Ying; Ghazi, Tinia Idaty Mohd	2011	365
9	Biodiesel production by simultaneous extraction and conversion of total lipids from microalgae, cyanobacteria, and wild mixed-cultures [40].	Wahlen, Bradley D.; Willis, Robert M.; Seefeldt, Lance C.	2011	319
10	The effects of catalysts in biodiesel production: A review [41].	Atadashi, I. M.; Aroua, M. K.; Aziz, A. R. Abdul; Sulaiman, N. M. N.	2013	308

The second most cited article on the list is *Non-edible vegetable oils: A critical evaluation of oil extraction, fatty acid compositions, biodiesel production, characteristics, engine performance, and emissions production* [33], which was cited by 645 papers within the period analyzed. This article discusses biodiesel production via non-edible vegetable oils, reviews recent related studies, and analyzes general aspects, advantages, and disadvantages of the commodities regarded as second-generation feedstocks. Among them are *Jatropha curcas*, *Pongamia pinnata* (Karanja), *Moringa oleifera*, and *Aleurites moluccanas*.

The visualization and analysis done on the keywords of the articles from our database, illustrated in Figure 8, revealed that, during the period spanning from January 2020 to December 2021, cooking oil residues showed the highest relevance (29 keyword occurrences), which may indicate a strong interest in this material, considering that it stands out even among other key words that are not necessarily raw materials. Palm oil and soy oil also appear in this list, with nine and six occurrences, respectively.

It is important to highlight that the journals that occupy the first and second positions in the ranking shown in Table 1 (Renewable & Sustainable Energy Reviews and Biotechnology Advances) are responsible for 8 of the 10 most cited in the area of feedstocks for biodiesel production (Table 3). This indicates that there is a large concentration of studies carried out by specific scientific groups. In addition, they constitute almost 15% of all publications in the collated database, among the 154 journals analyzed. Regarding the number of citations, the first journal earned 8257 (43% of the total), and the second, 1790 (9.3% of the total). This points to the high relevance of the publications in these journals and the wide dispersion of studies with less prominence across a great array of periodicals.

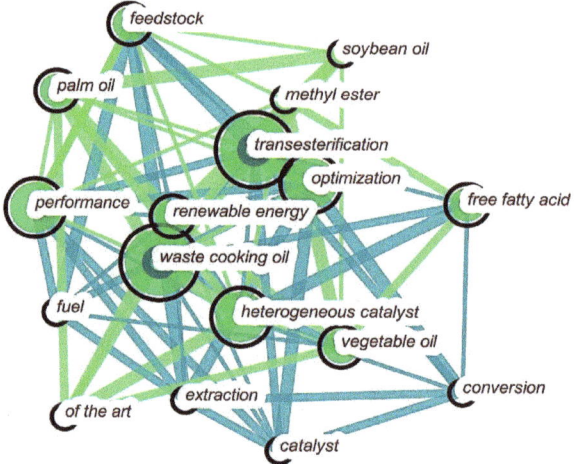

Figure 8. Keyword visualization network (January 2020 to December 2021).

The article entitled Patent landscape review on biodiesel production: Technology updates [32], despite its smaller number of citations comparatively to the articles in Table 3, shows great relevance owing to its very recent publication date (February 2020), having already gathered 152 citations. It appears to be more relevant, for example, than the Fractional characterization of jatropha, neem, moringa, trisperma, castor and candlenut seeds as potential feedstocks for biodiesel production in Cuba [42], which is almost 10 years older and, comparatively, only received 114 citations. Mahlia is also a co-author in other articles, as shown in Figure 7 (this author is in a prominent cluster), which attests to the relevance of their research methodology. The article in question is a review of 1660 patents related to biodiesel production. The paper organized the patents into five categories: raw materials, pretreatment methods, catalysts, reactors, and processing methodologies.

3.1.5. Research Areas

From the database compiled, we could observe that the 330 selected documents belonged to 25 research areas related to feedstocks for biodiesel production, from 2010 to 2021. Figure 9 shows that the most prominent them of was energy fuel, as 26.4% (163 out of 618) of the number of occurrences were from this area. Engineering was the second largest research area, with 16% of the total number of entries (99 occurrences). However, it can be seen that, despite the large number of occurrences in the area of energy fuels, there is not a high concentration in one single research area. The percentages across areas are similar, and the number of occurrences is fairly well distributed. The prominent relationship between energy fuels and engineering reveals that the market seeks scientific research that creates innovative solutions for solving existing problems related to the raw materials used for biodiesel production.

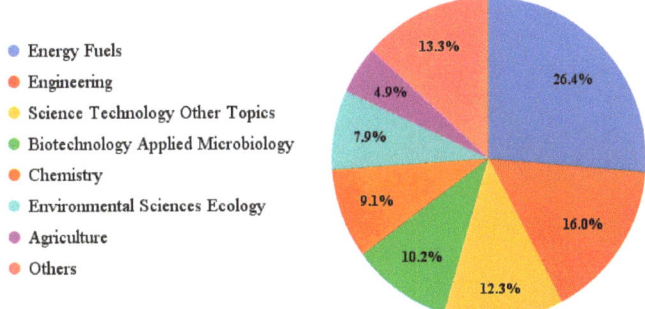

Figure 9. Distribution of research areas related to feedstock for biodiesel production.

When analyzing the database keywords, which totaled 1536 occurrences, it was possible to observe that the most used terms were "biodiesel" (225 occurrences), followed by "transesterification" (145 occurrences), and "waste cooking oil" (91 occurrences). This shows that this raw material is the most significant in the studies related to biodiesel production, given its prominence in the documents. In Figure 8, the keyword analysis was limited to the years 2020 and 2021, in order to enable the observation of the direction of the scientific research and the emerging feedstock trends for the future of biodiesel production.

There seem to not yet exist highly relevant studies on feedstocks that have great potential for the future of the market, as is the case of sewage sludge. From the search carried out on the Web of Science Core Collection site, there are only 60 documents that mention this material. The article with the most relevance is entitled Recent development on sustainable biodiesel production using sewage sludge [43], which gives an overview of biodiesel production via the use of municipal sewage sludge as a highly available and economical waste, since it is produced on a large scale due to increased urbanization worldwide. It also has zero added value. This article only has 13 citations, suggesting that the raw material is still underexplored, in comparison to soybean and waste cooking oil, for example.

It is essential to point out that a refinement was made in the database search by discarding the works that are not of interest for the visualization of trends in the proposed theme. Therefore, the limitation of inclusion to only research articles, conference papers, and review articles enabled us to obtain more precise and relevant conclusions. Furthermore, the search for documents related to feedstocks in general and for sewage sludge as a raw material made it necessary for us to search for these terms specifically, i.e., the terms "Sewage Sludge" and "Feedstock" were entered in the search field "All fields".

3.2. Feedstocks for Biodiesel Production

3.2.1. Classification of Feedstocks

The feedstocks used for biodiesel production are varied and can be classified into the four following categories: edible oils, inedible oils (e.g., inedible vegetable oils, residual oils), oils from microorganisms, and animal fats (Figure 10). The figure shows some of the feedstocks most used in the production process of biofuel, along with their classification. It is necessary to highlight the importance of those that are not edible, given that the biodiesel market does not intend to compete with the food industry. Edible oils are highly valued, which leads to price competition and causes these products to become more expensive [44].

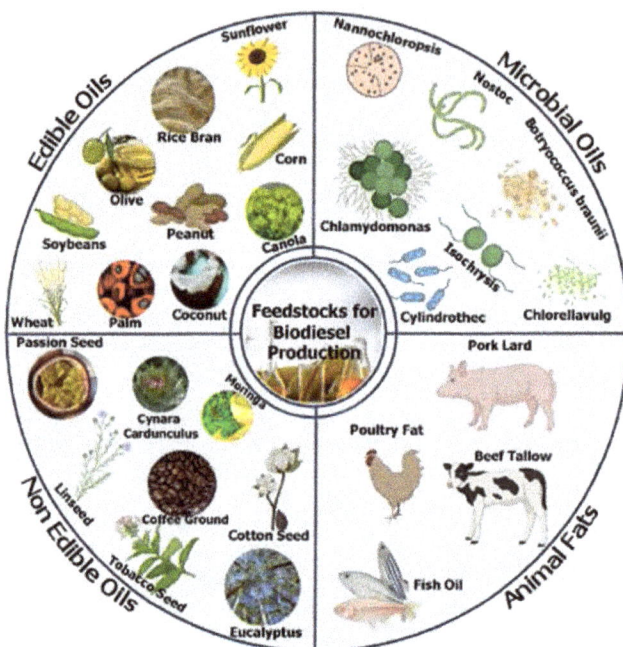

Figure 10. Various feedstocks used in the production of biodiesel. These are divided into four classes: edible oils, inedible oils, microbial oils, and animal fats.

The rapid growth of the world population and the extensive human consumption of edible oils can cause serious problems, such as starvation, in developing countries. Therefore, a category that becomes very promising as an alternative material for biodiesel production is non-edible oils. Due to the high demand for edible oils in the food industry, their prices are higher than fossil fuels, and this fact alone highlights the importance of the lower cost of cultivation of non-edible vegetable oils [45]. Oils classified as non-edible can be obtained from native plants distributed in different geographical areas of the globe. Some examples are babassu (*Orbignya* sp.), mahua (*Madhuca indica*), castor oil plant (*Ricinus communis*), jatropha (*Jatropha curcas* L.), macaúba (*Acrocomia aculeata*), andiroba (*Carapa guianensis*), karanja (*Pongamia pinnata*), seed oil from *Pistacia chinensis bge* and crambe (*Crambe abyssinica Hochst*), pinnai (*Calophyllum inophyllum*), rubber tree seed (*Hevea brasiliensis*), and coconut (*Cocos nucifera*) oil [31].

Among the raw materials reported, microalgae and waste cooking oil are highly relevant. Among the 332 publications in the database, 20 and 19 publications present the terms "waste cooking oil" or "microalgae" in their title, respectively. Together, they represent 11.7% of the documents. Accordingly, it is important to highlight that the evaluation criteria in this case maintained the same refinements of the initial search, but with the added logical operator "AND" with the following text: "waste cooking oil" or "microalgae". The articles reporting research that employ microalgae as the main object of study can be seen in Table 3, occupying prominent positions. The article with the highest number of citations that uses waste cooking oil as raw material is entitled A magnetically separable SO_4/Fe-Al-TiO(2) solid acid catalyst for biodiesel production from waste cooking oil [46]. It presents a new magnetic solid acid catalyst that was synthesized and used in the production of biodiesel through the transesterification of waste raw material. This scientific output merited 115 citations during the analyzed period.

Performing the same analysis on sewage sludge, it was possible to identify only two documents that present this raw material in their title—Biodiesel Production from Sewage

Sludge: New Paradigm for Mining Energy from Municipal Hazardous Material [47], with 83 citations, and Efficient extraction of lipids from primary sewage sludge using ionic liquids for biodiesel production [48], with 25 citations. The first article discusses the economic feasibility of biodiesel production using the lipids extracted from sewage sludge owing to the high oil yields and the lower value of this feedstock compared to those conventionally used. The second describes a new lipid extraction method from wet primary sludge for biodiesel production with ionic liquids.

3.2.2. Sewage Sludge

Studies involving biodiesel production through sewage sludge are scarce due to there being currently little interest in this research. However, according to the publications collected for this analysis, there are many advantages to using this raw material for manufacturing biofuels. Feedstocks that fall into the edible group (such as sunflower oil, rapeseed oil, and soybean) show low cultivation costs and land use requirements, while recyclable waste feedstocks (such as cooking oils and yellow grease) have limited availability. Thus, interest in using sewage sludge as a feedstock for biodiesel production has grown significantly [49] due to its excellent economic viability and a promising future intensification in its research.

In Europe, the country with the highest annual production of sewage sludge is Germany (1.85 million tons of dry solids), followed by the UK (1.14 million tons), and Spain (1.03 million tons). It can also be mentioned that more sewage sludge is produced in Japan than in the member countries in the EU. Despite scarce data on the USA, it is believed that its sludge production is the second largest in the world. China, being the country with the largest population (about 1.3 billion), has been experiencing a considerable continuous increase in its sewage sludge generation since 2011, and it surpassed the 12-million-ton mark in 2017. In that year, 45 million tons of this waste were produced worldwide [50]. Therefore, due to the huge economic potential and growth of municipal sewage sludge production, this constitutes a raw material that may become a strong trend in the future in this area.

Since about 85% of the total biodiesel production value is linked to feedstock, lipid-rich sewage sludge is also highly interesting for biodiesel production from an economic standpoint [51]. As stated in the European Union Report (Commission, 2008b), about 10 million tons of dry solids of sewage sludge were produced in the 26 European Union Member States in 2008, of which 36% (or 3.7 million tons of dry solids) was recycled for use in agriculture [52]. This shows that the material has great potential for production increase, given the urbanization rates seen worldwide.

Sewage sludge comprises organic materials (lipids, proteins, carbohydrates, etc.) and inorganic materials (heavy metals and ash, for example) from alternative waste treatment processes. It is an unavoidable waste, its production is constantly increasing, and it causes great concern regarding the risks imposed to human health and its significant potential for environmental damage. For this reason, strict regulations have limited the disposal of sewage sludge, causing alternative methods to be sought, including technological processes that employ principles of thermochemistry, such as gasification, pyrolysis, and direct combustion [53]. Using sludge for biodiesel production is an ecologically positive methodology for reusing and recycling this waste in the obtainment of versatile products via fermentation technology, with the potential aid of microorganisms. These approaches can significantly curb the amount of sludge disposal and provide high-value-added products with a lower cost of manufacture [54].

Figure 11A,B shows the number and ratio of occurrences among some of the raw materials used in the production of biodiesel and their presence in the titles of the 4586 analyzed documents. By analyzing this graph, microalgae and cooking oil residue, as previously mentioned, occupy a prominent place among the documents. Sewage sludge is the feedstock with the lowest relevance among these, as this is still in the research development stage and may be an emerging trend in the future of the scientific literature in this area.

Biodiesel production using sewage sludge as a feedstock is surrounded by complex challenges that must be addressed before it can see a breakthrough in the market. Some of these challenges are not unique to biodiesel production, but to the biofuel industry in general. A few of these challenges include sludge waste collection, product quality management, regulatory concerns, suboptimal yields, process economics, pharmaceutical and chemical contaminants, soap formation, and product separation, among others [35].

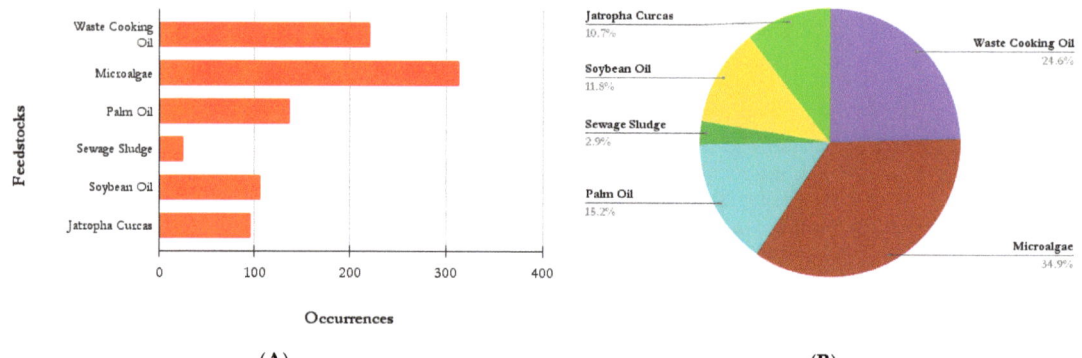

Figure 11. (**A**) Number of occurrences mentioning different feedstocks in the titles of their respective articles. (**B**) Ratio showing the percentage of articles on specific feedstocks that mention them in their respective titles.

Other advantages of sewage sludge are the fact that it is abundant and can deliver a continuous yield. The higher the value of food commodities in the future, the more competitive sewage sludge is set to become as a biodiesel feedstock [55]. Considering the facts discussed so far, it is clear that a future viability of sewage sludge is highly probable, given the drive towards ecological change for global sustainability. Municipal sewage sludge has a high potential to be a reliable and high-energy feedstock for future use in biodiesel production [56]. Increasingly more studies have been carried out on this topic, as shown in the graph related to Figure 1, which points to a sustainable interest in more intense biodiesel production using innovative feedstocks that contribute to this purpose.

Figure 12 illustrates the occurrence of the keywords from the 60 articles collected, and it can be seen that other related terms are also found, such as "lipid extraction" and "transesterification method", which are topics highly discussed in the content of the above documents. In addition, it is possible to notice the presence of discussions around microalgae that, although not the main focus of the articles in question, is clearly seen as a raw material of extreme importance in the literature on biofuel production. Figure 12 is a quantitative representation in the form of a density map, which reveals the prominence of some keywords over others, and enables the observation of a hierarchical formation of the organization of the keywords according to the number of occurrences.

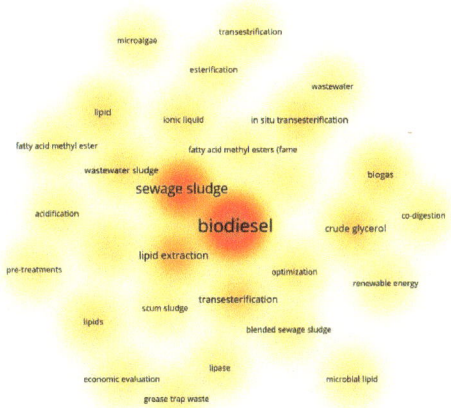

Figure 12. Density map of the database keywords relating to biodiesel production from sewage sludge. More intense color points indicate a higher number of keyword occurrences.

3.2.3. Microalgae

As discussed earlier in this article, microalgae are an extremely important raw material in most of the articles analyzed, given the fact they are the main subject of articles with very large numbers of citations. Again, the most cited article is entitled Microalgae for biodiesel production and other applications: A review [27], which has accumulated 3160 citations. Microalgae are composed of many photosynthetic microorganisms that can use carbon dioxide from the atmosphere to produce biomass faster and more efficiently than land plants. They are considered promising raw materials in various industry segments, such as feed, biofuels, food, pharmaceuticals, and nutraceuticals [57]. It is estimated that there are more than 100,000 species of algae. These plants can be used as biofertilizers for aquaculture, animal feed, and food, among other applications [58].

Microalgae cultivation can be carried out in environments considered unsuitable for the cultivation of other plants, such as in saltwater, brackish water, fresh water, or non-arable land that is unsuitable for conventional farming methods. In addition, cultivation can also be done in farms or even in bioreactors. Due to their non-selective, per hectare growth and development, microalgae can deliver higher production yields with better ecological performance [34]. Harvesting microalgae can cost as little as 20% of the total biodiesel production cost, depending on the farming method used [59].

Three main problems in algae-based biofuel production have been highlighted: supporting algae cultivation in different climates, the high water demand, and the technology deficit for commercialization. Nevertheless, the market is predicted to create future local and regional partnerships and collaborations with a view to maximize adoption and enable technologies for manufacturing them at a larger scale [60]. We can then conclude that this continues to be a raw material of extreme importance in the studies, and for the future, of the biofuel industry.

The potential of microalgae through high lipid productivity using small land areas, as well as the ability to use unproductive lands to this end, justify the recent investments in their culturing for fuels [61]. Therefore, given that there are significant economic advantages to the cultivation of this raw material and that the studies involving it are on the rise, microalgae are strong players among the raw materials currently used for biodiesel production, especially from a sustainability standpoint.

3.2.4. Waste Cooking Oil

After performing occurrence and relevance analyses, we also identify waste cooking oil as a feedstock of high interest in the research on biofuel production. Among the documents collected on this area from 2010 to 2021, waste cooking oil occupies the second place in the ranking of occurrences of articles with this specific feedstock in their title, with 221 documents (4.8% of the total). Moreover, according to Figure 8, which shows the keywords for the analysis of the emerging future trends in the research related to biodiesel production in 2020 and 2021, used cooking oil is repeatedly shown even among other terms related to methods and reactions inherent in different research topics (such as "transesterification", "heterogeneous catalyst", and "conversion").

Waste cooking oil is obtained from edible oils that have been used for frying food [62]. These types of food wastes are harmful to human health and to the environment when they are improperly disposed of by not submitting them to any treatment processes [63]. Waste cooking oil is seen as a promising feedstock for biodiesel production due to its low cost and abundance in several countries [64]. The employment of previously used cooking oils as raw materials does not create controversial issues, such as discussions about the clash between the food and fuel industries, or environmental reservations [63]. The use of this raw material for the production of biofuels was first reported a long time ago, and research on the topic started as early as the 1970s [65]. Waste cooking oil can be obtained from homes, hotels, restaurants, and food businesses that utilize frying operations and other similar food preparation processes [66]. Waste oils generated from household, commercial, and industrial sectors can be easily converted into biodiesel [67]. The production of biodiesel using used cooking oil is environmentally friendly and is a recognized solution in waste management practices [68].

Besides these raw materials, several others can be highlighted, especially in the non-edible group of materials. An example is beef tallow, which in Brazil is the second main raw material used in the production of biofuels after soybean oil. Biodiesel that is produced from tallow only generates between 17% and 35% of the total impact caused by low-sulfur diesel, thus showing an environmental advantage of using the former [69]. Another material is J. hieronymi, an endemic species in the semi-arid and arid northwestern region of Argentina, whose oil is not edible. It is a non-conventional oilseed species that does not compete with the food industry and hence has great economic potential as an alternative oil [70].

This work reveals several aspects of the biodiesel research that are clearly present across previously published review articles, such as the interest in researching feedstocks for biodiesel production in Asia, in view of the needs and economic specificities of this region. Additionally, new emerging trends are evidenced from the subsequent analysis of the articles published within the years delimited in this research, which also shows a constant search for sustainability approaches that are coupled with the maintenance of economic viability. These research trends are directed towards raw materials that do not create market competition with other industries, which is the case of the food sector, which in many cases utilizes a range of raw materials that can also be used for biodiesel production. Feedstocks such as sewage sludge, waste oils, and non-edible vegetable oils are the research foci in many of the papers reviewed earlier in this work.

4. Trendy Research Topics

4.1. Quantitative Analysis of Frequent Keywords

To better understand the development of the research on a specific area, it is necessary to carry out an analysis of the keywords in the published documents, as they reveal essential information about the topic, such as applications, trends, relevance of documents, discussions, and other general research characteristics. Table 4 lists the 24 most prominent keywords mentioned in the articles analyzed. The top six keywords are biodiesel (225), transesterification (145), residual cooking oil (91), optimization (62), soybean raw materials (60), and esterification (50).

Table 4. Ranking of the 24 most prominent keywords mentioned in the analyzed articles.

Rank	Keyword	Frequency	TLS	Rank	Keyword	Frequency	TLS
1	biodiesel	225	978	13	performance	31	170
2	transesterification	145	714	14	fuel production	31	183
3	waste cooking oil	91	475	15	lipase	27	136
4	optimization	62	323	16	*Jatropha curcas*	26	130
5	feedstocks	61	285	17	heterogeneous catalyst	26	137
6	soybean oil	60	336	18	free fatty acids	25	158
7	esterification	50	272	19	seed oil	24	133
8	vegetable oil	47	252	20	extraction	22	100
9	oil	40	149	21	in situ transesterification	20	104
10	microalgae	40	189	22	rapeseed oil	20	111
11	palm oil	35	205	23	vegetable oil	20	111
12	fuel	31	159	24	kinetics	20	127

Note: TLS: Total Link Strength.

Figure 13 illustrates the visualization network map generated from the VOSviewer program of the 60 most important keywords with at least eight occurrences in the collated database. The "biodiesel" keyword is clearly the largest group, belonging to the yellow cluster. This keyword is linked to "feedstocks", "microalgae," and "oil", which define the raw materials used in the research and are interconnected. The keyword "transesterification" is the second most reported. It is linked to "waste cooking oil" and "vegetable oil", which are also part of the blue cluster. Transesterification is the procedure that replaces the organic alkyl groups of a vegetable/plant oil (an ester) with a methyl alcohol group [71]. The cluster of optimization and general optimization encompasses some highly relevant topics in the search, such as emission characteristics, heterogeneous optimization, process optimization, and engine performance, giving us a good view into the important elements in the field of optimization. Regarding the methodologies and materials, it can be seen that this specific cluster illustrates the relationship of some keywords with "conversion", "esterification", "methane", and "acid".

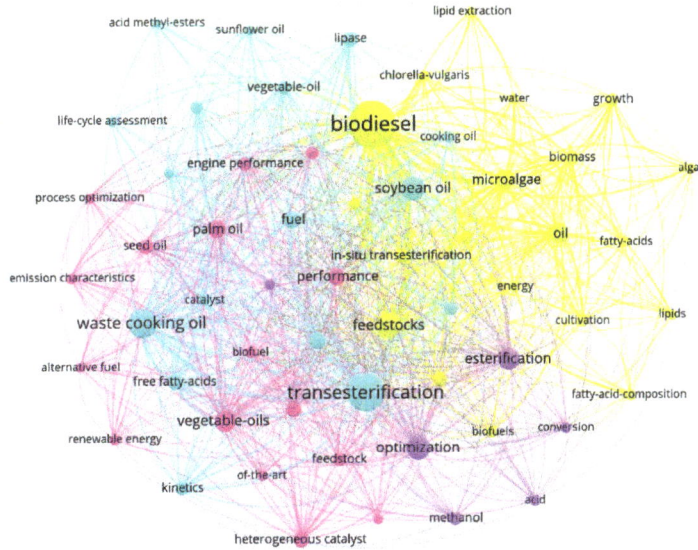

Figure 13. Visualization map of the co-citation network of keywords in the research related to raw materials for biodiesel production.

4.2. Research Areas

The CiteSpace software was used to organize the collated database and to analyze the emerging trends in the area. As mentioned earlier in this article, CiteSpace allows the visualization of knowledge development in a given area of research [72]. Analyzing a cluster enables the identification of study themes and data of greater relevance to the specific research [73]. Keywords can determine the future research paths related to biodiesel production and its raw materials. Table 5 illustrates the six primary sets of co-citations among the articles linked to this topic.

Table 5. Top six co-citation research clusters on raw materials for biodiesel production based on the CiteSpace analysis.

CID	Label	NS	Mean	Top Five Terms	Representative Articles
#0	oleaginous yeast	70	2015	oleaginous yeast; lipid extraction; wild strain; Egyptian freshwater habitat; different species.	(CHTOUROU, 2015) and (EL-SHEEKH, 2018)
#1	ionic liquid	63	2014	ionic liquid; ultrasound-assisted transesterification; using calcium oxide catalyst; economic variable; acyl acceptor.	(ZHANG, 2010) and (MANSIR, 2018)
#2	process design	54	2013	process design; heterogeneous catalyst; subcritical water; packed bed reactor; soybean soap stock acid oil.	(ZENG, 2014) and (SOARES, 2013)
#3	oil extraction	47	2014	oil extraction; plant seed; *Ricinus communis*; *Hevea brasiliensis*; *Calophyllum inophyllum* L.	(KENENI, 2017) and (SILITONGA, 2016)
#4	enzymatic biodiesel production	42	2013	enzymatic biodiesel production; low-cost feedstock; carbohydrate-derived solid acid catalyst; brown rice; ethanol fermentation.	(ADACHI, 2016) and (LOKMAN, 2014)
#5	calcium oxide	27	2016	calcium oxide; current state; technological progress; different type; LCA studies.	(MAZAHERI, 2021) and (FERNANDEZ, PENARRUBIA, 2017)

Note: CID = Cluster ID, NS = Node Size.

4.2.1. Research Fields

Cluster #0 has "oleaginous yeast" as its main keyword. These yeasts have gained significant prominence worldwide in metabolic engineering due to the fact they offer facilitated pathway manipulation, fatty acid- or oleochemical-derived metabolite enhancements, and simplified cultivation strategies [74]. The article that represents the cluster was published by Chtourou [75], and the main objective of their research was to perform the investigation of lipid accumulation in, and growth of, a new isolated marine microalgae strain. The optimization of the composition of the culture medium and the application of different stressful environmental conditions were used to this end. El-Sheekh [76] performed research on different species of isolated Scenedesmus, comparing their efficiency as feedstock for biodiesel production. The third reference article of the cluster was written by Abomohra [77], in which ten macroalgae were collected and selected as a biodiesel feedstock. The research confirmed that macroalgae are desirable potential alternatives as renewable feedstocks for biodiesel production.

Cluster #1 is represented by the keyword "ionic liquid". Ionic liquids are chemical compounds with one cation and one anion, defined by melting points of below 100 °C. Each of the above ions allows for the insertion of a unique property or function into a molecule [78]. Mansir [79] conducted research whose main objective was to demonstrate the current status of using heterogeneous bifunctional acid/base catalysts for biodiesel production from green and non-edible waste cooking oil. Zhang [80] was responsible for conducting a study that proposed a practical method used for biodiesel production from high FFA feedstocks with a high reaction rate, fewer environmental problems, less toxicity, and minimal corrosion. Petchsoongsakul [81], who published an article that showed great prominence in cluster #1, presented a novel hybridization of transesterification and esterification processes in a single reactive distillation column to be used in the production of biodiesel from waste cooking oil.

4.2.2. Emerging Trends

Cluster #2 shows the process designs involved in the various methodologies for producing biodiesel from feedstocks. One work [82] reviewed the transesterification process of low-content feedstocks for conversion to biodiesel via supercritical fluid technology, which is an environmentally friendly technique that shows higher process efficiency. Soares [83], in turn, performed an investigation of a new strategy used for a hydroesterification-based biodiesel production from low-cost oil feedstocks. This involves the complete hydrolysis of the feedstock to fatty acids in subcritical water, followed by the use of a packed bed reactor, which contains a fermented solid with lipase that performs the conversion of fatty acids into their ethyl esters. Jain [84] conducted a review of the kinetics of biodiesel production and reveals the results obtained from a two-step kinetic study of the acid–base-catalyzed transesterification process performed at preset temperatures of 65 and 50 °C in the esterification and transesterification process, respectively, under an optimal methanol-to-oil condition.

Cluster #3 depicts oil extraction processes, listing the studies covering different methodologies to this end. [85] carried out a study that presented, compared, and discussed several potential feedstocks for biofuel production, along with several oil extraction methods, and the advantages and disadvantages of using the different methodologies. [86], the second most relevant paper in cluster #3, deals with the biodiesel production from non-edible seeds, specifically by using *Hevea brasiliensis* (HB) and *Ricinus communis* (RC) as potential feedstocks. An esterification–neutralization–transesterification (ENT) process was used for biodiesel production. [87] investigates the potential of the promising feedstock *Calophyllum inophyllumas* for biodiesel production. The author assessed many crucial aspects of this process, such as the chemical and physical properties of the *Calophyllum inophyllum* crude oil and methyl ester, the mixture and engine performances, fatty acid composition, and the emissions of the *Calophyllum inophyllum* methyl ester.

In cluster #4, the focus is on the production of enzymatic biodiesel. Adachi [88] attempted to carry out the integration of a lipase-catalyzed ethanolysis and a fermentative bioethanol production process. Lokman's work [89], on the other hand, covers and explores the joint use of feedstocks that are considered to be of lower quality in tandem with carbon-based catalysts to perform the conversion of a refinery crude palm oil residue that contains a high percentage of free fatty acids. The production and characterization of the carbohydrate-derived solid acid catalysts were critically discussed, also with a view to measure their physicochemical properties. Another key article in this cluster is that of Aransiola [38], which reviews the various technologies used for biodiesel production to date, with a view to compare the commercial conformation of these methods based on feedstock availability. It was noted that there is a strong emphasis on using microalgae oil sources. The economic viability of the process is still a point that needs further discussion.

In cluster #5, calcium oxide is emphasized. CaO is an inexpensive material that can be found in abundant quantities in the Earth's lower mantle and a material that does not cause harm to the environment [90]. The prominent paper by Mazaheri [91] shows an overview of the advances in the use of calcium oxide-based catalysts in biodiesel production. The paper highlights the various factors involved in the synthesis of calcium oxide-based catalysts, and, furthermore, the effect of reaction parameters on their yield for biodiesel production is assessed. Another central article produced by Penarrubia [92] shows an industrial-scale simulation performed to compare the traditional alkali-catalyzed process via sodium hydroxide catalysis to an enzyme-catalyzed process that was developed by the research group of one of the authors involved. Finally, Moser's work [93] investigates the fuel characteristics that are highly dependent on the fatty acid composition of the feedstocks used in biodiesel production. Thus, the fatty acid profile was defined as a powerful screening tool to select feedstocks rich in monounsaturated fatty acids for further evaluation.

4.2.3. Two Key Insights

One issue with systematic reviews is that they may present data and conclusions within a confidence level that does not always reflect the reality, or that is feasible. One example is when systematic reviews for a particular emerging field of study do not yet exist. In addition, reviews that have been conducted within a certain time period may be out of date or may not have included all the scientific advances as comprehensively as it should. The CiteSpace tool was created to provide a potential solution to these challenges by enabling the use of customized datasets to answer questions about a field of knowledge that is changing rapidly [94]. It can also help extract valuable information from articles published on a specific subject matter. However, there is a need to consider its limitations concerning the optimization tools and the accuracy of the analysis of information from the database.

Node size is one of the key analysis points verified by CiteSpace. In Table 5, the data regarding the node size of each cluster in the collected database are measured, thus allowing for the quantification of the relevance of each cluster. It is also possible to observe that, among those listed, the cluster with the most recent node size (2016) is the one in position #5 and has "calcium oxide" as its most prominent keyword.

The first perspective pertains to studies related to the feedstocks used for biodiesel production. The selection of raw materials is a significant point dictating the quality and cost of the biodiesel produced [95]. There is some reported competition between the biodiesel and the edible oil markets, as many feedstocks are of use to both. Many studies have recently been performed with a view to try to find alternative feedstocks from non-edible sources that are low-cost and sustainable [44]. Performing a shelf-life analysis of feedstocks is crucial in biodiesel production [96], and many important aspects define the assessments made in the research related to these feedstocks. Attention to environmental degradation is becoming increasingly more important, and the scientific community sees the urgent need to improve the production process of renewable alternatives to petroleum-based fuels, and also to develop new ones. There is a high possibility that renewable fuels are to become an essential product in bio-based economies [97].

The second important insight from this work regards the production of biodiesel from sewage sludge, which can certainly provide a more sustainable angle to such processes, given the environmental need for this. Using sewage sludge as a feedstock for biodiesel production and therefore offering options to the resolution of the food–fuel debate can also help solve some of the difficulties in treating sludge [49]. Biodiesel from sludge can be a significantly superior alternative to other materials used as food-based feedstocks in biodiesel production. As mentioned earlier, the material has very low or zero added cost and is sustainably generated in wastewater treatment plants. The production value of the resulting biodiesel would be much lower when compared to that produced from other sources. This would eliminate the cost of feedstock materials, which is the most expensive element in biodiesel production and can account for almost 90% of the total cost of manufacture [53].

5. Conclusions

This article provided a comprehensive view into the literature related to feedstocks for biodiesel production, the emerging trends that are the focus of current work, and the promising alternatives to be explored in the future. In this work, a search methodology was devised to refine the search for, and the selection of, a set of articles, which started with a database of 4586 documents. From this initial database, two more targeted databases of 330 and 60 articles were created, respectively, in order to assess the direction of travel in this research area. The period analyzed (2010 to 2021) reveals a particular emergence of meaningful research relationships and collaborations, demonstrating a more sustainable vision among researchers for developing their work. Important conclusions from this review include:

- One of the key raw materials highlighted in this work, sewage sludge, has received little attention in the literature, considering that the number of cited articles is still small and despite the economic and sustainability advantages cited above. This is mostly due to the fact it is still a recent research theme and that raw materials that have been explored for longer are more attractive owing to the vast knowledge repertoire.
- Waste cooking oil and microalgae are the raw materials of most significant presence in the academic outputs analyzed. These are feedstocks that have been extensively reported on in the literature, mainly due to their long-standing reputation in the area. However, there has been a noticeable reduction in the volume of cited articles over the years.
- China, Malaysia, and India are the countries with the greatest research outputs relating to feedstocks for biodiesel production. It can be concluded, therefore, that Asia shows a great interest in this area. One factor that can explain this interest may be the sheer number of people concentrated in this region, since China, India, and Malaysia account for more than 35% of the world's total population.
- For a more specific analysis of this research, keywords collected by the CiteSpace program were used. From a systematic verification of the terms, it was possible to observe that the main research topics in this area include oleaginous yeasts, ionic liquids, process design, oil extraction, enzymatic biodiesel production, and calcium oxide.
- Two broad perspectives related to this research area have been emphasized. The first is the generalized view of the articles that engage in the topic of feedstocks for biodiesel production. It is concluded that there is a great deal of discussion regarding the economy versus sustainability dilemma, and researchers have sought practical solutions to the problems that arise from this. The second perspective concerns the recent emergence of academic interest in studying sewage sludge for biodiesel production. It is understood that this area will be further explored in the near future due to the solution that this raw material represents to the conundrum above.

Future development prospects in the biofuel market are mostly linked to a vision of sustainable change, as it was possible to observe with the analysis of the growth in publication numbers on the topic. Importantly, there is a clear concern from the countries that are active contributors to the global environmental degradation observed over the years to seek to reduce the impacts caused by them. The most influential researchers in this area are those who seek to fly the sustainability flag, as exemplified by Ahmad's article and also by the research topics underlying hundreds of papers in our database. In addition, a great economic interest and concern was observed in most of the above works, where authors described processes on the basis of financial costs and time or productivity. This was the case for works that had sewage sludge as their main theme, for example, which categorically sought to explain the advantages and productivity hurdles of this raw material and ways in which it could become a key trend within the biodiesel market.

The direction of the literature concerning the research topic under study is constantly being shaped due to the worldwide need for sustainability and a more environmentally centered economy. A very pertinent problem is the alleged competition between the biodiesel and the food industries for edible raw materials, in which, due to scale reasons, the latter industry will always have a competitive advantage over the former. The solution to this is in the exploitation of raw materials that are derived from non-edible oils. The environmental concern of researchers has shaped the methodological advances for using these oils in a way that can reconcile sustainability with the cost of manufacture, given that, even if a feedstock is exceptionally sustainable, it can be economically unviable.

The research contained in this review proposed to elucidate and explore the directions of travel of the research that will define the future approach to biodiesel production, highlighting the concerns regarding environmental degradation. Nevertheless, there are some limitations in this work, a major one being the fact that the Web of Science was the only database used to create our own databases. Apart from the refinements made (language and type of documents), other articles were certainly inadvertently excluded by not being in this specific database.

Furthermore, it is necessary to emphasize that the evaluation methodology used in this work was developed from the individual vision of the research described throughout the text. This should be considered, since several ways of approaching this research area have led to significantly different results. It could also be highlighted that the research domain encompassing feedstocks for biodiesel production is highly diverse due to the wide variety of research approaches undertaken on the topic. There is a clear evolving trend towards linking academic work and social compliance for sustainability. Some aspects of collaborative work were verified, showing that some countries still opt for internalized cooperation in their research development, in which authors mostly collaborate with others of the same nationality. The scientific literature on feedstocks for biodiesel production is extremely vast and can be analyzed further to allow for a comprehensive understanding of its specific characteristics that are of interest to researchers, driven mainly by a pressing, worldwide need for studies in this area.

Author Contributions: Conceptualization, M.A.d.S.R. and J.C.S.d.S.; methodology, A.A.S.L., R.K.C.d.L., M.A.d.S.R., J.C.S.d.S.; software, M.B.S., P.T.B., M.N.R.F., L.R.M.d.S., A.P.C.; validation, A.A.S.L., R.K.C.d.L., M.A.d.S.R., J.C.S.d.S.; formal analysis, M.B.S., P.T.B., M.N.R.F., L.R.M.d.S., A.P.C.; investigation, M.B.S., P.T.B., M.N.R.F., L.R.M.d.S., A.P.C.; resources, M.B.S., P.T.B., M.N.R.F., L.R.M.d.S., A.P.C.; data curation, M.B.S., P.T.B.; M.N.R.F., L.R.M.d.S., A.P.C.; writing—original draft preparation, M.B.S., P.T.B., M.N.R.F., L.R.M.d.S., A.P.C., A.A.S.L., R.K.C.d.L., M.A.d.S.R., J.C.S.d.S.; writing—review and editing, M.B.S., P.T.B., M.N.R.F., L.R.M.d.S., A.P.C., A.A.S.L., R.K.C.d.L., M.A.d.S.R., J.C.S.d.S.; visualization, A.A.S.L., R.K.C.d.L., M.A.d.S.R., J.C.S.d.S.; supervision, A.A.S.L., R.K.C.d.L., M.A.d.S.R., J.C.S.d.S.; project administration, J.C.S.d.S.; funding acquisition, A.A.S.L., R.K.C.d.L., M.A.d.S.R., J.C.S.d.S. All authors have read and agreed to the published version of the manuscript.

Funding: This research received no external funding.

Institutional Review Board Statement: Not applicable.

Informed Consent Statement: Not applicable.

Data Availability Statement: The data presented in this study are available on request from the corresponding author.

Acknowledgments: We gratefully acknowledge the following Brazilian Agencies for Scientific and Technological Development: Fundação Cearense de Apoio ao Desenvolvimento Científico e Tecnológico (FUNCAP) (PS1-0186-00216.01.00/21; PS1-00186-00255.01.00/21), Conselho Nacional de Desenvolvimento Científico e Tecnológico (CNPq) (311062/2019-9; 308280/2017-2; 313647/2020-8), and Coordenação de Aperfeiçoamento de Ensino Superior (CAPES) (finance code 001).

Conflicts of Interest: The authors declare no conflict of interest.

References

1. De Man, R.; German, L. Certifying the sustainability of biofuels: Promise and reality. *Energy Policy* **2017**, *109*, 871–883. [CrossRef]
2. Adnan, A.I.; Ong, M.Y.; Nomanbhay, S.; Chew, K.W.; Show, P.L. Technologies for biogas upgrading to biomethane: A review. *Bioengineering* **2019**, *6*, 92. [CrossRef] [PubMed]
3. Soleymani, M.; Rosentrater, K. Techno-economic analysis of biofuel production from macroalgae (seaweed). *Bioengineering* **2017**, *4*, 92. [CrossRef]
4. Kang, A.; Lee, T.S. Converting sugars to biofuels: Ethanol and beyond. *Bioengineering* **2015**, *2*, 184–203. [CrossRef]
5. Awogbemi, O.; Kallon, D.V.V. Application of tubular reactor technologies for the acceleration of biodiesel production. *Bioengineering* **2022**, *9*, 347. [CrossRef] [PubMed]
6. Rai, A.K.; Al Makishah, N.H.; Wen, Z.; Gupta, G.; Pandit, S.; Prasad, R. Recent developments in lignocellulosic biofuels, a renewable source of bioenergy. *Fermentation* **2022**, *8*, 161. [CrossRef]
7. Zhao, X.; Wei, L.; Cheng, S.; Julson, J. Review of heterogeneous catalysts for catalytically upgrading vegetable oils into hydrocarbon biofuels. *Catalysts* **2017**, *7*, 83. [CrossRef]

8. Moreira, K.S.; Moura Júnior, L.S.; Monteiro, R.R.C.; de Oliveira, A.L.B.; Valle, C.P.; Freire, T.M.; Fechine, P.B.A.; de Souza, M.C.M.; Fernandez-Lorente, G.; Guisan, J.M.; et al. Optimization of the Production of Enzymatic Biodiesel from Residual Babassu Oil (*Orbignya* sp.) via RSM. *Catalysts* **2020**, *10*, 414. [CrossRef]
9. de Oliveira, A.L.B.; Cavalcante, F.T.T.; da Silva Moreira, K.; de Castro Monteiro, R.R.; Rocha, T.G.; da Silva Souza, J.E.; da Fonseca, A.M.; Lopes, A.A.S.; Guimarões, A.P.; de Lima, R.K.C.; et al. Lipases immobilized onto nanomaterials as biocatalysts in biodiesel production: Scientific context, challenges, and opportunities. *Rev. Virtual Química* **2021**, *13*, 875–891. [CrossRef]
10. Monteiro, R.R.C.; Arana-Peña, S.; da Rocha, T.N.; Miranda, L.P.; Berenguer-Murcia, Á.; Tardioli, P.W.; dos Santos, J.C.S.; Fernandez-Lafuente, R. Liquid lipase preparations designed for industrial production of biodiesel. Is it really an optimal solution? *Renew. Energy* **2021**, *164*, 1566–1587. [CrossRef]
11. Cavalcante, F.T.T.; Neto, F.S.; Rafael de Aguiar Falcão, I.; Erick da Silva Souza, J.; de Moura Junior, L.S.; da Silva Sousa, P.; Rocha, T.G.; de Sousa, I.G.; de Lima Gomes, P.H.; de Souza, M.C.M.; et al. Opportunities for improving biodiesel production via lipase catalysis. *Fuel* **2021**, *288*, 119577. [CrossRef]
12. Rocha, T.G.; de L Gomes, P.H.; de Souza, M.C.M.; Monteiro, R.R.C.; dos Santos, J.C.S. Lipase cocktail for optimized biodiesel production of free fatty acids from residual chicken oil. *Catal. Lett.* **2021**, *151*, 1155–1166. [CrossRef]
13. Mota, G.F.; de Sousa, I.G.; de Oliveira, A.L.B.; Cavalcante, A.L.G.; da Silva Moreira, K.; Cavalcante, F.T.T.; da Silva Souza, J.E.; de Aguiar Falcão, Í.R.; Rocha, T.G.; Valério, R.B.R.; et al. Biodiesel production from microalgae using lipase-based catalysts: Current challenges and prospects. *Algal Res.* **2022**, *62*, 102616. [CrossRef]
14. Athar, M.; Zaidi, S. A review of the feedstocks, catalysts, and intensification techniques for sustainable biodiesel production. *J. Environ. Chem. Eng.* **2020**, *8*, 104523. [CrossRef]
15. Ma, X.; Gao, M.; Gao, Z.; Wang, J.; Zhang, M.; Ma, Y.; Wang, Q. Past, current, and future research on microalga-derived biodiesel: A critical review and bibliometric analysis. *Environ. Sci. Pollut. Res.* **2018**, *25*, 10596–10610. [CrossRef]
16. Jiang, W.; Aishan, T.; Halik, Ü.; Wei, Z.; Wumaier, M. A Bibliometric and Visualized Analysis of Research Progress and Trends on Decay and Cavity Trees in Forest Ecosystem over 20 Years: An Application of the CiteSpace Software. *Forests* **2022**, *13*, 1437. [CrossRef]
17. Li, N.; Han, R.; Lu, X. Bibliometric analysis of research trends on solid waste reuse and recycling during 1992–2016. *Resour. Conserv. Recycl.* **2018**, *130*, 109–117. [CrossRef]
18. Donthu, N.; Kumar, S.; Mukherjee, D.; Pandey, N.; Lim, W.M. How to conduct a bibliometric analysis: An overview and guidelines. *J. Bus. Res.* **2021**, *133*, 285–296. [CrossRef]
19. Kargbo, D.M. Biodiesel production from municipal sewage sludges. *Energy Fuels* **2010**, *24*, 2791–2794. [CrossRef]
20. Callegari, A.; Hlavinek, P.; Capodaglio, A.G. Production of energy (biodiesel) and recovery of materials (biochar) from pyrolysis of urban waste sludge. *Rev. Ambiente Água* **2018**, *13*, 1. [CrossRef]
21. Tamala, J.K.; Maramag, E.I.; Simeon, K.A.; Ignacio, J.J. A bibliometric analysis of sustainable oil and gas production research using VOSviewer. *Clean. Eng. Technol.* **2022**, *7*, 100437. [CrossRef]
22. Sweileh, W.M.; Al-Jabi, S.W.; AbuTaha, A.S.; Zyoud, S.H.; Anayah, F.M.A.; Sawalha, A.F. Bibliometric analysis of worldwide scientific literature in mobile—Health: 2006–2016. *BMC Med. Inform. Decis. Mak.* **2017**, *17*, 72. [CrossRef] [PubMed]
23. Chen, C. CiteSpace II: Detecting and visualizing emerging trends and transient patterns in scientific literature. *J. Am. Soc. Inf. Sci. Technol.* **2006**, *57*, 359–377. [CrossRef]
24. Veillette, M.; Chamoumi, M.; Nikiema, J.; Faucheux, N.; Heitz, M. Production of biodiesel from microalgae. In *Advances in Chemical Engineering*; InTech: Rijeka, Croatia, 2012.
25. Shah, S.H. Sustainable biodiesel production. In *Encyclopedia of Renewable and Sustainable Materials*; Elsevier: Amsterdam, The Netherlands, 2020; pp. 347–355.
26. Muthukumaran, C.; Sharmila, G.; Manojkumar, N.; Gnanaprakasam, A.; Sivakumar, V.M. Optimization and kinetic modeling of biodiesel production. In *Encyclopedia of Renewable and Sustainable Materials*; Elsevier: Amsterdam, The Netherlands, 2020; pp. 193–201.
27. Mata, T.M.; Martins, A.A.; Caetano, N.S. Microalgae for biodiesel production and other applications: A review. *Renew. Sustain. Energy Rev.* **2010**, *14*, 217–232. [CrossRef]
28. Aslam, M.M.; Khan, A.A.; Cheema, H.M.N.; Hanif, M.A.; Azeem, M.W.; Azmat, M.A. Novel mutant camelina and jatropha as valuable feedstocks for biodiesel production. *Sci. Rep.* **2020**, *10*, 21868. [CrossRef]
29. Ghani, N.; Iqbal, J.; Sadaf, S.; Nawaz Bhatti, H.; Asgher, M. Comparison of photo-esterification capability of bismuth vanadate with reduced graphene oxide bismuth vanadate (RGO/BiVO$_4$) composite for biodiesel production from high free fatty acid containing non-edible oil. *ChemistrySelect* **2020**, *5*, 9245–9253. [CrossRef]
30. Mureed, K.; Kanwal, S.; Hussain, A.; Noureen, S.; Hussain, S.; Ahmad, S.; Ahmad, M.; Waqas, R. Biodiesel production from algae grown on food industry wastewater. *Environ. Monit. Assess.* **2018**, *190*, 271. [CrossRef]
31. da Silva Dutra, L.; Costa Cerqueira Pinto, M.; Cipolatti, E.P.; Aguieiras, E.C.G.; Manoel, E.A.; Greco-Duarte, J.; Guimarães Freire, D.M.; Pinto, J.C. How the biodiesel from immobilized enzymes production is going on: An advanced bibliometric evaluation of global research. *Renew. Sustain. Energy Rev.* **2022**, *153*, 111765. [CrossRef]
32. Mahlia, T.M.I.; Syazmi, Z.A.H.S.; Mofijur, M.; Abas, A.E.P.; Bilad, M.R.; Ong, H.C.; Silitonga, A.S. Patent landscape review on biodiesel production: Technology updates. *Renew. Sustain. Energy Rev.* **2020**, *118*, 109526. [CrossRef]

33. Atabani, A.E.; Silitonga, A.S.; Ong, H.C.; Mahlia, T.M.I.; Masjuki, H.H.; Badruddin, I.A.; Fayaz, H. Non-edible vegetable oils: A critical evaluation of oil extraction, fatty acid compositions, biodiesel production, characteristics, engine performance and emissions production. *Renew. Sustain. Energy Rev.* **2013**, *18*, 211–245. [CrossRef]
34. Ahmad, A.L.; Yasin, N.H.M.; Derek, C.J.C.; Lim, J.K. Microalgae as a sustainable energy source for biodiesel production: A review. *Renew. Sustain. Energy Rev.* **2011**, *15*, 584–593. [CrossRef]
35. Halim, R.; Danquah, M.K.; Webley, P.A. Extraction of oil from microalgae for biodiesel production: A review. *Biotechnol. Adv.* **2012**, *30*, 709–732. [CrossRef] [PubMed]
36. Tan, T.; Lu, J.; Nie, K.; Deng, L.; Wang, F. Biodiesel production with immobilized lipase: A review. *Biotechnol. Adv.* **2010**, *28*, 628–634. [CrossRef] [PubMed]
37. Karmakar, A.; Karmakar, S.; Mukherjee, S. Properties of various plants and animals feedstocks for biodiesel production. *Bioresour. Technol.* **2010**, *101*, 7201–7210. [CrossRef] [PubMed]
38. Aransiola, E.F.; Ojumu, T.V.; Oyekola, O.O.; Madzimbamuto, T.F.; Ikhu-Omoregbe, D.I.O. A review of current technology for biodiesel production: State of the art. *Biomass Bioenergy* **2014**, *61*, 276–297. [CrossRef]
39. Koh, M.Y.; Ghazi, T.I.M. A review of biodiesel production from *Jatropha curcas* L. oil. *Renew. Sustain. Energy Rev.* **2011**, *15*, 2240–2251. [CrossRef]
40. Wahlen, B.D.; Willis, R.M.; Seefeldt, L.C. Biodiesel production by simultaneous extraction and conversion of total lipids from microalgae, cyanobacteria, and wild mixed-cultures. *Bioresour. Technol.* **2011**, *102*, 2724–2730. [CrossRef]
41. Atadashi, I.M.; Aroua, M.K.; Abdul Aziz, A.R.; Sulaiman, N.M.N. The effects of catalysts in biodiesel production: A review. *J. Ind. Eng. Chem.* **2013**, *19*, 14–26. [CrossRef]
42. Martín, C.; Moure, A.; Martín, G.; Carrillo, E.; Domínguez, H.; Parajó, J.C. Fractional characterisation of jatropha, neem, moringa, trisperma, castor and candlenut seeds as potential feedstocks for biodiesel production in Cuba. *Biomass Bioenergy* **2010**, *34*, 533–538. [CrossRef]
43. Srivastava, N.; Srivastava, M.; Gupta, V.K.; Manikanta, A.; Mishra, K.; Singh, S.; Singh, S.; Ramteke, P.W.; Mishra, P.K. Recent development on sustainable biodiesel production using sewage sludge. *3 Biotech* **2018**, *8*, 245. [CrossRef]
44. Mohiddin, M.N.B.; Tan, Y.H.; Seow, Y.X.; Kansedo, J.; Mubarak, N.M.; Abdullah, M.O.; Chan, Y.S.; Khalid, M. Evaluation on feedstock, technologies, catalyst and reactor for sustainable biodiesel production: A review. *J. Ind. Eng. Chem.* **2021**, *98*, 60–81. [CrossRef]
45. Banković-Ilić, I.B.; Stamenković, O.S.; Veljković, V.B. Biodiesel production from non-edible plant oils. *Renew. Sustain. Energy Rev.* **2012**, *16*, 3621–3647. [CrossRef]
46. Gardy, J.; Osatiashtiani, A.; Céspedes, O.; Hassanpour, A.; Lai, X.; Lee, A.F.; Wilson, K.; Rehan, M. A magnetically separable SO_4/Fe-Al-TiO_2 solid acid catalyst for biodiesel production from waste cooking oil. *Appl. Catal. B Environ.* **2018**, *234*, 268–278. [CrossRef]
47. Kwon, E.E.; Kim, S.; Jeon, Y.J.; Yi, H. Biodiesel production from sewage sludge: New paradigm for mining energy from municipal hazardous material. *Environ. Sci. Technol.* **2012**, *46*, 10222–10228. [CrossRef] [PubMed]
48. Olkiewicz, M.; Plechkova, N.V.; Fabregat, A.; Stüber, F.; Fortuny, A.; Font, J.; Bengoa, C. Efficient extraction of lipids from primary sewage sludge using ionic liquids for biodiesel production. *Sep. Purif. Technol.* **2015**, *153*, 118–125. [CrossRef]
49. Liu, X.; Zhu, F.; Zhang, R.; Zhao, L.; Qi, J. Recent progress on biodiesel production from municipal sewage sludge. *Renew. Sustain. Energy Rev.* **2021**, *135*, 110260. [CrossRef]
50. Chang, Z.; Long, G.; Zhou, J.L.; Ma, C. Valorization of sewage sludge in the fabrication of construction and building materials: A review. *Resour. Conserv. Recycl.* **2020**, *154*, 104606. [CrossRef]
51. Kumar, M.; Ghosh, P.; Khosla, K.; Thakur, I.S. Biodiesel production from municipal secondary sludge. *Bioresour. Technol.* **2016**, *216*, 165–171. [CrossRef]
52. Kacprzak, M.; Neczaj, E.; Fijałkowski, K.; Grobelak, A.; Grosser, A.; Worwag, M.; Rorat, A.; Brattebo, H.; Almås, Å.; Singh, B.R. Sewage sludge disposal strategies for sustainable development. *Environ. Res.* **2017**, *156*, 39–46. [CrossRef]
53. Arazo, R.O.; de Luna, M.D.G.; Capareda, S.C. Assessing biodiesel production from sewage sludge-derived bio-oil. *Biocatal. Agric. Biotechnol.* **2017**, *10*, 189–196. [CrossRef]
54. Balasubramanian, S.; Tyagi, R.D. Value-added bio-products from sewage sludge. In *Current Developments in Biotechnology and Bioengineering*; Elsevier: Amsterdam, The Netherlands, 2017; pp. 27–42.
55. Wu, X.; Zhu, F.; Qi, J.; Zhao, L. Biodiesel production from sewage sludge by using alkali catalyst catalyze. *Procedia Environ. Sci.* **2016**, *31*, 26–30. [CrossRef]
56. Navia, R.; Mittelbach, M. Could sewage sludge be considered a source of waste lipids for biodiesel production? *Waste Manag. Res.* **2012**, *30*, 873–874. [CrossRef] [PubMed]
57. Branco-Vieira, M.; Mata, T.M.; Martins, A.A.; Freitas, M.A.V.; Caetano, N.S. Economic analysis of microalgae biodiesel production in a small-scale facility. *Energy Rep.* **2020**, *6*, 325–332. [CrossRef]
58. Reza Talaghat, M.; Mokhtari, S.; Saadat, M. Modeling and optimization of biodiesel production from microalgae in a batch reactor. *Fuel* **2020**, *280*, 118578. [CrossRef]
59. Mubarak, M.; Shaija, A.; Suchithra, T. Flocculation: An effective way to harvest microalgae for biodiesel production. *J. Environ. Chem. Eng.* **2019**, *7*, 103221. [CrossRef]

60. Dev Sarkar, R.; Singh, H.B.; Chandra Kalita, M. Enhanced lipid accumulation in microalgae through nanoparticle-mediated approach, for biodiesel production: A mini-review. *Heliyon* **2021**, *7*, e08057. [CrossRef] [PubMed]
61. Enwereuzoh, U.; Harding, K.; Low, M. Microalgae cultivation using nutrients in fish farm effluent for biodiesel production. *S. Afr. J. Chem. Eng.* **2021**, *37*, 46–52. [CrossRef]
62. Amenaghawon, A.N.; Obahiagbon, K.; Isesele, V.; Usman, F. Optimized biodiesel production from waste cooking oil using a functionalized bio-based heterogeneous catalyst. *Clean. Eng. Technol.* **2022**, *8*, 100501. [CrossRef]
63. Suzihaque, M.U.H.; Syazwina, N.; Alwi, H.; Ibrahim, U.K.; Abdullah, S.; Haron, N. A sustainability study of the processing of kitchen waste as a potential source of biofuel: Biodiesel production from waste cooking oil (WCO). *Mater. Today Proc.* **2022**, *63*, S484–S489. [CrossRef]
64. Falowo, O.A.; Oladipo, B.; Taiwo, A.E.; Olaiya, A.T.; Oyekola, O.O.; Betiku, E. Green heterogeneous base catalyst from ripe and unripe plantain peels mixture for the transesterification of waste cooking oil. *Chem. Eng. J. Adv.* **2022**, *10*, 100293. [CrossRef]
65. Cordero-Ravelo, V.; Schallenberg-Rodriguez, J. Biodiesel production as a solution to waste cooking oil (WCO) disposal. Will any type of WCO do for a transesterification process? A quality assessment. *J. Environ. Manag.* **2018**, *228*, 117–129. [CrossRef] [PubMed]
66. Chen, C.; Chitose, A.; Kusadokoro, M.; Nie, H.; Xu, W.; Yang, F.; Yang, S. Sustainability and challenges in biodiesel production from waste cooking oil: An advanced bibliometric analysis. *Energy Rep.* **2021**, *7*, 4022–4034. [CrossRef]
67. Aghbashlo, M.; Tabatabaei, M.; Amid, S.; Hosseinzadeh-Bandbafha, H.; Khoshnevisan, B.; Kianian, G. Life cycle assessment analysis of an ultrasound-assisted system converting waste cooking oil into biodiesel. *Renew. Energy* **2020**, *151*, 1352–1364. [CrossRef]
68. Gouran, A.; Aghel, B.; Nasirmanesh, F. Biodiesel production from waste cooking oil using wheat bran ash as a sustainable biomass. *Fuel* **2021**, *295*, 120542. [CrossRef]
69. Rezania, S.; Oryani, B.; Park, J.; Hashemi, B.; Yadav, K.K.; Kwon, E.E.; Hur, J.; Cho, J. Review on transesterification of non-edible sources for biodiesel production with a focus on economic aspects, fuel properties and by-product applications. *Energy Convers. Manag.* **2019**, *201*, 112155. [CrossRef]
70. Ferrero, G.O.; Sánchez Faba, E.M.; Rickert, A.A.; Eimer, G.A. Alternatives to rethink tomorrow: Biodiesel production from residual and non-edible oils using biocatalyst technology. *Renew. Energy* **2020**, *150*, 128–135. [CrossRef]
71. Patel, N.K.; Shah, S.N. Biodiesel from plant oils. In *Food, Energy, and Water*; Elsevier: Amsterdam, The Netherlands, 2015; pp. 277–307.
72. Wang, H.; Zhang, W.; Zhang, Y.; Xu, J. A bibliometric review on stability and reinforcement of special soil subgrade based on CiteSpace. *J. Traffic Transp. Eng. Engl. Ed.* **2022**, *9*, 223–243. [CrossRef]
73. Wilks, D.S. Cluster Analysis. In *International Geophysics*; Elsevier: Amsterdam, The Netherlands, 2011; Volume 100, pp. 603–616. [CrossRef]
74. Chattopadhyay, A.; Mitra, M.; Maiti, M.K. Recent advances in lipid metabolic engineering of oleaginous yeasts. *Biotechnol. Adv.* **2021**, *53*, 107722. [CrossRef]
75. Chtourou, H.; Dahmen, I.; Jebali, A.; Karray, F.; Hassairi, I.; Abdelkafi, S.; Ayadi, H.; Sayadi, S.; Dhouib, A. Characterization of *Amphora* sp., a newly isolated diatom wild strain, potentially usable for biodiesel production. *Bioprocess Biosyst. Eng.* **2015**, *38*, 1381–1392. [CrossRef]
76. El-Sheekh, M.; Abomohra, A.E.-F.; Eladel, H.; Battah, M.; Mohammed, S. Screening of different species of scenedesmus isolated from Egyptian freshwater habitats for biodiesel production. *Renew. Energy* **2018**, *129*, 114–120. [CrossRef]
77. Abomohra, A.E.-F.; El-Naggar, A.H.; Baeshen, A.A. Potential of macroalgae for biodiesel production: Screening and evaluation studies. *J. Biosci. Bioeng.* **2018**, *125*, 231–237. [CrossRef] [PubMed]
78. Kosiński, S.; Rykowska, I.; Gonsior, M.; Krzyżanowski, P. Ionic liquids as antistatic additives for polymer composites—A review. *Polym. Test.* **2022**, *112*, 107649. [CrossRef]
79. Mansir, N.; Teo, S.H.; Rashid, U.; Saiman, M.I.; Tan, Y.P.; Alsultan, G.A.; Taufiq-Yap, Y.H. Modified waste egg shell derived bifunctional catalyst for biodiesel production from high FFA waste cooking oil. A review. *Renew. Sustain. Energy Rev.* **2018**, *82*, 3645–3655. [CrossRef]
80. Zhang, J.; Chen, S.; Yang, R.; Yan, Y. Biodiesel production from vegetable oil using heterogenous acid and alkali catalyst. *Fuel* **2010**, *89*, 2939–2944. [CrossRef]
81. Petchsoongsakul, N.; Ngaosuwan, K.; Kiatkittipong, W.; Aiouache, F.; Assabumrungrat, S. Process design of biodiesel production: Hybridization of ester-and transesterification in a single reactive distillation. *Energy Convers. Manag.* **2017**, *153*, 493–503. [CrossRef]
82. Zeng, D.; Li, R.; Wang, B.; Xu, J.; Fang, T. A review of transesterification from low-grade feedstocks for biodiesel production with supercritical methanol. *Russ. J. Appl. Chem.* **2014**, *87*, 1176–1183. [CrossRef]
83. Soares, D.; Pinto, A.F.; Gonçalves, A.G.; Mitchell, D.A.; Krieger, N. Biodiesel production from soybean soapstock acid oil by hydrolysis in subcritical water followed by lipase-catalyzed esterification using a fermented solid in a packed-bed reactor. *Biochem. Eng. J.* **2013**, *81*, 15–23. [CrossRef]
84. Jain, S.; Sharma, M.P. Biodiesel Production from *Jatropha curcas* Oil. *Renew. Sustain. Energy Rev.* **2010**, *14*, 3140–3147. [CrossRef]
85. Gonfa Keneni, Y.; Mario Marchetti, J. Oil extraction from plant seeds for biodiesel production. *AIMS Energy* **2017**, *5*, 316–340. [CrossRef]

86. Silitonga, A.S.; Masjuki, H.H.; Ong, H.C.; Yusaf, T.; Kusumo, F.; Mahlia, T.M.I. Synthesis and optimization of hevea brasiliensis and ricinus communis as feedstock for biodiesel production: A comparative study. *Ind. Crops Prod.* **2016**, *85*, 274–286. [CrossRef]
87. Atabani, A.E.; César, A.d.S. *Calophyllum inophyllum* L.—A prospective non-edible biodiesel feedstock. Study of biodiesel production, properties, fatty acid composition, blending and engine performance. *Renew. Sustain. Energy Rev.* **2014**, *37*, 644–655. [CrossRef]
88. Adachi, D.; Koda, R.; Hama, S.; Yamada, R.; Nakashima, K.; Ogino, C.; Kondo, A. An integrative process model of enzymatic biodiesel production through ethanol fermentation of brown rice followed by lipase-catalyzed ethanolysis in a water-containing system. *Enzym. Microb. Technol.* **2013**, *52*, 118–122. [CrossRef]
89. Lokman, I.M.; Rashid, U.; Yunus, R.; Taufiq-Yap, Y.H. Carbohydrate-derived solid acid catalysts for biodiesel production from low-cost feedstocks: A review. *Catal. Rev.* **2014**, *56*, 187–219. [CrossRef]
90. Zul, N.A.; Ganesan, S.; Hamidon, T.S.; Oh, W.-D.; Hussin, M.H. A review on the utilization of calcium oxide as a base catalyst in biodiesel production. *J. Environ. Chem. Eng.* **2021**, *9*, 105741. [CrossRef]
91. Mazaheri, H.; Ong, H.C.; Amini, Z.; Masjuki, H.H.; Mofijur, M.; Su, C.H.; Anjum Badruddin, I.; Khan, T.M.Y. An overview of biodiesel production via calcium oxide based catalysts: Current state and perspective. *Energies* **2021**, *14*, 3950. [CrossRef]
92. Peñarrubia Fernandez, I.A.; Liu, D.-H.; Zhao, J. LCA studies comparing alkaline and immobilized enzyme catalyst processes for biodiesel production under brazilian conditions. *Resour. Conserv. Recycl.* **2017**, *119*, 117–127. [CrossRef]
93. Moser, B.R.; Vaughn, S.F. Efficacy of fatty acid profile as a tool for screening feedstocks for biodiesel production. *Biomass Bioenergy* **2012**, *37*, 31–41. [CrossRef]
94. Chen, C. *CiteSpace: A Practical Guide for Mapping Scientific Literature*; Nova Science: New York, NY, USA, 2016.
95. Maheshwari, P.; Haider, M.B.; Yusuf, M.; Klemeš, J.J.; Bokhari, A.; Beg, M.; Al-Othman, A.; Kumar, R.; Jaiswal, A.K. A review on latest trends in cleaner biodiesel production: Role of feedstock, production methods, and catalysts. *J. Clean. Prod.* **2022**, *355*, 131588. [CrossRef]
96. Ambat, I.; Srivastava, V.; Sillanpää, M. Recent advancement in biodiesel production methodologies using various feedstock: A review. *Renew. Sustain. Energy Rev.* **2018**, *90*, 356–369. [CrossRef]
97. Bergmann, J.; Tupinambá, D.; Costa, O.Y.; Almeida, J.R.; Barreto, C.; Quirino, B. Biodiesel production in brazil and alternative biomass feedstocks. *Renew. Sustain. Energy Rev.* **2013**, *21*, 411–420. [CrossRef]

Review

Bioengineering to Accelerate Biodiesel Production for a Sustainable Biorefinery

Dheeraj Rathore [1], Surajbhan Sevda [2], Shiv Prasad [3], Veluswamy Venkatramanan [4], Anuj Kumar Chandel [5], Rupam Kataki [6], Sudipa Bhadra [2], Veeranna Channashettar [7], Neelam Bora [6] and Anoop Singh [8,*]

[1] School of Environment and Sustainable Development, Central University of Gujarat, Gandhinagar 382030, Gujarat, India
[2] Environmental Bioprocess Laboratory, Department of Biotechnology, National Institute of Technology, Warangal 506004, Telangana, India
[3] Division of Environment Science, ICAR—Indian Agricultural Research Institute, New Delhi 110012, Delhi, India
[4] School of Interdisciplinary and Transdisciplinary Studies, Indira Gandhi National Open University, New Delhi 110068, Delhi, India
[5] Department of Biotechnology, Engineering School of Lorena (EEL), University of São Paulo (USP), Estrada Municipal do Campinho, Lorena 12602-810, SP, Brazil
[6] Department of Energy, Tezpur University, Napaam, Tezpur 784028, Assam, India
[7] Environmental and Industrial Biotechnology Division, The Energy and Resources Institute, Lodhi Road, New Delhi 110003, Delhi, India
[8] Department of Scientific and Industrial Research (DSIR), Ministry of Science and Technology, Government of India, Technology Bhawan, New Mehrauli Road, New Delhi 110016, Delhi, India
* Correspondence: apsinghenv@gmail.com

Citation: Rathore, D.; Sevda, S.; Prasad, S.; Venkatramanan, V.; Chandel, A.K.; Kataki, R.; Bhadra, S.; Channashettar, V.; Bora, N.; Singh, A. Bioengineering to Accelerate Biodiesel Production for a Sustainable Biorefinery. *Bioengineering* **2022**, *9*, 618. https://doi.org/10.3390/bioengineering9110618

Academic Editors: Indra Neel Pulidindi and Aharon Gedanken

Received: 30 September 2022
Accepted: 20 October 2022
Published: 27 October 2022

Publisher's Note: MDPI stays neutral with regard to jurisdictional claims in published maps and institutional affiliations.

Copyright: © 2022 by the authors. Licensee MDPI, Basel, Switzerland. This article is an open access article distributed under the terms and conditions of the Creative Commons Attribution (CC BY) license (https://creativecommons.org/licenses/by/4.0/).

Abstract: Biodiesel is an alternative, carbon-neutral fuel compared to fossil-based diesel, which can reduce greenhouse gas (GHGs) emissions. Biodiesel is a product of microorganisms, crop plants, and animal-based oil and has the potential to prosper as a sustainable and renewable energy source and tackle growing energy problems. Biodiesel has a similar composition and combustion properties to fossil diesel and thus can be directly used in internal combustion engines as an energy source at the commercial level. Since biodiesel produced using edible/non-edible crops raises concerns about food vs. fuel, high production cost, monocropping crisis, and unintended environmental effects, such as land utilization patterns, it is essential to explore new approaches, feedstock and technologies to advance the production of biodiesel and maintain its sustainability. Adopting bioengineering methods to produce biodiesel from various sources such as crop plants, yeast, algae, and plant-based waste is one of the recent technologies, which could act as a promising alternative for creating genuinely sustainable, technically feasible, and cost-competitive biodiesel. Advancements in genetic engineering have enhanced lipid production in cellulosic crops and it can be used for biodiesel generation. Bioengineering intervention to produce lipids/fat/oil (TGA) and further their chemical or enzymatic transesterification to accelerate biodiesel production has a great future. Additionally, the valorization of waste and adoption of the biorefinery concept for biodiesel production would make it eco-friendly, cost-effective, energy positive, sustainable and fit for commercialization. A life cycle assessment will not only provide a better understanding of the various approaches for biodiesel production and waste valorization in the biorefinery model to identify the best technique for the production of sustainable biodiesel, but also show a path to draw a new policy for the adoption and commercialization of biodiesel.

Keywords: biodiesel; bioengineering; biorefinery; waste; valorization; life cycle assessment; sustainability

1. Introduction

The increasing GHG emissions and depleting fossil-based energy resources require potential and environmentally sound sustainable energy alternatives to overcome these global problems. Surging energy demand owing to rapid population growth, industrial and economic development, accelerated urbanization, and technological advancement requires more energy harvesting from all available sources. Energy consumption was increased from 109,583 terawatt hours in 2000 to 162,194 terawatt hours in 2019 [1] and it is predicted to be 50% higher than the present consumption by 2050 [2]. Simultaneously, the increase in CO_2 emissions needs to be controlled to prevent climate change. Achieving CO_2 emission within the range of a 'safe zone', i.e., 450 ppm, requires an emission reduction of 50–85% by 2050 [3]. The report of Renewable Energy Policy Network (REN) indicated that around 80% of primary energy comes from the fossil-based resources [4], which is the major cause of GHG emissions. Thus, the emissions of carbon dioxide (CO_2) from energy-related applications is predicted to continue to increase globally [5]. Biobased energy alternatives are showing potential and gaining significant global attention to replace fossil fuels and resolve concerns regarding climate change. Biodiesel is the most tested and prominent biofuel with similar a composition and combustion properties to fossil diesel. It can be blended or directly used without modifying the engine [6,7]. Various first, second and third-generation feedstocks for biodiesel production have led to promising results [8] and biodiesel produced from various edible and non-edible (energy) crops has already been commercialized [4,9]. Globally, the European Union, USA, Brazil, Argentina, Indonesia and other countries constitute 43%, 15%, 13%, 13%, 6%, and 10% of biodiesel production, respectively [10]. Figure 1 showed the global status of crops used for biodiesel production. However, the debate over food or fuel has been a major concern to overcome for the development of biodiesel [11,12]. Further, land use changes, monocropping for energy crops that caused an impact on land fertility, converting grassland and deforestation are often associated with biodiesel [8]. Thus, it is essential to explore new approaches, including feedstock and those of technology to advance biodiesel production and maintain its sustainability.

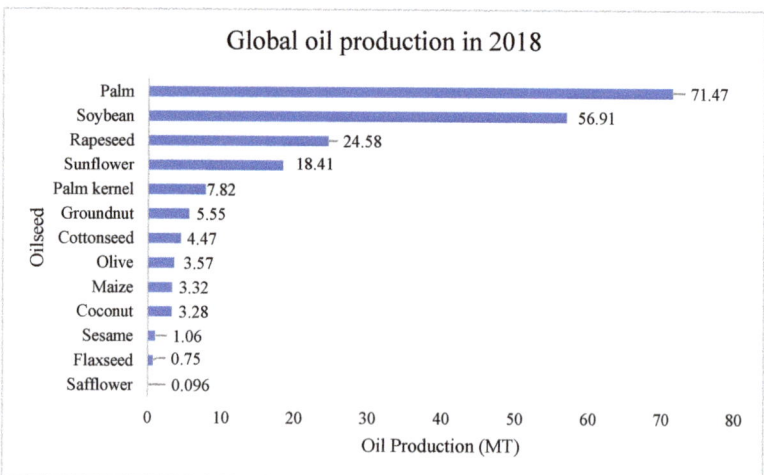

Figure 1. Global oil production in 2018 from oilseed crops. Adapted with permission from Nehmeh et al. [13].

Several potential feedstocks including vegetable oil, animal fat, lipid and fatty acids from algae, yeast and other microbes and pathways have been identified to produce biodiesel without affecting the present agricultural system. However, their production potential necessitates bringing it to a commercial level. The high production cost of biodiesel is a major constraint to substituting corresponding fossil diesel [14]. Approximately 75–80% of the total manufacturing cost of biodiesel is attributed to vegetable oil feedstock, which leads to biodiesel production costs reaching almost double that of commercial diesel costs [6,15].

Further, the feedstock with higher free fatty acids, which declines the quality and yield, requires treatment that involves added costs [6]. An economic model study of *Jatropha curcas* biodiesel identified that the expenses would always be greater than income for an annual production capacity of 10,000 m^3 year^{-1} [16]. Analysis by Sun et al. [17] found significantly higher production costs of microalgae-based biodiesel. This study also identified that the microalgae yield, system operational days are the major obstructions to economical feasible biodiesel production using microalgae. The literature survey demonstrates the need for high lipid production through advancements in technology for cost reduction and popularization of biodiesel.

Genetic engineering methods to produce biodiesel from various sources including plant, yeast, algae, and agricultural or other waste is one of the recent technologies, which could be a promising alternative for creating truly sustainable, technically feasible, and cost-competitive biodiesel [18–20]. The higher production of microbial lipids with lower production input may develop a commercial scalable biodiesel production system. Significant progress in both cellular and bioprocess engineering has been achieved in past decades [20]. The metabolic engineering approach includes improved carbon assimilation, limiting fatty acid flux, higher precursors such as acetyl-CoA availability, higher activity of lipid synthesis enzymes, added gene expression towards lipid synthesis, down- regulating the catabolism of fatty acids by inhibiting β-oxidation or lipase hydrolysis and transcription factor engineering [18,21]. Research showed that the metabolically engineered *Yarrowia lipolytica* can accumulate 70–90% lipids of biomass from glucose only [22–24]. The *Y. lipolytica* yeast was also successfully engineered to directly use starch [25] or use both C5 and C6 sugars derived from lignocellulosic biomass [26] for oil production. Similarly, gene overexpression was another bioengineered approach adopted for enhanced lipid accumulation in cells. Overexpression of acetyl-CoA carboxylase (ACC) and acyl-CoA synthetases (ACSs), which are responsible to form malonyl-CoA from Acetyl-CoA and thioesterification of fatty acids with coenzyme A to form activated intermediates, respectively, is another strategy to enhance the fatty acid synthesis [21,27]. This approach is successfully demonstrated in several organisms including plants, bacteria, and fungi [28–30]. Enhanced lipid synthesis in microalgae was reported by Reik et al. [31] through overexpression of the zinc-finger protein transcription factor.

As identified previously [26,32], the cost of the biofuel should be competitive with corresponding fossil fuels to achieve maximum success. Major production costs come from the raw material, thus, using cheaper feedstock, such as waste cooking oil, inexpensive sugars or sugars obtained from lignocellulosic biomass or animal fat, is a feasible option and can reduce the total cost of bulk chemicals [14]. Focusing on second and third- generation feedstock with the adoption of a biorefinery approach for biodiesel production could be an economically viable option to enhance its sustainability [33]. Moreover, a sustainability assessment for efficient GHGs saving and energy balance is also needed to develop environment-friendly transportation fuel. This article shall henceforth be a review of state-of-the-art research on accelerating biodiesel production using bioengineering approaches and identifying the gap to make biodiesel a sustainable and cost-effective alternative to fossil fuel through a sustainable biorefinery model.

2. Scientometric Analysis

Scientometrics is becoming a leading tool for measuring the value of research activities. It has extensive applications in understanding the structure of a discipline, research trends, impact and networks, growth of knowledge and potentiality of cross-disciplinary/cross-boundary work. It also helps in the decision-making for maximum visibility, the introduction of a new policy, tracking emerging trends and finding niche research areas [34].

The present scientometric analysis is conducted using the SCOPUS database using the keywords 'biodiesel' and 'biodiesel and bioengineering'. Figure 2 shows that researchers are highly focused on biodiesel research, as about 3000 articles have been published every year since the last decade and researchers have started applying bioengineering tools in biodiesel production. It has been observed that more than 45,000 documents have been published on biodiesel out of which about 200 documents have been published on biodiesel research involving bioengineering (Figure 3). Researchers have published patents for more than 25,000 findings on biodiesel, out of which about 1200 patents covered the bioengineering aspect. The maximum number of patents, more than 20,000 on biodiesel research, out of which about 1000 patents have bioengineering aspect, were published with the United States Patent and Trademark Office (Figure 4). The majority of documents published on biodiesel are articles (70%) followed by conference papers (16%) and reviews (6%), while this trend gets shifted for biodiesel along with bioengineering, maximum documents are article (60%), followed by reviews (31%) (Figure 5). The research on biodiesel along with bioengineering aspects are majorly conducted under energy, chemical engineering, environmental sciences, engineering, chemistry, agricultural and biological sciences, biochemistry, genetics and molecular biology, immunology and microbiology disciplines (Figure 6).

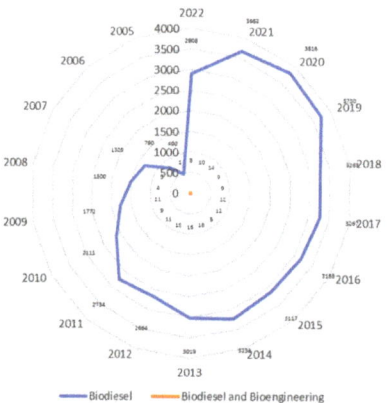

Figure 2. Number of documents published with keywords 'biodiesel' and 'biodiesel and bioengineering'.

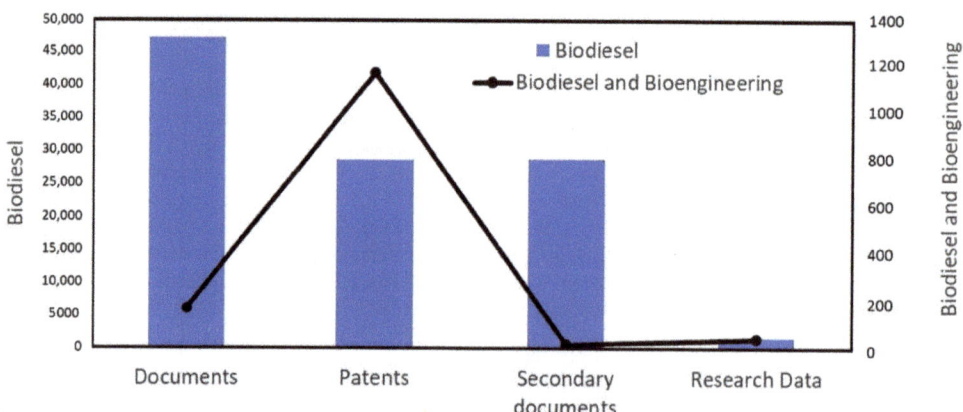

Figure 3. Different types of publications with keywords 'biodiesel' and 'biodiesel and bioengineering'.

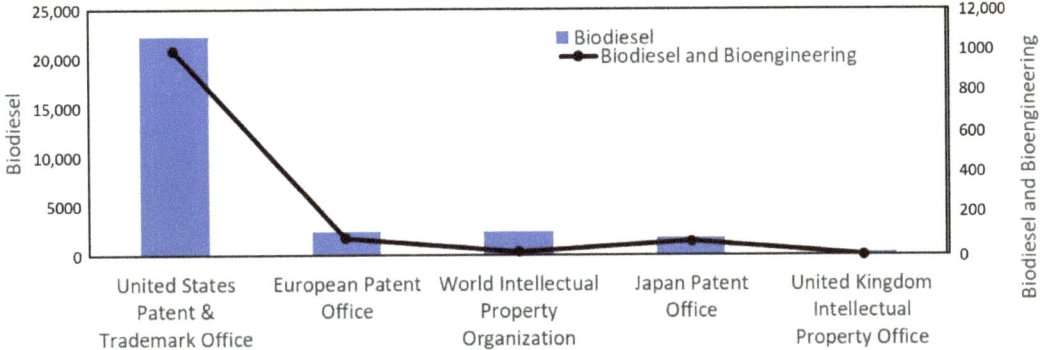

Figure 4. Patents published with different patent offices with keywords 'biodiesel' and 'biodiesel and bioengineering'.

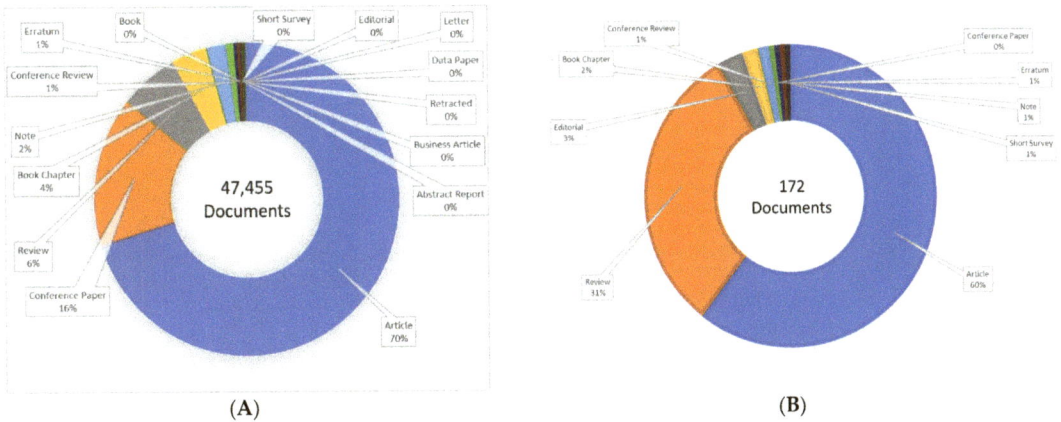

Figure 5. Different types of documents published with keywords (**A**) 'biodiesel' and (**B**) 'biodiesel and bioengineering'.

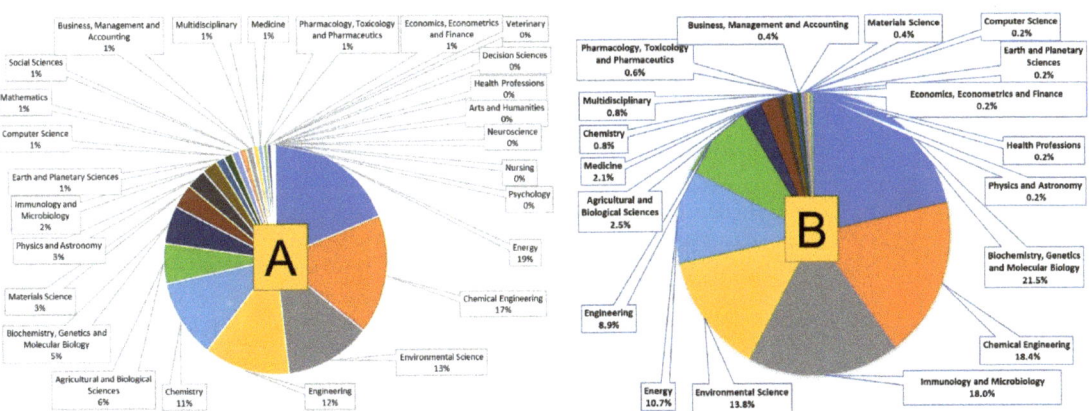

Figure 6. Documents published in different discipline with keywords (**A**) 'biodiesel' and (**B**) 'biodiesel and bioengineering'.

3. Approaches to Accelerate Biodiesel Production

Plants, microalgae, microbes-derived oil, fat, and lipids-based biodiesel are considered the most promising and sustainable feedstock for fossil fuels [25,35]. Biodiesel, as an alternative to fossil diesel, is produced via transesterification mainly by three typical routes [20]: (i) by using vegetable/crop seed oil, (ii) microbial conversion of carbohydrate/sugars to lipid, and (iii) usage of microalgal oil/lipids as shown in Figure 7.

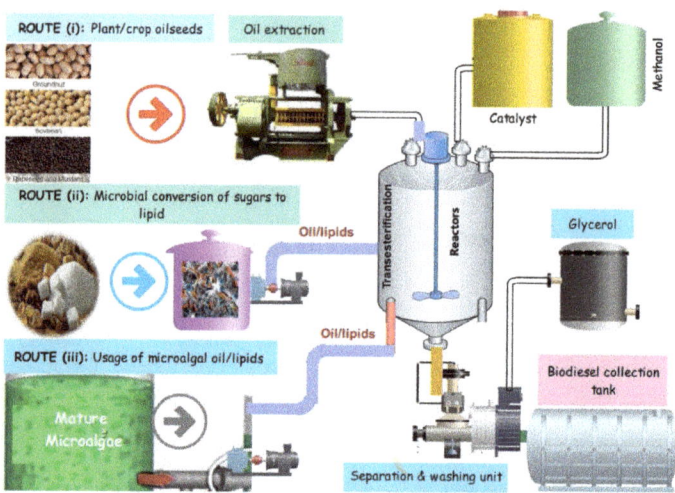

Figure 7. Biodiesel production via the transesterification process through three important routes.

Plant oils, i.e., various edible and non-edible seed oils and many other oil/lipid-bearing materials, are used for biodiesel production [36]. Significant progress has been made so far in biodiesel production from microalgae [37]. However, microalgal oil production using current technologies is still too expensive to commercialize due to efficient photo-bioreactor designs, contamination control methods, and downstream processing. Microalgae biomass cultivation contributes 60–65% of the total production cost for 20–30% of biomass recovery [37,38]. These challenges must be resolved to increase efficiency by incorporating modern tools and bioengineering techniques.

Conversion of sugars/starch/lignocellulose or other C-containing biomass and its bioprocess to lipids through microbial strain bioengineering have several advantages over conventional biodiesel feedstocks. Microbes-based lipids/oils have also been identified as more advantageous in terms of cost-effectiveness, process flexibility, and industrial biotechnology platform for biodiesel production [20,39].

3.1. Biomass Selection in a Carbon-Neutral Manner

Biodiesel production with competitive pricing is among the key factor in the bio-based transportation fuels eventually developing the carbon neutral economy. Additionally, biodiesel production contributes to a resource-efficient value-chain in the low carbon fuels and chemicals production. While the dwindling oil prices and political conflicts constituted some constraints in recent years for biofuel production, the bio-diesel industry still seems to have the momentum to continue its production [40,41]. Although a variety of potential feedstock has been evaluated, current biodiesel production processes are not economically and sustainably viable at a large scale, owing to the higher feedstock cost, round-the-year surplus availability of feedstock and energy/cost-intensive processing steps required for biodiesel production. It is evident from some studies that feedstock cost contributes approximately 80% of the total expenditure of biodiesel production [42].

The selection of the right biomass plays a decisive factor in cost-efficient biodiesel production in a carbon-neutral manner. Vegetable oil, non-agricultural (animal fat), Jatropha and biomass-based feedstock are principal resources for biodiesel production in the world. Among the vegetal feedstock, rapeseed, palm, soya, and sunflower are commonly utilized for biodiesel production in the world. Tallow oil (animal fat) and recycled cooking oil are also considered as a sizeable feedstock for the biodiesel production. Rapeseed is the principal feedstock in the world, representing approximately 65% share of global biodiesel production. Palm oil is another major feedstock for the production of biodiesel in countries like Malaysia, Thailand and Indonesia, among others. It constitutes approximately 6% share of biodiesel production in the world. Soybean is the principal feedstock for biodiesel production in the USA and Brazil, constituting around 15% fraction of total biodiesel production [43]. In Brazil, the total production of biodiesel was 5.8 billion liters in 2019, showing an eight percent increase in production relative to 2018. In Brazil, soya alone represents an almost 81.36% share for biodiesel production, followed by animal fat (13.36%), cotton (4.11%), and other fatty material (1.17%) [44]. The use of food resources for biodiesel production developed concerns over food vs. fuel and energy vs. environment. This also caused high food costs and deforestation in turn ameliorating the greenhouse gas emissions [45].

The regular supply of feedstock in large amounts is pivotal to the successful deployment of biodiesel production at a commercial scale. The oil yield in the feedstock (gallons/acre), biodiesel productivity from the feedstock (metric tons/hectare/year), and energy content are the pivotal factors for the selection of an appropriate substrate for biodiesel production. These are trinomial characteristics for commercially successful biodiesel production. The surplus availability of feedstock throughout the year and the cost of feedstocks are the main influencing factors for the successful deployment of biodiesel production in long term. Readily available oil-rich feedstock, for example, soybean or rapeseed looks more viable but in the long term when the concern arises for the selection of food or fuel, then the non-edible feedstock such as *Jatropha curcas*, Karanja or Pongamia oil, Neem oil, Jojoba oil, among others seems more sustainable [46]. Large populous countries such as India and China mainly need to address these feedstocks for biodiesel production. Waste cooking oil is also a substitute feedstock, but impurities and degree of saturation are the possible constraints for using these oils in the transesterification process. The most emerging feedstock for biodiesel production is algal feedstock, which looks more promising in terms of biomass productivity (140–255 Mt/ha/year), high oil content (35–65%), biodiesel productivity (50–100 Mt/ha/year) and high energy content (1150–2000 barrel of oil equivalent/1000 ha/day) [47]. Algal feedstock seems even more financially viable since they are not only high oil-yielding substrates but also high value-added products such as nutraceuticals (biopigments, amino acids, and vitamins, etc.) can also be derived conforming within sustainability guidelines in 3G biorefineries [48]. Lignocellulose feedstocks are also considered important for biodiesel production indirectly. There are plenty of microorganisms which produce oils and lipids and can be grown on lignocellulosic sugars. These microorganisms can be harvested for oil recovery followed by oil transesterification into biodiesel [49].

Considering the current debates on food vs. fuel, energy vs. environment, and the burgeoning demand for transportation fuels, there is a primary concern over the selection of appropriate feedstock for biodiesel production. The use of edible oil as feedstock of biodiesel production may not be a practically viable option. Instead, feedstock such as microalgae, lignocellulosic biomass and used cooking oil are viable options for biodiesel production reducing dependency on conventional diesel eventually contributing to the development of a low-carbon economy.

3.2. Constraints

Growing concerns for climate change and increasing energy demand requires the adoption of bioenergy, particularly biofuels. Sustainable alternative energy sources, such as biofuels, limit greenhouse gas emissions and enable a carbon-neutral economy [50]. Nevertheless, the production and utilization of biofuels are influenced by socio-economic environmental and political factors. The primary concern for biofuel production is the choice of feedstock.

Biofuel feedstocks include crops, agricultural by-products, vegetable oils and organic wastes. Based on the types of feedstocks and technology options, the biofuels are grouped into different generations of biofuels. Generally, carbohydrate-rich plant biomass is used as a feedstock for the production of "first-generation biofuels". However, first-generation biofuels are constrained by socio-economic and environmental issues such as "food security" and greenhouse gas emissions. Lignocellulosic biomass considered second-generation accounts for a major portion (50%) of the total available biomass on Earth [51]. Availability, renewability and cost-effectiveness make the lignocellulosic biomass a valuable feedstock for biofuel production. The biochemical conversion of lignocellulosic biomass involves processes like partial depolymerization of biomass (pre-treatment), formation of simple sugars through the action of enzymes, fermentation of sugars and distillation. Lignin, due to its complex and stable aromatic structures, is recalcitrant to degradation. The pre-treatment methods including the biochemical and physical methods are important to augment the conversion efficiency of biomass [52]. The biological methods are reported to be cost-effective. Huge numbers of microbes including fungi, bacteria and actinobacteria exhibited hemicellulolytic and cellulolytic capabilities [51]. The lignocellulolytic enzymes involved in the biochemical transformation include the hydrolytic enzymes and ligninolytic enzymes [53]. The cost of lignocellulolytic enzymes is a cause of concern for sustainable biofuel production. However, advances in biochemical processes have the potential to increase the conversion efficiency of lignocellulose into simple sugars. Bioengineering of ligninolytic enzymes can augment biofuel production from lignocellulosic biomass [51]. Microbes can be employed for the production of second-generation biofuels by widening the substrate range, increasing their productivity, increasing their tolerance abilities and enabling the production of value-added biochemicals. Metabolic engineering can improve the efficiency of the microbial strain. Metabolic engineering aims to "design native or entirely new metabolic pathways in a cell" [54]. Metabolic engineering improves cellular activities by manipulating the metabolic, transport system and regulatory functions of the cells [50]. Recently, metabolic engineering has been employed to augment biofuel production. Studies have documented the engineering of metabolic pathways for biofuel generation [51,55,56].

The third-generation biofuel feedstock includes cyanobacteria, algae and seaweeds which are a significant source of the production of triglycerides, fatty acids, and lipids. These feedstocks have advantages such as large biomass production and a shorter harvesting cycle [55]. Microalgae, by utilizing sunlight and atmospheric carbon dioxide as energy and carbon sources, respectively, synthesize lipids. The high cost of microalgal production and lipid extraction limits the use of microbially derived lipids for biodiesel production and industrial applications. Challenges in the domain of strain improvement to increase lipid production, growth and substrate requirements, utilization of non-expensive substrates and lipid extraction need to be addressed to augment biodiesel production [55]. The sustainability of algal biofuel production rests on the cultivation system, choice of algal species, source of nutrients, harvesting and downstream processing [57,58]. The production cost of biodiesel using microalgae can be minimized by effective resource utilization, water recycling, and adopting a biorefinery approach [59,60]. To increase the biodiesel production from microalgae, Ranjbar and Malcata [56] suggested genetic engineering measures such as improving lipid yield through manipulation of lipid biosynthesis pathway, increasing metabolic flow towards lipid biosynthesis, lipid secretion, increasing the biomass yield, "transcription factor engineering" and "transporter engineering".

3.3. Bioengineering

Biodiesel contains fatty acid methyl esters (FAME), and it is considered a carbon-neutral fuel as it is made from vegetable oils, hence reducing carbon emission as compared to conventional fossil fuel. Glycerol is widely used in various bioprocess due to its availability and it is also a sub-product of biodiesel production. Khan et al. [61] optimized process parameters for biodiesel production from microalgae using response surface methodology (RSM) and genetic engineering (GA). Carbon dioxide is converted into carbon-rich lipids by microalgae in the presence of sunlight and micronutrients. Microalgae, a non-edible feedstock that can be found in the fresh water, ponds and marine habitats, is advantageous for biodiesel production compared to other high lipid content sources. Process parameter optimization is a very important aspect to produce a high quantity of carbon-rich lipids. The RSM and GA are important techniques to optimize the processes such as molar ratio, reaction time, operating pressure, catalyst concentration, carbon dioxide concentration, pH and temperature.

Singh et al. [62] investigated the thermal pretreatment of bagasse of genetically engineered sorghum for the high recovery of glucose and xylose. Due to advancements in genetic engineering, lipid production is also enhanced in the cellulosic crop and it can be used for bioethanol and biodiesel generation. They also reported that new enzymes can be induced in the cellulosic crops to store more lipids. They used engineered sorghum bagasse as a substrate for lipid and sugar recovery. They used liquid hot water pretreatment for enhancing lipid and sugar recovery and reported that liquid hot water pretreatment at 170 °C for 20 min increases the recovery of lipids and glucose by two-fold [62].

Due to the increase in human population and decrease in cropland and resources, metabolic engineering is required to boost the lipid content in seeds and other plant tissue. In general, more lipid is stored in oil seeds while in other tissue, such as leaves, there is a very minor amount (0.04% to 0.20%) of neutral lipid triacylglycerol (TAG). In the last two decades, metabolic engineering approaches have been studied on various plants to increase the lipid content in plants' vegetative tissue [63]. In general, three different strategies are used: (i) push: carbon quantity flux is increased through fatty acid and glycolysis pathways; (ii) pull: optimizing TAG assembly; and (iii) protect: reducing the turnover of resulting oil bodies. Vanhercke et al. [64] explained all three different genes that influence stored lipid accumulation in the plant's vegetative tissues.

Waste cooking oil (WCO) is another alternative for biodiesel generation as it is cheaper and non-edible. The WCO contains a high amount of free fatty acids (FFAs) and its direct use in the transesterification process forms soap in the presence of a base catalyst and reduces the overall efficiency of biodiesel generation. To avoid its soap formation, an acid catalyst can be used to convert FFAs into FAME before the addition of a base catalyst to form TAG. Adding one more step increased the overall cost of the process. To avoid this additional step, lipases can be used to convert both TAGs and FFAs into the FAME in the mild condition of temperature and pH. Heater et al. [65] developed a single step, genetically engineered immobilized lipase for the higher production rate of biodiesel from WCO. They showed that genetic fusion of the *Proteus mirabilis* lipase to Cry3Aa allowed for the production of immobilized lipase crystals (Cry3Aa–PML) directly in bacterial cells. The novel approach showed a 4.3-fold higher enzyme efficiency compared to the conventional lipase enzyme methods. Heater et al. [65] also showed the high activity of Cry3Aa–PML catalyst for 15 cycles for the conversion of WCO to biodiesel. Takeshita et al. [66] genetically muted *Parachlorella kessleri* using heavy-ion beam irradiation for the production of high levels of both starch and lipids. The muted strain is named PK4 and compared to the wild strain it accumulates more lipid, at 1.75 g/L compared to the wild strain of 1.17 g/L. Advanced genetic engineering and metabolic engineering approaches will be used to improve the lipid content in vegetable tissues and can also be used for further biodiesel production. Various researchers reported that the target gene varied with the vegetative part of the plant. Table 1 shows the details of targeted genes responsible for the increase in the lipid content in the plant's vegetative tissues.

Table 1. Details of targeted genes responsible for the increase in the lipid content in the plant's vegetative tissues.

Species	Tissue	Target Gene (s)	Total Fatty Acid (TFA) and Triacylglycerol (TGA) Level	Reference
Arabidopsis thaliana	Leaf	LEC2	TAG not quantified	Santos Mendoza et al. [67]
Nicotiana tabacum	Leaf	WRI1, DGAT1, L-OLEOSIN	17.7% TFA (DW), 15.8% TAG (DW)	Vanhercke et al. [64]
A. thaliana	Leaf	PDAT1	2.6% TAG (DW)	Fan et al. [68]
A. thaliana	Leaf	*tgd1*	Not quantified	Xu et al. [69]
Solanum tuberosum	Tuber	ACCase	0.03% TAG (DW)	Klaus et al. [29]
A. thaliana	Seedling	WRI1, AGPase RNAi	5.8-fold increase	Sanjaya et al. [70]
Nicotiana benthamiana	Seedling	MGAT2	6.2-fold TAG increase	Petrie et al. [71]

The production cost of biofuels can be significantly reduced using less biomass expensive materials. However, bacteria or yeasts can convert biomass-derived sugars to lipids/fatty acids that transform into biodiesel via transesterification [20]. Advances in synthetic and metabolic pathway bioengineering have expanded the substrate ranges. *E. coli* was the first microbe studied to produce fatty acid ethyl esters (FAEEs) or methyl esters (FAMEs) directly, which can be used to produce biodiesel. An overview of the synthesis of lipid pathways to produce biodiesel in *E. coli* is shown in Figure 8. The expressions: fatty acyl thioesterases (FATs) lead to the synthesis of free fatty acids (FFAs) converted to fatty acid (FAMEs) by FAMT using AdoMet as FAEEs by acyl-CoA synthase (FadD) and wax synthase (WS). Acyl-ACP is converted to fatty aldehyde by acyl-ACP reductase (ACR) and then to alkanes/alkenes by ADC or fatty alcohols by fatty aldehyde reductase (ALR). Acyl-CoA reductase (ACAR1) can facilitate acyl-CoA conversion to fatty alcohols.

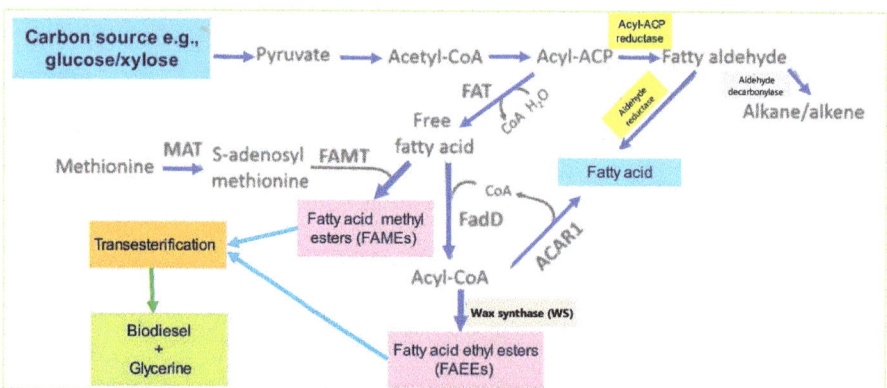

Figure 8. Lipid synthesis pathways to produce biodiesel by *E. coli*.

Kalscheuer et al. [72] investigated *E. coli* biosynthesis for FAEEs, and they achieved 1.28 g/L FAEE production in a 2-L fed-batch fermentation. Further, Steen et al. [73] produced biodiesel by engineered *E. coli* and *Z. mobilis*. The engineered *E. coli* was capable of producing FAEEs directly from hemicellulose. Elbahloul and Steinbuechel [74] reported that an engineered *E. coli* can produce FAEEs at a pilot scale with glucose and oleic acid as feeding substances. The classic yeast *S. cerevisiae* was employed by de Jong et al. [75] to produce FAEEs from ethanol and fatty acyl-CoAs by heterologous expression of a wax ester synthase (WS2) and reported the production of 10 mg/L FAEEs. They suggested examining the direct FAEE production using the oleaginous yeast Yarrowia lipolytica in the future, as S. cerevisiae is not a typical fatty acid producer. Microbial strain engineering, particularly in yeast Y. lipolytica, produces higher fatty acids from various industrial wastes and bio-derived sugars (glucose and xylose) to produce biodiesel. High lipid was produced from xylose by engineered xylose use pathway from lignocellulosic biomass [76].

4. Waste Valorization

4.1. Current Status

Energy is available in numerous forms, however, population upsurge in conjunction with economic development has imposed a major strain on conventional fuels [77]. To address its repercussion against adverse climatic change, efforts have been directed towards sustainable practices. Sustainable development goals (SDGs) aim to orient economic development with environmental protection owing to the close association of environmental and ecological degradation with extensive economic activities. This has prompted a shift from fossil-based fuels to non-conventional energy sources which includes biomass, solar, geothermal and wind. The energy potential of biomass is considerably high owing to its high availability, lesser emissions along with advanced technologies for its efficient conversion. However, its market share in the energy sector is considerably less [78]. There are a few bottlenecks that exist to obtain the entire energy potential of biomass, thereby, a holistic waste management strategy is needed to recover energy as well as essential nutrients from waste.

Biodiesel is a biodegradable and potential alternative fuel which is derived from renewable sources; however, its large-scale production generates different types of residues such as oil cake and seed kernels in huge quantities after oil extraction. In this context, various research initiatives have been carried out in the waste valorisation of biodiesel residues which offers an excellent opportunity to derive value-added products for different applications. The various findings regarding the valorization of waste generated during biodiesel production are summarized in Table 2. Oilseed cakes or deoiled cakes and seed kernels are the leftovers generated by extracting oils from them. According to the Food and Agricultural Organisation (FAO)'s Food Outlook November 2020, the worldwide production of oilseed cakes is predicted to be 158.3 million tonnes in 2020–2021 [79]. It includes edible oil cake (e.g., sunflower, mustard, peanuts, soybean) having high protein content along with vitamins and antioxidants that are generally used as supplement feed for cattle, and non-edible oil cake (e.g., castor, neem, mahua, karanja cakes) used in the production of bio-based products such as biofuel, biogas, chemicals, organic fertilizers, pesticides, biopolymer, etc. [80]. Oilseed cakes are an effective way of utilizing agro-waste with an integrated biorefinery approach with the co-production of protein and vitamin-added value products, enzymes, bioethanol, bioplastics and bioelectricity.

De-oiled cake as a biosorbent is studied for decontamination of wastewater or dye in terms of its adsorption behavior. Jatropha oil cake is used as an effective biosorbent in treating aqueous solution containing reactive red dye. The adsorption process was dependent on pH, concentration, temperature, contact time and dose. It was found that at normal conditions (T = 30 ± 2 °C; pH = 7 and 6 h adsorption period), the highest dye adsorption capacity was obtained which was best represented by Redlich-Peterson and Sip isotherms [81]. Hydrolyzed olive cake also exhibited good adsorption-desorption cycles for the removal of copper (II) contaminated fertilizer industry wastewater, with the highest adsorption capacity of 7.32 mg·g^{-1} and the desorption yield changed from 86% to 67.1% [82]. Equilibrium sorption of Cu (II) from synthetic solution by *Jatropha curcas* deoiled cake was higher than Cr (VI) in terms of pH, adsorbent dosage, initial metal concentration and dosage time, which was best fitted by Freundlich isotherm model. Desorption involves the use of chemical reagents such as HNO$_3$ [83] or HCl [82] are used for maximum metal recovery. Activation of carbonized oil palm decanter cake (OPDC) exhibited higher adsorption capacities on Cu (II), Pb (II) and Zn (II), but was not found to be suitable for Cd (II) and Cr (VI) adsorption. The adsorption capacities were Pb (II)

(128.51 mg/g) > Cu (II) (45.01 mg/g) > Zn (II) (39.21 mg/g), while raw OPDC were more effective in adsorbing Cd (II) and Cr (VI) [84]. Neem oil cake (NOC) finds its wide usage in organic farming as a novel biopesticide and biofertilizer. Additionally, NOC exhibits high adsorptivity for Pb (II) (98%) at low pH (pH = 4) with breakthrough capacity of Pb (II) (30 mg/g) > Cd (II) (15 mg/g) > Cu (II) (10 mg/g) [85]. The occurrence of Ni (II) concentration in the environment from several sources such as metal finishing, tableware plating, forging as well as mine drainage is a serious concern as it may cause several health issues. Jatropha oil cake possesses high affinity to sorb Ni (II) species via different mechanisms, which include ion exchange, chemisorption and physical forces, chelation, complexation and entrapment in the capillaries and pores of the polysaccharide network. Ni (II) adsorption by Jatropha oil cake in its natural form exhibited a removal efficiency of 62% within an hour and 63% in its immobilized form within 90 min [86]. Hydroxyl, carbonyl and carboxyl groups are primarily involved in the biosorption process of Ni (II) [87]. The addition of sodium dodecyl sulfate (SDS) favored the adsorption process of Ni (II) and Zn (II) on carbon derived from mustard oil cake. As revealed by Reichenberg equation, along with pore diffusion, other processes like film diffusion were also the rate-determining steps that were involved during the adsorption process [88]. CO_2 adsorption was studied using *Pongamia pinnata* seed cake that was processed by hydrothermal and extraction treatments. Breakthrough curves revealed that hydrothermally treated curves exhibited enhanced adsorption capacity, easy desorption as well as good recyclability, thereby proving it to be a promising adsorbent [89].

Table 2. Different applications of biodiesel waste through valorization.

Waste from Biodiesel	Type of Feedstock	Applications	Outcomes	Reference
Jatropha curcas Deoiled Cake	Non-edible	Adsorption of Cr(VI) and Cu (II) from wastewater	• Optimum contact time between adsorbate and adsorbent were 15 min and 60 min for Cr(VI) and Cu(II) respectively • Recommended pH of the absorbate were 2 and 6 for Cr(VI) and Cu(II) respectively	Rawat et al. [83]
Oil Palm decanter cake (OPDC)	Edible	Adsorption of heavy metals such as Cu (II), Pb (II) and Zn (II) from waste streams	• Maximum adsorption capacities of activated carbon prepared from OPDC were Pb(II) (128 mg/g) > Cu (II) (45.01 mg/g) > Zn (II) (39.21 mg/g) • Maximum adsorption capacities of activated OPDC were higher than those of the raw OPDC	Yusoff et al. [84]
Mustard Oil Cake	Edible	Adsorption of Ni (II) from aqueous solution	• Optimum pH for biosorption: 8 • Highest breakthrough and exhaustive capacities for 10 mg/L Ni (II) concentration were 4.5 and 9.5 mg/g respectively	Khan et al. [87]
Carbon derived from mustard oil cake (CMOC)	Edible	Adsorption of Zn (II) and Ni (II) from aqueous solution	• Optimum adsorption capacity of Zn (II) was Ni (II) were 45.8 mg/g and 47.2 mg/g respectively • Recovery of Zn (II) and Ni (II) were 75% and 78.97%	Rao et al. [88]
Neem Oil Cake	Edible	Removal and recovery of Cu (II), Cd (II) and Pb (II) from wastewater	• Highest adsorptivity was found for Pb (II) (98%) at pH 4 • Breakthrough capacities: Pb (II) (30 mg/g) > Cd (II) (10 mg/g) > Cu (II) (10 mg/g)	Rao and Khan [85]
Olive Oil Cake	Edible	Biogas Production	• Cumulative yield of biogas: 1226 mL (inoculum ratio: 0.64)	Sarkar [90]
Cotton Oil Cake	Non-edible	Biogas Production	• Highest Methane production of 78 mL from 1 g of cotton oil cake	Isci and Demirer [91]

Table 2. Cont.

		Preparation of Spray-dried functional powders for food applications as emulsion stabilizers			
Flaxseed Oil Cake	Edible		•	Highest stability of the emulsions prepared with the powder was at 200 °C	Drozlowska et al. [92]
Neem Oil Cake	Edible	Evaluation of the effect on plant growth, yield, and management of Alternaria tenuissima leaf spot disease, and rhizosphere microorganisms in chilli crop	•	Effectiveness in improving plant growth and reducing leaf spot disease: Simarouba > Madhuca > Neem	VasudhaUdupa et al. [93]
Madhuca Oil Cake	Non-edible				
Simarouba	Edible				
Coconut kernel cake	Edible	Used as substrate for lipase production	•	Highest lipase production: 698 U/g Dry Substrate	Venkatesagowda et al. [94]
Neem Cake	Edible	Used for soil amendment against root knot nematode (*Meloidogyne incognita*) infecting Black gram (*Vigna mungo*)	•	Effectiveness in plant growth enhancement: Neem > Mustard > Castor > Linseed	Rehman et al. [95]
Mustard Cake	Edible				
Castor Cake	Non-edible		•	Neem oil cake was found to be most effective in controlling root knot nematode (*Meloidogyne incognita*)	
Linseed Cake	Edible				

De-oiled cakes contain micronutrients or different chemicals that might serve as growth enhancers and bio-control agents of beneficial micro-organisms, such as soil bacteria or fungi to antagonize crop-based pathogens and mitigate the use of synthetic agricultural inputs. In a study, four different de-oiled cakes, namely mahua, neem, jatropha and karanja were used as substrate to examine the survival, mass multiplication and population dynamics of *Trichoderma harzianum*. Out of the four substrates, it was found that neem cake was the best substrate in terms of longevity and population dynamics, supporting the survival and growth of *T. harzianum* for more than 105 days, while Jatropha, karanja and mahua cakes could support the growth and longevity of *T. harzianum* for up to 90 days [96]. Eight different oil-seed cakes were used as substrates for lipase production by five different fungi, namely *Chalaropsis thielavioides*, *Aspergillus niger*, *Phoma glomerata*, *Colletotrichum gloeosporioides* and *Lasiodiplodia theobromae*. Out of these, coconut kernel cake as substrate for *Lasiodiplodia theobromae* exhibited maximum lipase productivity, which was further optimized in terms of different operating conditions and the lipase productivity was increased to 698 U/g Dry Substrate [94]. Biopotency of oilcakes, namely neem, mustard, castor, linseed, cotton, olive, flax, soybean, sesame, madhuca and simarouba were also found to increase the yield and growth of *Vigna mungo* [95], tomato [97] and chilli [93] and aided towards direct toxicity and protective action against *Meloidogyne incognita* and *Alternaria tenuissima*, respectively. A reduction in crop productivity and tropical soil fertility causes hindrances to attaining food security. Traditional approaches include the uneconomical usage of mineral fertilizers that cause eutrophication, soil acidity and leaching of phosphates and nitrates that adversely affect the whole ecosystem. Jatropha cake has an abundance of phosphorus, nitrogen, potassium and other organic nutrient sources that improves water, air infiltration and allows for deeper root lengths. Jatropha cake supplemented with compost at different optimized treatment conditions helped to improve the available P, soil pH, exchangeable K, thereby enriching the soil fertility [98]. Nevertheless, Hirota et al. [99] addressed biosafety issues with respect to cake application as soil amendment or biofertilizer. The biopesticidal activity was exhibited by a non-edible oil cake produced from karanja that was used as a substrate for the growth of the fungus *Paecilomyces*. At a C/N ratio of 40:1 and pH = 7, potent termite mortality was observed [100].

Oil seed cakes are used for the preparation of functional powders for use in food industries as emulsion stabilizers. Flaxseed oil cake extract was used as a substrate for the preparation of powder having emulsifying activity. The study investigated the effect of spray-drying process inlet temperature on physicochemical features, regarding oil binding capacity, water holding capacity, solubility, antioxidant activity, chemical composition, surface morphology, color and water activity. It was found that the inlet temperature played a crucial role in the functional and physicochemical properties of the powders in such a way that with the rise in inlet temperature, the antioxidant activity and solubility reduced, but the oil-binding capacity, water-holding capacity, as well as emulsifying activity increased. The highest stability was achieved with emulsions prepared with the powder at 200 °C [92]. Energy production from oil-seed cakes has also been utilized for biogas and hydrogen generation. Biogas production from sunflower oil cakes was in the range of 186–215 mL CH_4/g volatile solids which reflects its low conversion efficiency that is attributed to the lesser bioavailability of hollocellulose that is interlinked as an intricate polymer of lignin, hemicellulose and cellulose to microbial action. Pre-treatment methods aid to disintegrate these complex polysaccharide linkages to enhance the exposure of fermentable sugars to microbial action. Pre-treatment with 1% H_2SO_4 and at 170 °C further enhanced the methane yield to 302 ± 10 mL CH_4/g volatile solid [101]. Hydrothermal pretreatment enhances cellulose accessibility by increasing the surface area and minimizing the crystallinity of lignocellulosic biomass. Under batch fermentation in mesophilic conditions of sunflower oil cake at 25, 100, 150 and 200 °C, highest methane yield of 310 ± 4 mL CH_4/g COD_{added} for the liquid fraction and 105 ± 7 mL CH_4/g COD_{added} for solid fraction was obtained [102]. Hydrogen is an emerging and alternative fuel due to its unique characteristics such as high energy yield and emission of water vapor upon combustion that altogether represents its carbon neutral property. De-oiled jatropha waste was subjected to direct, semicontinuous hydrogen fermentation and the obtained hydrogen yield was 8.7 mL H_2/g and hydrogen production rate was 1.48 L/L-d when the operational parameters of the reactor were as follows: hydraulic retention time: 2 days, concentration of de-oiled jatropha cake: 200 g/L, pH: 6.5 and temperature: 55 °C [103].

4.2. Opportunities

The composition and quantity of oil cake obtained mainly vary with the type of feedstock used, plant growing and processing conditions. Oil cake can be edible or inedible. Edible oil cakes are rich in proteins which adds nutritional value to their use as an animal feed supplement. Oil cakes compose of different nutrients and minerals which makes them a valuable source of nitrogenous fertilizers. Oil cake/meal can be utilized as a substrate for growing microorganisms, and they have been widely used for the production of essential nutrients and chemicals such as amino acids, enzymes, ethanol, organic acids, antibiotics, antioxidants, vitamins, and other bio-chemicals which can be utilized in various foods and pharmaceutical industries [104,105]. Oil cakes and oil seeds are also being investigated as promising sources for the production of biochar, bio-oil and syngas which have many useful applications. Biochar produced can be used as adsorbents for the removal of dyes [106], ion adsorption [107], and many others which highlights the significance of the feedstock. The bio-oil obtained can be upgraded to advanced biofuels and it is a potential storehouse of different chemicals. Oil cakes and oil seeds can be utilized for the generation of hydrogen and biogas and proper integration of biodiesel and non-edible seeds can be used to produce biogas for bioenergy generation which is an economically viable technique [80].

The development of an integrated biodiesel refinery with the simultaneous valorization of its residues and by-products presents opportunities to lower biodiesel production costs as well as the exploitation of biodiesel wastes to derive value-added products is beneficial which increases the overall economic value with reduced emissions and addresses waste disposal issues make biodiesel a sustainable energy resource.

4.3. Challenges

As discussed, earlier oil cakes are rich in dietary fibers, proteins and compounds with antioxidant properties, that can be used in bakeries, infant products and supplements [108]. In addition, substrates for producing vitamins, amino acids, antibiotics, enzymes, flavors, pigments, surfactants and bioactive compounds can also be derived from it [109]. Although the production of oil cake is increasing as a result of seed oil production, its application is limited, resulting in low yields from abundant resources. Varieties of applications of the by-products are possible with suitable procedures. Significant pollution is caused by improper management of the oil cakes. The extraction, utilization, and incorporation of dietary fiber and antioxidants in food products need further investigation.

Due to their high nutritional value and moisture content, oil cakes are prone to deterioration. The natural drying is ineffective owing to unpredictable weather and time requirements, causing the product to rot before drying [110]. Hence, effective drying methods with respect to energy, time, cost, and acceptable product quality need to be addressed. Oil cakes also include anti-nutrients such as tannins, phytic acid, antivitamins, saponins and trypsin inhibitors [111]. Although multiple techniques are available for anti-nutrient removal, a common and cost-effective method is yet to be developed.

The high nutritional value and functional benefits of oil cake come with allergenicity due to plant proteins, which stand as a barrier to human consumption. The allergens need to be identified before the introduction of a new protein source as a food ingredient. In addition to that, the digestive aspects of the dose of protein, physicochemical properties and immune response need to be studied in detail. Another challenge with the novel food product is acceptance by consumers. Despite higher nutritional value and added health benefits, not all individuals are willing to try new products, which is known as food neophobia [112].

Biogas production from oil cake is another approach towards value generation from waste. Several factors need to be considered to make it sustainable. Land protection laws in different countries hinder the growth of biogas projects. This is due to increasing population density in certain countries. Thus, biogas production from oil press-cakes needs an in-depth study of social acceptance, related policies and techno-economic feasibility. The supply chain of press cake, including collection, storage and transportation must be clearly identified.

Alkaline-hydrolysis of Jatropha press cake results in a nitrogen source for growing fungi and the production of lipase [113]. Glucose and maltodextrin as carbon sources stimulate fungal biomass formation but decreases lipase production due to catabolite repression. The utilization of alternative carbon sources to overcome catabolite repression and maximization of fungal biomass as well as lipase production needs further research. Oil cakes are not given importance with respect to the amount of bioenergy and valuable products that can be generated. The valorization of oil cakes may help to solve the environmental issue related to waste disposal and support the zero-waste concept. Moreover, to popularize the concept of oil cake valorization, multiple awareness and training programs needs to be conducted. The farmers and industries dealing with seed oil production should be encouraged to learn more about waste valorization and resultant benefits.

As oil cake valorization is a relatively new technology, identification of the supply chain is a major challenge. Technical knowledge towards efficient oil cake generation and its economic value determination is lacking. This will reduce the chances of investment by various stakeholders like farmers, oil millers, traders, power plants and the food industry. Well-defined policies and incentives are also required for the optimal development of this sector. Agri-based products are facilitated by the cooperative expansion of associated organizations [80]. The lack of subsidy, legal issues and lack of cooperative culture in many countries is a serious drawback to the oil cake valorization sector.

5. Lifecycle Assessment and Sustainability

Biofuels should be environmentally and economically advantageous to become a sustainable alternative to petroleum [114]. The sustainability of biofuel production systems incorporates energy saving and reduction in greenhouse gas (GHG) emissions and cleaner environmental and social acceptability. A life cycle assessment (LCA) is an internationally recognized tool for determining the associated environmental impacts with all the life cycle stages of biofuel production, processing and utilization, making biofuels more sustainable [8]. LCA is a methodology customarily employed to assess the impact of the product on the environment induced by industrial operations and services, from acquisition, manufacturing, and usage of raw material and its maintenance until the final disposal of products or utilization of services [115]. LCA can also suggest alternative sub-processes to make the process sustainable [35]. With the help of the LCA study, Foteinis et al. [116] explained that the transportation of feedstock is the primary cause of environmental burden for used cooking oil biodiesel, while overall, its environmental sustainability is higher with a 40% reduction of GHGs and ecological footprint. Further, gaining value-added by-products (glycerol, potassium sulfate, and other by-produce) has extra monetary and environmental benefits. Correspondingly, Arguelles-Arguelles et al. [115] conducted an LCA study to produce renewable diesel from palm oil, and reported around a 110% decrease in CO_2 emissions compared to fossil diesel combustions. Nevertheless, this investigation also reported the impact of palm oil-based biodiesel on the environment and its toxicity to humans due to high agrochemical use in palm plantations.

The LCA of biofuel production systems needs a meticulous design to define the study's goals, scope, functional units, system boundaries, reference system, comprehensive inventory establishment and detailed information on emissions from products and by-products [117,118]. For example, Larson [119] represented four input parameters that cause more significant variation and uncertainties in LCA results of energy production, which include climate-active plant species (species capability to adjust under altered climate change), N_2O emission assumption, method allocation for co-product, credits, and dynamics of soil organic carbon. Therefore, in the LCA study, the defining of a goal and scope are two basic steps to determine the system boundary of the study.

Biorefinery is an industrial establishment that sustainably transforms biomass waste materials into biofuels and valuable biochemicals. It is much more complicated than a single biofuel generation system. It is a multi-functional system that creates multiple biochemicals similar to traditional biorefineries. However, it needs a different assessment to determine the goal and scope of such a multi-functional system. In a LCA, various allocation techniques, such as physical or economic allocation are employed to separate the environmental burden of a process or product when numerous functions reflect the same process. Therefore, allocation methods varied by system expansion, monetary value, mass (wet or dry), energy and C-content [114], which may influence the results of LCA studies. Karka et al. [120] assessed the effect of various allocation strategies on LCA results of multi-product biofuel refineries. They figured deviation in the results according to the allocation method followed. Many investigators have recommended adopting an expanded system strategy to compare the environment-related burden of biofuels with fossil fuels [121,122]. In the case when an allocation cannot be avoided, inputs and outputs of the system may be partitioned between various products and by-products [123], which may be performed based on the mass, volume, energy, or C-content of the co-products [124]. The selection of the allocation process dramatically affects the results [121,125]. Allocation based on C-content may be preferred in the investigations pertaining to bioenergy generation because such investigations are targeted to reduce emissions by substituting the traditional fuel resource.

Sheehan et al. [126] conducted a study on LCA to assess energy consumption. They reported that the effectiveness of fossil energy resources using biodiesel produced 3.2 units of fuel product energy per unit of fossil fuel use in LCA. In comparison, fossil diesel LCA produced only 0.83 units of fuel product energy per unit of fossil fuel consumed. Biodiesel use also decreases net CO_2 emissions by approximately 78.5%, corresponding to diesel per unit of work performed by a bus engine. Such measures confirm the renewable nature of biodiesel. Sieira et al. [127] conducted an LCA study and estimated that biodiesel production generates about 174 times less CO_2 than diesel production. Nguyen et al. [128] conducted 'a multi-product landscape life-cycle assessment approach for evaluating local climate mitigation potential'. They reported that the GHG mitigation potential was higher than the C-sequestration value of leaving the corn stover in place for biofuel generation. While intensified and realistic solutions are usually considered through the lens of a stark trade-off, the landscape–LCA results suggest synergies can be achieved when these strategies are combined. Co-adoption of corn-stover collection in fields, no-till system and cover cropping enhances SOC levels, fetch farm revenues, and total landscape production compared to business-as-usual (BAU) management. Malik et al. [129] indicated that the integration of multi-product algal biorefineries and wastewater treatment approach is more sustainable for producing green products in a circular bioeconomy paradigm while keeping the water-energy-environment nexus sustainable. They found a novel and sustainable algal biorefinery route that is more efficient in achieving better biomass valorization by adaptation and use of wastewater for biomass production and efficient removal of N, P, COD, and BOD in the wastewater, sedimentation-based harvesting system, production of alfa-amylase and mycoproteins from residual algal biomass along with recovery of most of the solvents utilized in the process, making it a zero-waste approach.

Singh and Olsen [130], in a LCA study of microalgal biodiesel production considering cultivation of *Nannochloropsis* sp. in a flat-panel photobioreactor (FPPBR), compared six different biodiesel production pathways (three different harvesting techniques, i.e., aluminum as flocculent, lime flocculent and centrifugation and two different oil extraction methods, i.e., supercritical CO_2 (sCO_2) and press & co-solvent extraction). They reported that harvesting with lime flocculation and press & co-solvent oil extraction scenarios of biodiesel production provides maximum savings on total impacts. They also reported that scenarios considered in this study also offer GHG savings over fossil diesel, but algal biodiesel is not better in terms of impacts on human health, ecosystem quality, and resources. They suggested that it can be improved by expanding the system boundaries to include the utilization of coproducts and by-products. Arguelles-Arguelles et al. [109] studied the environmental impact of renewable diesel production using an attributional life cycle assessment considering five production stages regarding palm cultivation and harvest, oil extraction, refining, diesel production, and its use and concluded that biodiesel significantly reduces GHG emissions (about 110%) compared to fossil ones. Gupta et al. [131] compared centralized and decentralized rapeseed-biorefinery using the life cycle assessment technique and concluded that centralized biorefinery emitted lesser while energy demand is higher than decentralized ones. They suggested that the system can be improved by the use of low nitrogen nutrients during cultivation, use of biodiesel in farm machineries, improved heating techniques, utilization of waste like glycerol, rape straw, rape cake, etc., and better oil extraction techniques.

LCA can be used as a tool to identify the best combination of various processes involved in the biodiesel refinery by comparing various pathways/techniques available. LCA may also indicate possibilities to improve the system by the adoption of alternate processes/techniques and/or by inclusion of waste valorization within the system boundaries to get sustainable biorefinery.

6. Conclusions and Future Prospects

In recent decades, serious efforts have been made to develop strategies to enhance renewable resource efficiency to produce sustainable and eco-friendly energy carriers. Top priority has been given to the compatibility of biofuels with internal combustion (IC) engines. Among the diverse renewable energy resources, biodiesel is reportedly more compatible with IC engines and more environment-friendly. Hence, biodiesel has the capability of meeting energy needs and is also helpful in the transition toward replacing conventional fossil-derived diesel fuels [132]. An increased production of global biodiesel of 19%, from 34.3 billion liters in 2018 to 40.9 billion liters in 2019 was reported in the Renewables 2020—Global Status Report [133]. However, its production is restricted by limited and inadequate raw materials, low economic benefits, lengthy life cycles, impact on food commodities prices, utilization of by-products, net energy ratio (NER) and eco-friendliness limit its application and industrialization.

The extent of these challenges demands several approaches and strategies to capture from the environment and employ bioengineering technologies for efficient waste valorization to establish and achieve circular economy goals [134]. Hossain [135] indicated that the food vs. fuel security conflict might arise due to the utilization of arable land resources for fuel raw materials production. As the oilseed crop remains the primary source of edible oil, its limited oil and lipid production capacity and uncertainty to cater sufficient oils to produce biodiesel in a country like India and China. The utilization of microalgal or lignocellulosic biomass for lipid production is an excellent alternative feedstock. However, lipid production cost and quantity are significant factors behind its applicability. Bioengineering intervention to produce lipids/fat/oil (TGA) and further their chemical or enzymatic transesterification to accelerate biodiesel production has a great future [18,20,62,136]. Microalgal oil production from current technologies is still too expensive to commercialize due to efficient photo-bioreactor designs, contamination control methods, and downstream processing. Such limitations and technological challenges must be tackled, leading to the development of new genetic and bioengineering strategies to improve biodiesel production. Recent advancements in process technology and bioengineering intervention provide promising results in terms of production potential and cost-effectiveness. Microbes, i.e., non-conventional yeast, algae, and bacterial strains, have enormous potential to produce oil/fat/lipids under aerobic fermentation conditions from various economical substrates. Metabolic engineering of microbial strains to synthesize higher lipids and fatty acids by efficiently utilizing substrates has a great future and excellent potential to lower operational costs, improve its economics and accelerate biodiesel production [136–138].

The biorefinery for biodiesel and other valuable by-products physically symbolizes the circular bioeconomy concept. While biorefineries are not equally economically strong compared to petroleum hydrocarbon refineries, adoption of a biorefinery model similar to a conventional refinery along with by-products utilization provides economical sustainability through additional income to compensate for the production costs and make it more eco-friendly as emission burdens will be allocated among biodiesel and its by-products. Therefore, biodiesel commercialization will require arduous and combined efforts of R&D work for its capacity for scalability [139]. The biorefinery concept of biodiesel production will also help in improving the NER as some energy inputs will also be apportioned to by-products. A comprehensive environmental sustainability analysis of biodiesel production based on lifecycle assessment (LCA) and various multi-dimensional decision-making techniques would help to develop and prioritize for the achievement of future goals and the aim of sustainable biodiesel biorefinery.

The following outcomes could be made from the present literature survey:

➢ The second and third generation feedstocks are latent resources that can overcome the feedstock restriction.
➢ Advancements in bioengineering approaches used in biodiesel production can enhance production capacity and reduce production costs.
➢ Adopting biorefinery approach provides additional benefit for commercial biodiesel production.
➢ The LCA could be employed as the best tool to identify the best combination of various processes involved in the biodiesel refinery by comparing various pathways/techniques available for the sustainable biodiesel production system.

Author Contributions: Conceptualization, D.R., S.P. and A.S.; writing—original draft preparation, D.R., S.S., S.P., V.V., A.K.C., R.K., S.B., V.C., N.B. and A.S.; writing—review and editing, D.R., S.S., S.P., V.V., A.K.C., R.K., S.B., V.C., N.B. and A.S.; visualization, D.R., S.P. and A.S.; supervision, D.R., S.P. and A.S.; project administration, A.S. All authors have read and agreed to the published version of the manuscript.

Funding: This research received no external funding.

Institutional Review Board Statement: Not applicable.

Informed Consent Statement: Not applicable.

Data Availability Statement: Data sharing is not applicable to this article.

Conflicts of Interest: The authors declare no conflict of interest.

References

1. Primary Energy Consumption. 2019. Available online: https://ourworldindata.org/grapher/primary-energy-cons?tab=chart&time=2000..latest&country=~{}OWID_WRL (accessed on 15 August 2022).
2. USEIA. *Today in Energy*; U.S. Energy Information Administration: Washington, DC, USA, 2021.
3. Schenk, P.M.; Thomas-Hall, S.R.; Stephens, E.; Marx, U.C.; Mußgnug, J.; Posten, C.; Kruse, O.; Hankamer, B. Second generation biofuels: High-efficiency microalgae for biodiesel production. *Bioenerg. Res.* **2008**, *1*, 20–43. [CrossRef]
4. REN21. *Renewables Global Status*; Report: 2009 Update; Renewable Energy Global Status Report 2021; REN21: Paris, France, 2009.
5. Awogbemi, O.; Kallon, D.V.V. Application of Tubular Reactor Technologies for the Acceleration of Biodiesel Production. *Bioengineering* **2022**, *9*, 347. [CrossRef] [PubMed]
6. Elgharbawy, A.S.; Sadik, W.A.; Sadek, O.M.; Kasaby, M.A. A review on biodiesel feedstocks and production technologies. *J. Chil. Chem. Soc.* **2021**, *66*, 5098–5109. [CrossRef]
7. Linganiso, E.C.; Tlhaole, B.; Magagula, L.P.; Dziike, S.; Linganiso, L.Z.; Motaung, T.E.; Moloto, N.; Tetana, Z.N. Biodiesel Production from Waste Oils: A South African Outlook. *Sustainability* **2022**, *14*, 1983. [CrossRef]
8. Rathore, D.; Nizami, A.S.; Singh, A.; Pant, D. Key issues in estimating energy and greenhouse gas savings of biofuels: Challenges and perspectives. *Biofuel Res. J.* **2016**, *3*, 380–393. [CrossRef]
9. OECD; FAO. *OECD-FAO Agricultural Outlook 2015–2024*; OECD: Paris, France; FAO: Rome, Italy, 2015.
10. IEA. *Biofuel Production by Country/Region and Fuel Type, 2016–2022*; IEA: Paris, France, 2021.
11. Mukherjee, P.; Varshney, A.; Johnson, S.; Jha, T.B. Jatropha curcas: A review on biotechnological status and challenges. *Plant Biotechnol. Rep.* **2011**, *5*, 197–215. [CrossRef]
12. Sharma, V.; Ramawat, K.G.; Choudhary, B.L. Biodiesel Production for Sustainable Agriculture. In *Sustainable Agriculture Reviews*; Lichtfouse, E., Ed.; Springer: Berlin/Heidelburg, Germany, 2012; pp. 133–160.
13. Nehmeh, M.; Rodriguez-Donis, I.; Cavaco-Soares, A.; Evon, P.; Gerbaud, V.; Thiebaud-Roux, S. Bio-Refinery of Oilseeds: Oil Extraction, Secondary Metabolites Separation towards Protein Meal Valorisation—A Review. *Processes* **2022**, *10*, 841. [CrossRef]
14. Gebremariam, S.N.; Marchetti, J.M. Economics of biodiesel production: Review. *Energy Convers. Manag.* **2018**, *168*, 74–84. [CrossRef]
15. Phan, A.N.; Phan, T.M. Biodiesel Production from Waste Cooking Oils. *Fuel* **2008**, *87*, 3490–3496. [CrossRef]
16. Navarro-Pineda, F.S.; Ponce-Marbán, D.V.; Sacramento-Rivero, J.C.; Barahona-Pérez, L.F. An economic model for estimating the viability of biodiesel production from *Jatropha curcas* L. *J. Chem. Technol. Biotechnol.* **2017**, *92*, 971–980. [CrossRef]
17. Sun, J.; Xiong, X.; Wang, M.; Du, H.; Li, J.; Zhou, D.; Zuo, J. Microalgae biodiesel production in China: A preliminary economic analysis. *Renew. Sustain. Energy Rev.* **2019**, *104*, 296–306. [CrossRef]
18. Deniz, I.; Aslanbay, B.; Imamoglu, E. State of Art Strategies for Biodiesel Production: Bioengineering Approaches. In *Liquid Biofuel Production*; Sigh, L.K., Chaudhary, G., Eds.; Scrivener Publishing LLC: Beverly, MA, USA, 2019; pp. 319–350.
19. Hegde, K.; Chandra, N.; Sarma, S.J.; Brar, S.K.; Veeranki, V.D. Genetic Engineering Strategies for Enhanced Biodiesel Production. *Mol. Biotechnol.* **2015**, *57*, 606–624. [CrossRef]
20. Xie, D. Integrating Cellular and Bioprocess Engineering in the Non-Conventional Yeast Yarrowia lipolytica for Biodiesel Production: A Review. *Front. Bioeng. Biotechnol.* **2017**, *5*, 65. [CrossRef] [PubMed]
21. Lin, H.; Castro, N.M.; Bennett, G.N.; San, K.Y. Acetyl-CoA synthetase overexpression in *Escherichia coli* demonstrates more efficient acetate assimilation and lower acetate accumulation: A potential tool in metabolic engineering. *Appl. Microbiol. Biotechnol.* **2006**, *71*, 870–874. [CrossRef] [PubMed]
22. Liu, L.; Pan, A.; Spofford, C.; Zhou, N.; Alper, H.S. An evolutionary metabolic engineering approach for enhancing lipogenesis in Yarrowia lipolytica. *Metab. Eng.* **2015**, *29*, 36–45. [CrossRef] [PubMed]
23. Qiao, K.; Imam Abidi, S.H.; Liu, H.; Zhang, H.; Chakraborty, S.; Watson, N.; Ajikumar, P.K.; Stephanopoulos, G. Engineering lipid overproduction in the oleaginous yeast Yarrowia lipolytica. *Metab. Eng.* **2015**, *29*, 56–65. [CrossRef] [PubMed]
24. Qiao, K.; Wasylenko, T.M.; Zhou, K.; Xu, P.; Stephanopoulos, G. Lipid production in Yarrowia lipolytica is maximized by engineering cytosolic redox metabolism. *Nat. Biotechnol.* **2017**, *35*, 173–179. [CrossRef]
25. Ledesma-Amaro, R.; Dulermo, T.; Nicaud, J.-M. Engineering Yarrowia lipolytica to produce biodiesel from raw starch. *Biotechnol. Biofuels* **2015**, *8*, 148–159. [CrossRef] [PubMed]
26. Ledesma-Amaro, R.; Lazar, Z.; Rakicka, M.; Guo, Z.; Fouchard, F.; Coq, A.M.C.L.; Nicaud, J.-M. Metabolic engineering of Yarrowia lipolytica to produce chemicals and fuels from xylose. *Metab. Eng.* **2016**, *38*, 115–124. [CrossRef]
27. Black, P.N.; DiRusso, C.C. Yeast acyl-CoA synthetases at the crossroads of fatty acid metabolism and regulation. *Biochim. Biophys. Acta* **2007**, *1771*, 286–298. [CrossRef]
28. Davis, M.S.; Solbiati, J.; Cronan, J.E. Overproduction of acetyl-CoA carboxylase activity increases the rate of fatty acid biosynthesis in *Escherichia coli*. *J. Biol. Chem.* **2000**, *275*, 28593–28598. [CrossRef]
29. Klaus, D.; Ohlrogge, J.B.; Neuhaus, H.E.; Dörmann, P. Increased fatty acid production in potato by engineering of acetyl-CoA carboxylase. *Planta* **2004**, *219*, 389–396. [CrossRef]
30. Ruenwai, R.; Cheevadhanarak, S.; Laoteng, K. Overexpression of acetyl-CoA carboxylase gene of Mucor rouxii enhanced fatty acid content in Hansenula polymorpha. *Mol. Biotechnol.* **2009**, *42*, 327–332. [CrossRef]
31. Reik, A.; Zhou, Y.; Collingwood, T.N.; Warfe, L.; Bartsevich, V.; Kong, Y.; Henning, K.A.; Fallentine, B.K.; Zhang, L.; Zhong, X.; et al. Enhanced protein production by engineered zinc finger proteins. *Biotechnol. Bioeng.* **2007**, *97*, 1180–1189. [CrossRef]

32. Liu, H.-H.; Ji, X.-J.; Huang, H. Biotechnological applications of Yarrowia lipolytica: Past, present and future. *Biotechnol. Adv.* **2015**, *33*, 1522–1546. [CrossRef]
33. Doornbosch, R.; Steenblik, R. *Biofuels: Is the Cure Worse than the Disease?* OECD: Paris, France, 2007.
34. Jayasree, A.V.; Baby, M.D. Scientometrics: Tools, techniques and software for analysis. *Indian J. Inf. Sources Serv.* **2019**, *9*, 116–121. [CrossRef]
35. Prasad, S.; Singh, A.; Korres, N.E.; Rathore, D.; Sevda, S.; Pant, D. Sustainable utilization of crop residues for energy generation: A Life Cycle Assessment (LCA) perspective. *Bioresour. Technol.* **2020**, *303*, 122964. [CrossRef]
36. Tan, T.; Lu, J.; Nie, K.; Deng, L.; Wang, F. Biodiesel production with immobilized lipase: A review. *Biotechnol. Adv.* **2010**, *28*, 628–634. [CrossRef]
37. Dickinson, S.; Mientus, M.; Frey, D.; Amini-Hajibashi, A.; Ozturk, S.; Shaikh, F.; Sengupta, D.; El-Halwagi, M.M. A review of biodiesel production from microalgae. *Clean Technol. Environ. Policy* **2017**, *19*, 637–668. [CrossRef]
38. Mata, T.M.; Martins, A.A.; Caetano, N.S. Microalgae for biodiesel production and other applications: A review. *Renew. Sustain. Energy Rev.* **2010**, *14*, 217–232. [CrossRef]
39. Beopoulos, A.; Cescut, J.; Haddouche, R.; Uribelarrea, J.L.; Molina-Jouve, C.; Nicaud, J.M. Yarrowia lipolytica as a model for bio-oil production. *Prog. Lipid Res.* **2009**, *48*, 375–387. [CrossRef]
40. Gude, V.; Grant, G.; Patil, P.; Deng, S. Biodiesel production from low cost and renewable feedstock. *Open Eng.* **2013**, *3*, 595–605. [CrossRef]
41. Chandel, A.K.; Garlapati, V.K.; Kumar, S.P.J.; Hans, M.; Singh, A.K.; Kumar, S. The role of renewable chemicals and biofuels in building a bioeconomy. *Biofuels Bioprod. Biorefin.* **2020**, *14*, 830–844. [CrossRef]
42. Heo, H.Y.; Heo, S.; Lee, J.H. Comparative techno-economic analysis of transesterification technologies for microalgal biodiesel production. *Ind. Eng. Chem. Res.* **2019**, *58*, 18772–18779. [CrossRef]
43. Maheshwari, P.; Haider, M.B.; Yusuf, M.; Klemeš, J.J.; Bokhari, A.; Beg, M.; Al-Othman, A.; Kumar, R.; Jaiswal, A.K. A review on latest trends in cleaner biodiesel production: Role of feedstock, production methods, and catalysts. *J. Clean. Prod.* **2022**, *355*, 131588. [CrossRef]
44. Bergmann, J.C.; Tupinambá, D.D.; Costa, O.Y.A.; Almeida, J.R.M.; Barreto, C.C.; Quirino, B.F. Biodiesel production in Brazil and alternative biomass feedstocks. *Renew. Sustain. Energy Rev.* **2013**, *21*, 411–420. [CrossRef]
45. Fargione, J.; Hill, J.; Tilman, D.; Polasky, S.; Hawthorne, P. Land Clearing and the Biofuel Carbon Debt. *Science* **2008**, *319*, 1235–1238. [CrossRef]
46. Hossain, A.K.; Davies, P.A. Plant oils as fuels for com-pression ignition engines: A technical review and life-cycle analysis. *Renew. Energy* **2010**, *35*, 1–13. [CrossRef]
47. Elgharbawy, A.S.; Sadik, W.A.; Sadek, O.M.; Kasaby, M.A. Glycerolysis treatment to enhance biodiesel production from low-quality feedstocks. *Fuel* **2021**, *284*, 118970. [CrossRef]
48. Bhatia, L.; Bachheti, R.K.; Garlapati, V.K.; Chandel, A.K. Third-generation biorefineries: A sustainable platform for food, clean energy, and nutraceuticals production. *Biomass Conv. Bioref.* **2022**, *12*, 4215–4230. [CrossRef]
49. Chintagunta, A.D.; Zuccaro, G.; Kumar, M.; Kumar, S.P.J.; Garlapati, V.K.; Postemsky, P.D.; Kumar, N.S.S.; Chandel, A.K.; Simal-Gandara, J. Biodiesel production from lignocellulosic biomass using oleaginous microbes: Prospects for integrated biofuel production. *Front. Microbiol.* **2021**, *12*, 658284. [CrossRef]
50. Shah, S.; Venkatramanan, V. Advances in microbial technology for upscaling sustainable biofuel production. In *New and Future Developments in Microbial Biotechnology and Bioengineering: Microbial Secondary Metabolites Biochemistry and Applications*; Gupta, V.K., Pandey, A., Eds.; Elsevier: Amsterdam, The Netherlands, 2019; pp. 69–76. [CrossRef]
51. Lu, H.; Yadav, V.; Zhong, M.; Bilal, M.; Taherzadeh, M.J.; Iqbal, H.M.N. Bioengineered microbial platforms for biomass-derived biofuel production—A review. *Chemosphere* **2022**, *288*, 132528. [CrossRef]
52. Mankar, A.R.; Pandey, A.; Modak, A.; Pant, K.K. Pretreatment of lignocellulosic biomass: A review on recent advances. *Bioresour. Technol.* **2021**, *334*, 125235. [CrossRef]
53. Adegboye, M.F.; Ojuederie, O.B.; Talia, P.M.; Babalola, O.O. Bioprospecting of microbial strains for biofuel production: Metabolic engineering, applications, and challenges. *Biotechnol. Biofuels* **2021**, *14*, 5. [CrossRef]
54. Xu, P.; Koffas, M.A. Metabolic engineering of *Escherichia coli* for biofuel production. *Eng. Life Sci.* **2010**, *1*, 493–504. [CrossRef]
55. Saha, R.; Mukhopadhyay, M. Prospect of metabolic engineering in enhanced microbial lipid production: A review. *Biomass Convers. Biorefin.* **2021**, 1–22. [CrossRef]
56. Ranjbar, S.; Malcata, F.X. Challenges and prospects for sustainable microalga-based oil: A comprehensive review, with a focus on metabolic and genetic engineering. *Fuel* **2022**, *324*, 124567. [CrossRef]
57. Chia, S.R.; Ong, H.C.; Chew, K.W.; Show, P.L.; Phang, S.M.; Ling, T.C.; Nagarajan, D.; Lee, D.J.; Chang, J.S. Sustainable approaches for algae utilisation in bioenergy production. *Renew. Energy* **2018**, *129*, 838–852. [CrossRef]
58. Iniyakumar, M.; Venkatramanan, V.; Ramalakshmi, A.; Bobita, R.; Tharunkumar, J.; Jothibasu, K.; Rakesh, S. Overview on Advanced Microalgae-Based Sustainable Biofuel Generation and Its Life Cycle Assessment. In *Micro-Algae: Next-Generation Feedstock for Biorefineries: Contemporary Technologies and Future Outlook*; Verma, P., Ed.; Springer Nature: Singapore, 2022; pp. 53–71. [CrossRef]
59. Zhu, L.; Nugroho, Y.K.; Shakeel, S.R.; Li, Z.; Martinkauppi, B.; Hiltunen, E. Using microalgae to produce liquid transportation biodiesel: What is next? *Renew. Sustain. Energy Rev.* **2017**, *78*, 391–400. [CrossRef]

60. Koyande, A.K.; Show, P.L.; Guo, R.; Tang, B.; Ogino, C.; Chang, J.S. Bio-processing of algal bio-refinery: A review on current advances and future perspectives. *Bioengineered* **2019**, *10*, 574–592. [CrossRef]
61. Khan, T.A.; Khan, T.A.; Kumar Yadav, A. A Hydrodynamic Cavitation-Assisted System for Optimization of Biodiesel Production from Green Microalgae Oil Using a Genetic Algorithm and Response Surface Methodology Approach. *Environ. Sci. Pollut. Res.* **2022**, *29*, 49465–49477. [CrossRef]
62. Singh, R.; Liu, H.; Shanklin, J.; Singh, V. Hydrothermal Pretreatment for Valorization of Genetically Engineered Bioenergy Crop for Lipid and Cellulosic Sugar Recovery. *Bioresour. Technol.* **2021**, *341*, 125817. [CrossRef]
63. Vanhercke, T.; Petrie, J.R.; Singh, S.P. Energy Densification in Vegetative Biomass through Metabolic Engineering. *Biocatal. Agric. Biotechnol.* **2014**, *3*, 75–80. [CrossRef]
64. Vanhercke, T.; Tahchy, A.E.; Liu, Q.; Zhou, X.-R.; Shrestha, P.; Divi, U.K.; Ral, J.-P.; Mansour, M.P.; Nichols, P.D.; James, C.N.; et al. Metabolic Engineering of Biomass for High Energy Density: Oilseed-like Triacylglycerol Yields from Plant Leaves. *Plant Biotechnol. J.* **2014**, *12*, 231–239. [CrossRef]
65. Heater, B.S.; Chan, W.S.; Lee, M.M.; Chan, M.K. Directed Evolution of a Genetically Encoded Immobilized Lipase for the Efficient Production of Biodiesel from Waste Cooking Oil. *Biotechnol. Biofuels* **2019**, *12*, 165. [CrossRef]
66. Takeshita, T.; Ivanov, I.N.; Oshima, K.; Ishii, K.; Kawamoto, H.; Ota, S.; Yamazaki, T.; Hirata, A.; Kazama, Y.; Abe, T.; et al. Comparison of Lipid Productivity of Parachlorella Kessleri Heavy-Ion Beam Irradiation Mutant PK4 in Laboratory and 150-L Mass Bioreactor, Identification and Characterization of Its Genetic Variation. *Algal Res.* **2018**, *35*, 416–426. [CrossRef]
67. Santos Mendoza, M.; Dubreucq, B.; Miquel, M.; Caboche, M.; Lepiniec, L. LEAFY COTYLEDON 2 Activation Is Sufficient to Trigger the Accumulation of Oil and Seed Specific MRNAs in Arabidopsis Leaves. *FEBS Lett.* **2005**, *579*, 4666–4670. [CrossRef]
68. Fan, J.; Yan, C.; Zhang, X.; Xu, C. Dual Role for Phospholipid:Diacylglycerol Acyltransferase: Enhancing Fatty Acid Synthesis and Diverting Fatty Acids from Membrane Lipids to Triacylglycerol in Arabidopsis Leaves. *Plant Cell* **2013**, *25*, 3506–3518. [CrossRef]
69. Xu, C.; Fan, J.; Froehlich, J.E.; Awai, K.; Benning, C. Mutation of the TGD1 Chloroplast Envelope Protein Affects Phosphatidate Metabolism in Arabidopsis. *Plant Cell* **2005**, *17*, 3094–3110. [CrossRef]
70. Sanjaya; Durrett, T.P.; Weise, S.E.; Benning, C. Increasing the Energy Density of Vegetative Tissues by Diverting Carbon from Starch to Oil Biosynthesis in Transgenic Arabidopsis. *Plant Biotechnol. J.* **2011**, *9*, 874–883. [CrossRef]
71. Petrie, J.R.; Vanhercke, T.; Shrestha, P.; El Tahchy, A.; White, A.; Zhou, X.-R.; Liu, Q.; Mansour, M.P.; Nichols, P.D.; Singh, S.P. Recruiting a New Substrate for Triacylglycerol Synthesis in Plants: The Monoacylglycerol Acyltransferase Pathway. *PLoS ONE* **2012**, *7*, e35214. [CrossRef]
72. Kalscheuer, R.; Stoelting, T.; Steinbuechel, A. Microdiesel: *Escherichia coli* engineered for fuel production. *Microbiology* **2006**, *152*, 2529–2536. [CrossRef]
73. Steen, E.J.; Kang, Y.; Bokinsky, G.; Hu, Z.; Schirmer, A.; Mcclure, A.; del Cardayre, S.B.; Keasling, J.D. Microbial production of fatty-acid-derived fuels and chemicals from plant biomass. *Nature* **2010**, *463*, 559–562. [CrossRef] [PubMed]
74. Elbahloul, Y.; Steinbüchel, A. Pilot-scale production of fatty acid ethyl esters by an engineered *Escherichia coli* strain harboring the p (Microdiesel) plasmid. *Appl. Environ. Microbiol.* **2010**, *76*, 4560–4565. [CrossRef]
75. de Jong, B.W.; Shi, S.; Siewers, V.; Nielsen, J. Improved production of fatty acid ethyl esters in Saccharomyces cerevisiae through up-regulation of the ethanol degradation pathway and expression of the heterologous phosphoketolase pathway. *Microb. Cell Fact.* **2014**, *13*, 39. [CrossRef]
76. Dulermo, R.; Gamboa-Meléndez, H.; Ledesma-Amaro, R.; Thevenieau, F.; Nicaud, J.-M. Yarrowia lipolytica AAL genes are involved in peroxisomal fatty acid activation. *Biochim. Biophys. Acta* **2016**, *1861*, 555–565. [CrossRef]
77. Das, S.; Reshad, A.S.; Bhuyan, N.; Sut, D.; Tiwari, P.; Goud, V.V.; Kataki, R. Utilization of nonedible oilseeds in a biorefinery approach with special emphasis on rubber seeds. In *Waste Biorefinery: Integrating Biorefineries for Waste Valorisation*; Bhaskar, T., Pandey, A., Rene, E.R., Tsang, D.C.W., Eds.; Elsevier, B.V.: Amsterdam, The Netherlands, 2020; pp. 311–336.
78. Bilgili, F.; Koçak, E.; Bulut, U.; Kuşkaya, S. Can biomass energy be an efficient policy tool for sustainable development? *Renew. Sustain. Energy Rev.* **2017**, *71*, 830–845. [CrossRef]
79. FAO. *Food Outlook*; Food and Agricultural Organization (FAO): Rome, Italy, 2020.
80. Mohanty, A.; Rout, P.R.; Dubey, B.; Meena, S.S.; Pal, P.; Goel, M. A critical review on biogas production from edible and non-edible oil cakes. *Biomass Convers. Biorefin.* **2022**, *12*, 949–966. [CrossRef]
81. Das, A.; Busetty, S.; Goel, M.; Ram Kiran, B.; Sudharshan, K.; Kanapuram, R. Alternative usage of edible deoiled cake for decolonisation of Reactive Red Dye. *Desalin. Water Treat.* **2015**, *53*, 2720–2726. [CrossRef]
82. Fernandez-Gonzalez, G.; Martin-Lara, M.A.; Blazquez, G.; Tenorio, G.; Calero, M. Hydrolyzed olive cake as novel adsorbent for copper removal from fertilizer industry wastewater. *J. Clean. Prod.* **2020**, *268*, 121935. [CrossRef]
83. Rawat, A.P.; Rawat, M.; Rai, J.P.N. Toxic Metals Biosorption by Jatropha curcas Deoiled Cake: Equilibrium and Kinetic Studies. *Water Environ. Res.* **2013**, *85*, 733–742. [CrossRef]
84. Yusoff, M.; Mohamed, E.; Idris, J.; Zainal, N.H.; Ibrahim, M.F.; Abd-Aziz, S. Adsorption of heavy metal ions by oil palm decanter cake activated carbon. *Makara J. Technol.* **2019**, *23*, 2.
85. Rao, R.A.K.; Khan, M.A. Removal and recovery of Cu(II), Cd(II) and Pb(II) ions from single and multimetal systems by batch and column operation on neem oil cake (NOC). *Sep. Purif. Technol.* **2007**, *57*, 394–402. [CrossRef]
86. Mahajan, G.; Garg, U.; Sud, D.; Garg, V. Utilization Properties of Jatropha De-Oiled Cake for Removal of Nickel (II) from Aqueous Solutions. *Bioresources* **2013**, *8*, 5596–5611. [CrossRef]

87. Khan, M.A.; Ngabura, M.; Choong, T.S.Y.; Masood, H.; Chuah, L.A. Biosorption and desorption of Nickel on oil cake: Batch and column studies. *Bioresour. Technol.* **2012**, *103*, 35–42. [CrossRef] [PubMed]
88. Rao, R.A.K.; Khan, M.A.; Jeon, B.H. Utilization of carbon derived from mustard oil cake (CMOC) for the removal of bivalent metal ions: Effect of anionic surfactant on the removal and recovery. *J. Hazard. Mater.* **2010**, *173*, 273–282. [CrossRef]
89. Upendar, K.; Sagar, T.V.; Raveendra, G.; Lingaiah, N.; Rao, B.V.S.K.; Prasad, R.B.N.; Sai Prasad, P.S. Development of a low temperature adsorbent from karanja seed cake for CO_2 capture. *RSC Adv.* **2014**, *4*, 7142. [CrossRef]
90. Sarkar, S. Exploring biogas potential data of cattle manure and olive cake to gain insight into farm and commercial scale production. *Data Br.* **2020**, *32*, 106045. [CrossRef]
91. Isci, A.; Demirer, G.N. Biogas production potential from cotton wastes. *Renew. Energy* **2007**, *32*, 750–757. [CrossRef]
92. Drozłowska, E.; Łopusiewicz, L.; Mezynska, M.; Bartkowiak, A. Valorization of Flaxseed Oil Cake Residual from Cold-Press Oil Production as a Material for Preparation of Spray-Dried Functional Powders for Food Applications as Emulsion Stabilizers. *Biomolecules* **2020**, *10*, 153. [CrossRef]
93. VasudhaUdupa, A.; Balakrishna, G.; Shivanna, M.B. Influence of non-edible oil-cakes and their composts on growth, yield and Alternaria leaf spot disease in chilli. *Int. J. Recycl. Org. Waste Agric.* **2022**, *11*, 301–318.
94. Venkatesagowda, B.; Ponugupaty, E.; Barbosa, A.M.; Dekker, R.F.H. Solid-state fermentation of coconut kernel-cake as substrate for the production of lipases by the coconut kernel-associated fungus Lasiodiplodia theobromae VBE-1. *Ann. Microbiol.* **2015**, *65*, 129–142. [CrossRef]
95. Rehman, B.; Ganai, M.A.; Parihar, K.; Asif, M.; Siddiqui, M. A Biopotency of Oilcakes Against Meloidogyne incognita Affecting Vigna mungo. *Eur. J. Appl. Sci.* **2014**, *6*, 57–63. [CrossRef]
96. Singh, R.; Kumar, A.; Tomer, A. De-oiled Cakes of Neem, Jatropha, Mahua and Karanja: A New Substrate for Mass Multiplication of T. harzianum. *J. Plant Pathol. Microbiol.* **2015**, *6*, 1000288. [CrossRef]
97. Radwan, M.A.; El-Maadawy, E.K.; Kassem, S.I.; Abu-Elamayem, M.M. Oil cakes soil amendment effects on Meloidogyne incognita, root-knot nematode infecting tomato. *Arch. Phytopathol. Plant Prot.* **2009**, *42*, 58–64. [CrossRef]
98. Olowoake, A.A.; Osunlola, O.S.; Ojo, J.A. Influence of compost supplemented with jatropha cake on soil fertility, growth, and yield of maize (zea mays L.) in a degraded soil of Ilorin, Nigeria. *Int. J. Recycl. Org. Waste Agric.* **2018**, *7*, 67–73. [CrossRef]
99. Hirota, M.; Suttajit, M.; Suguri, H.; Endo, Y.; Shudo, K.; Wongchai, V.; Hecker, E.; Fujiki, H. A new tumor promoter from the seed oil of Jatropha curcas L., an intramolecular diester of 12-deoxy-16-hydroxyphorbol. *Cancer Res.* **1988**, *48*, 5800–5804. [PubMed]
100. Sharma, S.; Verma, M.; Sharma, A. Utilization of Non Edible Oil Seed Cakes as Substrate for Growth of Paecilomyces lilacinus and as Biopesticide Against Termites. *Waste Biomass Valoriz.* **2013**, *4*, 325–330. [CrossRef]
101. Monlau, F.; Latrille, E.; Da Costa, A.C.; Steyer, J.P.; Carrere, H. Enhancement of methane production from sunflower oil cakes by dilute acid pretreatment. *Appl. Energy* **2013**, *102*, 1105–1113. [CrossRef]
102. Fernandez-Cegri, V.; De la Rubia, M.A.; Raposo, F.; Borja, R. Effect of hydrothermal pre-treatment of sunflower oil cake on biomethane potential focusing on fibre composition. *Bioresour. Technol.* **2012**, *123*, 424–429. [CrossRef]
103. Kumar, G.; Lin, C.Y. Biogenic Hydrogen Conversion of De-Oiled Jatropha Waste via Anaerobic Sequencing Batch Reactor Operation: Process Performance, Microbial Insights, and CO_2 Reduction Efficiency. *Sci. World J.* **2014**, *2014*, 946503. [CrossRef] [PubMed]
104. Ramachandran, S.; Singh, S.K.; Larroche, C.; Soccol, C.R.; Pandey, A. Oil cakes and their biotechnological applications—A review. *Bioresour. Technol.* **2007**, *98*, 2000–2009. [CrossRef] [PubMed]
105. Vaštag, Ž.; Popović, L.; Popović, S.; Krimer, V.; Peričin, D. Production of enzymatic hydrolysates with antioxidant and angiotensin-I converting enzyme inhibitory activity from pumpkin oil cake protein isolate. *Food Chem.* **2011**, *124*, 1316–1321. [CrossRef]
106. Jose, S.; Roy, R.; Phukan, A.R.; Shakyawar, D.B.; Sankaran, A. Biochar from oil cakes: An efficient and economical adsorbent for the removal of acid dyes from wool dye house effluent. *Clean Technol. Environ. Policy* **2022**, *24*, 1599–1608. [CrossRef]
107. Kalinke, C.; Mangrich, A.S.; Marcolino-Junior, L.H.; Bergamini, M.F. Biochar prepared from castor oil cake at different temperatures: A voltammetric study applied for Pb^{2+}, Cd^{2+} and Cu^{2+} ions preconcentration. *J. Hazard. Mater.* **2016**, *318*, 526–532. [CrossRef]
108. Ancuța, P.; Sonia, A. Oil press-cakes and meals valorization through circular economy approaches: A review. *Appl. Sci.* **2020**, *10*, 7432. [CrossRef]
109. Nevara, G.A.; Giwa Ibrahim, S.A.; Syed Muhammad, S.K.; Zawawi, N.; Mustapha, N.A.; Karim, R. Oilseed meals into foods: An approach for the valorization of oilseed by-products. *Crit. Rev. Food Sci. Nutr.* **2022**, *28*, 1–14. [CrossRef]
110. Colletti, A.; Attrovio, A.; Boffa, L.; Mantegna, S.; Cravotto, G. Valorisation of by-products from soybean (Glycine max (L.) Merr.) processing. *Molecules* **2020**, *25*, 2129. [CrossRef]
111. Martín-Cabrejas, M.A.; Aguilera, Y.; Pedrosa, M.M.; Cuadrado, C.; Hernández, T.; Díaz, S.; Esteban, R.M. The impact of dehydration process on antinutrients and protein digestibility of some legume flours. *Food Chem.* **2009**, *114*, 1063–1068. [CrossRef]
112. Rabadán, A.; Álvarez-Ortí, M.; Martínez, E.; Pardo-Giménez, A.; Zied, D.C.; Pardo, J.E. Effect of replacing traditional ingredients for oils and flours from nuts and seeds on the characteristics and consumer preferences of lamb meat burgers. *LWT* **2021**, *136*, 110307. [CrossRef]
113. Ilmi, M.; Hidayat, C.; Hastuti, P.; Heeres, H.J.; Van Der Maarel, M.J.E.C. Utilisation of Jatropha press cake as substrate in biomass and lipase production from Aspergillus niger 65I6 and Rhizomucor miehei CBS 360.62. *Biocatal. Agric. Biotechnol.* **2017**, *9*, 103–107. [CrossRef]

114. Rathore, D.; Pant, D.; Singh, A. A comparison of life cycle assessment studies of different biofuels. In *Life Cycle Assessment of Renewable Energy Sources, Series: Green Energy and Technology*; Singh, A., Olsen, S.I., Pant, D., Eds.; Springer: London, UK, 2013. [CrossRef]
115. Arguelles-Arguelles, A.; Amezcua-Allieri, M.A.; Ramírez-Verduzco, L.F. Life cycle assessment of green diesel production by hydrodeoxygenation of palm oil. *Front. Energy Res.* **2021**, *9*, 690725. [CrossRef]
116. Foteinis, S.; Chatzisymeon, E.; Litinas, A.; Tsoutsos, T. Used-cooking-oil biodiesel: Life cycle assessment and comparison with first- and third-generation biofuel. *Renew. Energy* **2020**, *153*, 588–600. [CrossRef]
117. Singh, A.; Olsen, S.I. Key Issues in Life Cycle Assessment of Biofuels. In *Sustainable Bioenergy and Bioproducts: Value Added Engineering Applications, Green Energy and Technology*; Gopalakrishnan, K., Leeuwen, J.H.v., Brown, R.C., Eds.; Springer: London, UK, 2012; pp. 213–228.
118. Singh, A.; Pant, D.; Olsen, S.I.; Nigam, P.S. Key issues to consider in microalgae based biodiesel. *Energy Educ. Sci. Technol. Part A Energy Sci. Res.* **2012**, *29*, 687–700.
119. Larson, E.D. A review of life-cycle analysis studies on liquid biofuel systems for the transport sector. *Energy Sustain. Dev.* **2006**, *10*, 109–126. [CrossRef]
120. Karka, P.; Papadokonstantakis, S.; Hungerbühler, K.; Kokossis, A. Life Cycle Assessment of Biorefinery Products Based on Different Allocation Approaches. *Comput. Aided Chem. Eng.* **2015**, *37*, 2573–2578. [CrossRef]
121. Gnansounou, E.; Dauriat, A.; Villegas, J.; Panichelli, L. Life cycle assessment of biofuels: Energy and greenhouse gas balances. *Bioresour. Technol.* **2009**, *100*, 4919–4930. [CrossRef]
122. Cherubini, F.; Bird, N.D.; Cowie, A.; Jungmeier, G.; Schlamadinger, B.; Woess-Gallasch, S. Energy- and greenhouse gas-based LCA of biofuel and bioenergy systems: Key issues, ranges and recommendations. *Resour. Conserv. Recycl.* **2009**, *53*, 434–447.
123. SAIC. *Life Cycle Assessment: Principles and Practice*; EPA/600/R-06/060; National Risk Management Research Laboratory, Office of Research and Development, US Environmental Protection Agency: Cincinnati, OH, USA, 2006.
124. Singh, A.; Pant, D.; Korres, N.E.; Nizami, A.S.; Prasad, S.; Murphy, J.D. Key issues in life cycle assessment of ethanol production from lignocellulosic biomass: Challenges and perspectives. *Bioresour. Technol.* **2010**, *101*, 5003–5012.
125. Luo, L.; van der Voet, E.; Huppes, G.; Udo de Haes, H.A. Allocation issues in LCA methodology: A case study of corn stover-based fuel ethanol. *Int. J. Life Cycle Assess.* **2009**, *14*, 529–539. [CrossRef]
126. Sheehan, J.; Camobreco, V.; Duffield, J.; Graboski, M.; Shapouri, H. *An Overview of Biodiesel and Petroleum Diesel Life Cycles*; National Renewable Energy Laboratory: Golden, CO, USA, 1998.
127. Sieira, P.; Galante, E.B.F.; Mendes, A.J.B.; Haddad, A. Life Cycle Assessment of a Biodiesel Production Unit. *Am. J. Chem. Eng.* **2015**, *3*, 25–29. [CrossRef]
128. Nguyen, T.H.; Field, J.L.; Kwon, H.; Hawkins, T.R.; Paustian, K.; Wang, M.Q. A multi-product landscape life-cycle assessment approach for evaluating local climate mitigation potential. *J. Clean. Prod.* **2022**, *354*, 131691. [CrossRef]
129. Malik, S.; Shahid, A.; Betenbaugh, M.J.; Liu, C.-G.; Mehmood, M.A. A novel wastewater-derived cascading algal biorefinery route for complete valorization of the biomass to biodiesel and value-added bioproducts. *Energy Convers. Manag.* **2022**, *256*, 115360. [CrossRef]
130. Singh, A.; Olsen, S.I. Comparison of algal biodiesel production pathways using life cycle assessment tool. In *Life Cycle Assessment of Renewable Energy Sources, Green Energy and Technology Series*; Singh, A., Pant, D., Olsen, S.I., Eds.; Springer: London, UK, 2013; pp. 145–168.
131. Gupta, R.; McRoberts, R.; Yu, Z.; Smith, C.; Sloan, W.; You, S. Life cycle assessment of biodiesel production from rapeseed oil: Influence of process parameters and scale. *Bioresour. Technol.* **2022**, *360*, 127532. [CrossRef]
132. Vairaprakash, P.; Arumugam, A. Sustainability, Commercialization, and Future Prospects of Biodiesel Production. In *Biodiesel Production: Feedstocks, Catalysts, and Technologies*; Rokhum, S.L., Halder, G., Assabumrungrat, S., Ngaosuwan, K., Eds.; John Wiley & Sons Ltd.: Hoboken, NJ, USA, 2022; pp. 355–376. [CrossRef]
133. REN21. *Renewables 2020 Global Status Report*; REN21 Secretariat: Paris, France, 2020.
134. Pavan, M.; Reinmets, K.; Garg, S.; Mueller, A.P.; Marcellin, E.; Kopke, M.; Valgepea, K. Advances in systems metabolic engineering of autotrophic carbon oxide-fixing biocatalysts towards a circular economy. *Metab. Eng.* **2022**, *71*, 117–141. [CrossRef]
135. Hossain, S.M.Z.; Razzak, S.A.; Al-Shater, A.F.; Moniruzzaman, M.; Hossain, M.M. Recent Advances in Enzymatic Conversion of Microalgal Lipids into Biodiesel. *Energy Fuels* **2020**, *34*, 6735–6750. [CrossRef]
136. Wang, H.; Peng, X.; Zhang, H.; Yang, S.; Li, H. Microorganisms-promoted biodiesel production from biomass: A review. *Energy Convers. Manag. X* **2021**, *12*, 100137. [CrossRef]
137. Fan, J.; Andre, C.; Xu, C. A chloroplast pathway for the de novo biosynthesis of triacylglycerol in Chlamydomonas reinhardtii. *FEBS Lett.* **2011**, *585*, 1985–1991. [CrossRef]
138. Cagliari, A.; Margis, R.; Maraschin, F.S.; Turchetto-Zolet, A.C.; Loss, G.; Margis-Pinheiro, M. Biosynthesis of triacylglycerols (TAGs) in plants and algae. *Int. J. Plant Biol.* **2011**, *2*, 40–52. [CrossRef]
139. Severo, I.A.; Siqueira, S.F.; Deprá, M.C.; Maroneze, M.M.; Zepka, L.Q.; Jacob-Lopes, E. Biodiesel facilities: What can we address to make biorefineries commercially competitive? *Renew. Sustain. Energy Rev.* **2019**, *112*, 686–705. [CrossRef]

Review

Zr-Based Metal-Organic Frameworks for Green Biodiesel Synthesis: A Minireview

Qiuyun Zhang [1,2,*], Jialu Wang [1,3,*], Shuya Zhang [2], Juan Ma [2], Jingsong Cheng [2] and Yutao Zhang [1,2,*]

[1] College Rural Revitalization Research Center of Guizhou, Anshun University, Anshun 561000, China
[2] School of Chemistry and Chemical Engineering, Anshun University, Anshun 561000, China
[3] School of Resource and Environmental Engineering, Anshun University, Anshun 561000, China
* Correspondence: sci_qyzhang@126.com (Q.Z.); wangjl226@126.com (J.W.); zyt0516@126.com (Y.Z.)

Abstract: Metal–organic frameworks (MOFs) have widespread application prospects in the field of catalysis owing to their functionally adjustable metal sites and adjustable structure. In this minireview, we summarize the current advancements in zirconium-based metal–organic framework (Zr-based MOF) catalysts (including single Zr-based MOFs, modified Zr-based MOFs, and Zr-based MOF derivatives) for green biofuel synthesis. Additionally, the yields, conversions, and reusability of Zr-based MOF catalysts for the production of biodiesel are compared. Finally, the challenges and future prospects regarding Zr-based MOFs and their derivatives for catalytic application in the biorefinery field are highlighted.

Keywords: Zr-based MOFs; heterogeneous catalysis; esterification; transesterification; biofuels

Citation: Zhang, Q.; Wang, J.; Zhang, S.; Ma, J.; Cheng, J.; Zhang, Y. Zr-Based Metal-Organic Frameworks for Green Biodiesel Synthesis: A Minireview. *Bioengineering* 2022, *9*, 700. https://doi.org/10.3390/bioengineering9110700

Academic Editors: Chengfei Zhang, Ilaria Fratoddi, Indra Neel Pulidindi and Aharon Gedanken

Received: 12 October 2022
Accepted: 15 November 2022
Published: 17 November 2022

Publisher's Note: MDPI stays neutral with regard to jurisdictional claims in published maps and institutional affiliations.

Copyright: © 2022 by the authors. Licensee MDPI, Basel, Switzerland. This article is an open access article distributed under the terms and conditions of the Creative Commons Attribution (CC BY) license (https://creativecommons.org/licenses/by/4.0/).

1. Introduction

With rapid population growth and high industrial energy demand from fossil fuel resources, fossil fuel usage leading to the energy security crisis and climate change (e.g., greenhouse gas emissions) has been of great concern [1]. Based on this, it is necessary to search for alternative fuel resources. Currently, renewable biofuels derived from biomass have gained enormous attention, and one of these liquid biofuels is biodiesel [2]. Chemically, biodiesel (fatty acid alkyl ester, FAME) is generally produced via the transesterification of edible oils (e.g., rapeseed oil, palm oil, sunflower oil), non-edible oils (e.g., *Jatropha curcus*, *Euphorbia lathyris* L., *Monotheca buxifolia*, *Sinapis Arvensis*), microalgal oils, or waste cooking oil or via the esterification of free fatty acids (e.g., oleic acid, lauric acid, palmitic acid) and methanol or ethanol using acid/alkali as catalysts [3–8].

Traditionally, biodiesel production is carried out using liquid acid/alkali catalysts due to their high catalytic activity. Unfortunately, these homogeneous catalytic processes exhibit numerous disadvantages, such as high operating costs for steps such as product purification, catalyst neutralization, and a large amount of industrial wastewater that requires treatment [9]. With regard to this, the utilization of heterogeneous catalysts is becoming an efficient candidate for the production of biodiesel, because of their simple recovery, ease of reuse, insolubility in reaction solvents, and reduction in waste treatment [10]. According to Figure 1, various kinds of heterogeneous acid/alkali catalysts are available for different types of organic reactions, including metal oxides, ionic liquids, heteropoly acids, zeolites, sulfonic-acid-functionalized catalysts, etc. [11,12]. However, some problems such as leaching, lesser activity, low stability, and longer reaction time for some heterogeneous catalysts are found in the transesterification/esterification reaction process [13].

Quite recently, metal–organic frameworks (MOFs) have gained enormous attention due to their unique features, such as great specific surface area, uniformity in pore size, large porosity, adjustable properties, tunable structures, and controllable functional groups [14]. In addition, MOFs are suitable to be functionalized by coordinating acid/base functional groups, which have been widely studied in catalysis [15–17]. Among the numerous types of MOFs, zirconium-based MOFs (Zr-based MOFs) have been frequently applied as potential porous materials owing to the presence of Lewis and Brønsted acidity [18,19]. At present, several reviews have summarized the applications of Zr-MOFs in catalysis [20,21]. However, none has given a detailed study on the applications of Zr-MOFs and derivatives for green biodiesel synthesis. Thus, the present review focuses on the current development of Zr-based MOFs and derivatives for green biofuel production. More importantly, the catalytic performance and reusability of single Zr-based MOFs, modified Zr-based MOFs, and Zr-based MOF derivatives are systematically discussed. Finally, the conclusions and prospects are emphasized.

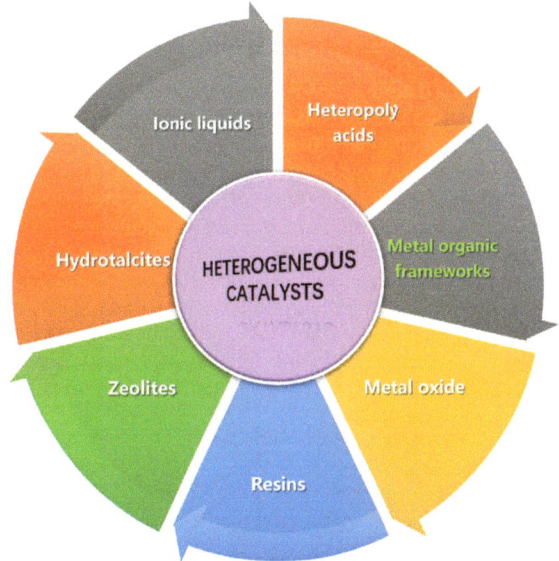

Figure 1. Various heterogeneous catalysts.

2. Zr-Based MOF Catalysts

In 2008, a zirconium-based inorganic building brick (Zr-MOFs) was first reported by Lillerud et al. [22], and the results showed that Zr-MOFs possess high stability due to the combination of strong Zr-O bonds, the inner Zr_6 cluster, and the addition of μ_3-OH groups. A large number of Zr-based MOF catalysts have been widely used for fuel synthesis, including UiO-66 (see Figure 2), UiO-67, MOF-801, MOF-808, UiO-66-NH_2, etc., owing to their excellent chemical and thermal stability under harsh conditions, large specific surface areas, smaller particle size, and strong acid sites by the tuning of structural defects [23,24].

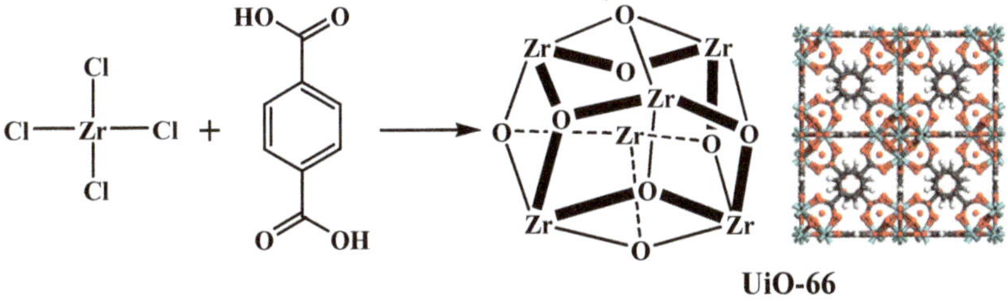

Figure 2. Schematic diagram of the structure and synthesis of UiO-66.

2.1. Single Zr-Based MOF Catalysts

Single Zr-based MOFs usually show fewer Lewis acid characteristics due to the saturated Zr atom, and they have exhibited less catalytic activity in acid-catalyzed organic reactions. In view of this, the synthesis of various zirconium-containing UiO-66 samples by varying the synthesis temperatures and terephthalic acid/$ZrCl_4$ ratios was reported by Zhou, et al. [25]. They found that UiO-66 catalysts with different amounts of defects could be synthesized under various synthesis conditions, and the catalytic activities of the UiO-66 depended on the defect amount. The obtained catalyst was used to catalyze the transesterification of soybean oil with methanol, and a conversion rate of 98.5% was acquired through a catalyst amount of 9%, with an oil/methanol molar ratio of 1:40, and at 140 °C for 5 h. UiO-66 with defects can be especially easily reused. Caratelli and his co-workers [26] also utilized UiO-66 MOFs as an acid catalyst for the production of ethyl levulinate via esterification from levulinic acid and ethanol. Various defective hydrated and dehydrated UiO-66 materials demonstrated excellent performance, and a maximum yield (>70%) of ethyl levulinate was achieved.

A similar study was conducted by Jrad, et al. [27], in which three isostructural Zr-based MOFs (UiO-66, UiO-66(COOH)$_2$, and UiO-66(NH$_2$)) were prepared (Figure 3). UiO-66(COOH)$_2$ demonstrated superior catalytic activity in the esterification of butyric acid and butanol, and a 90% conversion rate of butyl butyrate was achieved. This better catalytic activity was possibly related to the smaller particle size of the catalyst and the additional active acid functional groups grafted onto the original organic linker.

Figure 3. The suggested structures of UiO-66, UiO-66(COOH)$_2$, and UiO-66(NH$_2$).

Wei's group [28] also designed a series of defective UiO-66 catalysts for the esterification of levulinic acid with ethanol. The results showed that the synergistic effects between unsaturated Zr_6 nodes and hydroxyl groups can have a significant influence on catalytic activity. The synthesized defective UiO-66 catalyst possessed excellent stability and could retain 75% of its initial activity after five cycles. Similarly, Chaemchuen and co-workers [29] synthesized UiO-66 catalyst for the esterification reaction between oleic acid and methanol. According to the kinetic analysis, an activation energy of 54.9 ± 1.8 kJ/mol was obtained, and the desorption of methyl oleate was found to be irreversible. Desidery, et al. [30] tested MOF-808 for the conversion of dimethyl carbonate into ethyl methyl carbonate. The MOF-808 catalyst exhibited superior catalytic performance and could be recycled for up to four cycles without any major change in activity or structure. Shaik, et al. [31] developed the Zr-fumarate MOF (MOF-801) as a heterogeneous catalyst for biodiesel production. The characterization results demonstrated that MOF-801 possessed cubic structure, high crystallinity, good thermal stability, and moderate catalytic activity. Under optimal reaction conditions, the conversion rate of used vegetable oil was 60%; the activity of MOF-801 is probably due to the cationic Zr and anionic O_2 sites in the crystal structure.

Besides this, de la Flor, et al. [32] reported the synthesis of defective UiO-66(Zr) catalyst for the production of jet-fuel precursors via aldol-condensation, and total furfural conversion and selectivity (~100%) were obtained. Rapeyko, et al. [33] also reported that the as-synthesized UiO-66 could efficiently catalyze the selective ketalization of levulinic acid and 1,2-propanediol, and high selectivity (91–93%) was attained.

From the studied literature, it is derived that the single Zr-based MOFs can be considered as a catalyst for acid-catalyzed reactions. However, the activity of single Zr-based MOFs still needs to be further improved, and facilely tuning the defect density on nodes by introducing modulators may be a very interesting approach to developing highly active Zr-based MOF catalysts.

2.2. Modified Zr-Based MOF Catalysts

To improve the catalytic activity and chemical stability of Zr-based MOFs, several researchers have investigated various types of modification methods, including functional organic linkers, loading active components (e.g., lipase, ionic liquids, heteropoly acid, etc.), and the incorporation of metal ions into Zr-based MOFs (Table 1).

Table 1. Recent findings on green fuel production using modified Zr-based MOF catalysts.

Entry	Raw Material	Catalyst	Reaction Conditions (Time, Temperature, Catalyst Amount, Molar Ratio (Acid(Oil):Alcohol))	Yield (Y/%) or Conversion (C/%)	Reusability	E_a (KJ/mol)	Ref.
1	Lauric acid + Methanol	UiO-66-NH$_2$	2 h, 60 °C, 8%, 1:26	Y > 99	Not reported	\	[34]
2	Levulinic acid + n-butanol	UiO-66-NH$_2$	5 h, 120 °C, 1.8%, 1:6	Y = 99%	3 cycles, no significant loss	\	[35]
3	Levulinic acid + Ethanol	UiO-66-(COOH)$_2$	24 h, 78 °C, 0.39%, 1:20	Y = 97%	5 cycles, Y = 93.9%	\	[36]
4	Oleic acid + Methanol	UiO-66(Zr)-NH$_2$	4 h, 60 °C, 6%, 1:39	C = 97%	4 cycles, C > 50%	15.13	[37]
5	Oleic acid + Methanol	10SA/UiO-66(Zr)	4 h, 25 °C, 6%, 1:39	C = 94.5%	6 cycles, C = 83%	32.53	[38]
6	Levulinic acid + Ethanol	UiO66-SO$_3$H(100)	6 h, 80 °C, 0.4%, 1:10	Y = 87%	4 cycles, Y = 84%	\	[39]
7	Ricinus communis oil + Methanol	Lipase/Zr-MOF/PVP	12 h, 50 °C, 2 mg, 1:3	C = 83%	7 cycles, C = 66%	\	[40]
8	Oleic acid + Methanol	UiO-G	2 h, 70 °C, 8%, 1:12	C = 91.3%	4 cycles, C = 66.6%	28.61	[41]
9	Acetic acid + Isooctyl alcohol	UiO-67-CF$_3$SO$_3$	18 h, 90 °C, 0.2 g, 6:1	C = 98.6%	5 cycles, C = 95.9%	\	[42]
10	Tripalmitin + Methanol	UiO-66-[C$_3$NH$_2$][SO$_3$CF$_3$]	12 h, 85 °C, 0.025 g, 1:121.5	Y = 86.6–98.4%	Not reported	38.9	[43]
11	Jatropha oil + Methanol	PSH/UiO-66-NO$_2$	4 h, 70 °C, 4%, 1:25	C = 97.57%	3 cycles, C = 77.14%	\	[44]
12	Oleic acid + Methanol	AIL@NH$_2$-UiO-66	6 h, 75 °C, 5%, 1:14	C = 95.22%	6 cycles, C = 90.42%	\	[45]
13	Oleic acid + Methanol	Ca^{2+}/UiO-66(Zr)	4 h, 60 °C, 6%, 1:39	Y = 98%	5 cycles, Y = 84%	36.73	[46]
14	Oleic acid + Methanol	K-PW$_{12}$@UIO-66(Zr)	4 h, 75 °C, 5%, 1:20	C = 90%	10 cycles, no significant loss	\	[47]
15	Acetic acid + n-butanol	HPW@UiO-66	3 h, 120 °C, 3%, 1:2	C = 80.2%	4 cycles, C = 63%	\	[48]
16	Soybean oil + C8 + C10	Cs$_{2.5}$H$_{0.5}$PW$_{12}$O$_{40}$@UiO-66	10 h, 150 °C, 7%, 1:5:5	FA incorporation =20.3%	5 cycles, no significant loss	\	[49]
17	Soybean oil + Methanol	AILs/HPW/UiO-66-2COOH	6 h, 110 °C, 10%, 1:35	C = 95.8%	5 cycles, C > 80%	\	[50]
18	Euphorbia Lathyris L. oil + Methanol	HPW/UiO-66-NH$_2$	8 h, 180 °C, 3.5%, 1:40	Y = 91.2%	4 cycles, no significant loss	31.0	[51]

Table 1. Cont.

Entry	Raw Material	Catalyst	Reaction Conditions (Time, Temperature, Catalyst Amount, Molar Ratio (Acid(Oil):Alcohol))	Yield (Y/%) or Conversion (C/%)	Reusability	Ea (KJ/mol)	Ref.
19	Oleic acid + Methanol	FDCA/SA-UiO-66(Zr)	24 h, 60 °C, 6%, 1:40	Y = 98.4%	6 cycles, Y > 90%	\	[52]
20	Soybean oil + Methanol	PW$_{12}$@UIO-66	4 h, 75 °C, 0.2 g, 1 g:5.5 ml	C = 91.1%	4 cycles, no significant loss	\	[53]
21	Lauric acid + Methanol	HSiW-UiO-66	4 h, 160 °C, 7%, 1:20	C = 80.5%	4 cycles, C = 70.2%	27.5	[54]
22	Oleic acid + Methanol	ZrSiW/UiO-66	4 h, 150 °C, 8%, 1:20	C = 98.0%	4 cycles, C = 88.9%	\	[55]
23	Lauric acid + Methanol	Ag$_1$(NH$_4$)$_2$PW$_{12}$O$_{40}$/UiO-66	3 h, 150 °C, 10%, 1:15	C = 75.6%	4 cycles, C = 70.6%	35.2	[56]
24	Oleic acid + Methanol	Ce-BDC@HSiW@UiO-66	4 h, 130 °C, 0.2 g, 1:30	C = 81.5%	6 cycles, C = 76.9%	\	[57]

Cirujano's group [34,35] prepared an UiO-66-NH$_2$ catalyst to convert lauric acid to methyl laurate via an esterification reaction. The high activity of UiO-66-NH$_2$ with respect to UiO-66 is attributed to the occurrence of cooperative acid–base catalysis in the frame network. In addition, UiO-66-NH$_2$ has been successfully used for esterification of levulinic acid with various alcohols. A possible bifunctional acid–base catalyst mechanism for esterification was proposed, as displayed in Figure 4.

Likewise, Wang, et al. [36] employed UiO-66-(COOH)$_2$ as a heterogeneous catalyst for the esterification of levulinic acid, and Abou-Elyazed, et al. [37] also employed UiO-66(Zr)-NH$_2$ for the esterification of oleic acid. Meanwhile, Abou-Elyazed's group [38] also demonstrated the direct preparation of Ca^{2+}-doped UiO-66(Zr) under solvent-free conditions. In detail, the introduction of Ca^{2+} could greatly enhance the catalytic performance and stability in the esterification because of the existence of double active sites with the formation of more defects.

Desidery, et al. [39] investigated partially and fully sulfonated hydrated UiO66 catalysts prepared by one-step solvothermal synthesis. Compared to that of commercial Amberlyst 15, the activity of the fully sulfonated hydrated UiO66 afforded the highest yield of ethyl levulinate.

UiO-66(Zr) used as a support for p-toluenesulfonic acid (PTSA) through a defect coordination strategy was proposed by Li, et al. [41]. Their results indicated that the PTSA was successfully introduced into UiO-66(Zr), and the highest conversion rate of oleic acid to biodiesel of 91.3% was acquired under mild conditions. More specifically, a reusability study showed that the conversion was dramatically reduced from 91.3% to 76.65% after four cycles, and they verified a loss of Zr and S in the reaction system.

Recently, acidic or basic ionic liquids (ILs) have shown efficient catalytic activities in various organic reactions. However, they also suffer from several shortcomings, such as high viscosity, diffusion limitations, and difficulty in separation. In order to overcome these problems, the introduction of ILs into Zr-based MOFs has been studied [42–45]. Acidic ILs (AIL) were introduced into the NH$_2$-UiO-66 matrix (See Figure 5) via acid–base interaction by Lu, et al. [45]. Accordingly, the best mass ratio of AIL to NH$_2$-UiO-66 in 3AIL/NH$_2$-UiO-66 displayed excellent activity and reusability in the esterification of oleic acid; a conversion rate of 95.22% was achieved in 6 h, and it could still reach 90.42% conversion after six cycles. Moreover, it was concluded that the good conversion rate was attributed to the stimulating synergy between the -SO$_3$H group of AIL and the -NH$_2$ group of NH$_2$-UiO-66 on the MOFs.

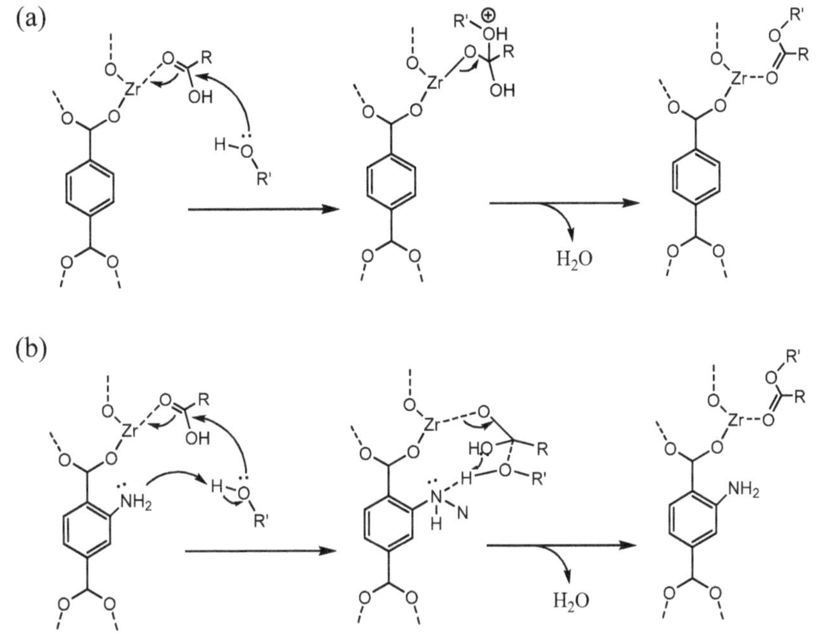

Figure 4. Plausible mechanisms for esterification: (**a**) UiO-66 catalyst; (**b**) UiO-66-NH$_2$ acid–base catalyst. (Adapted with permission from Ref. [34]. Copyright 2015, Elsevier.)

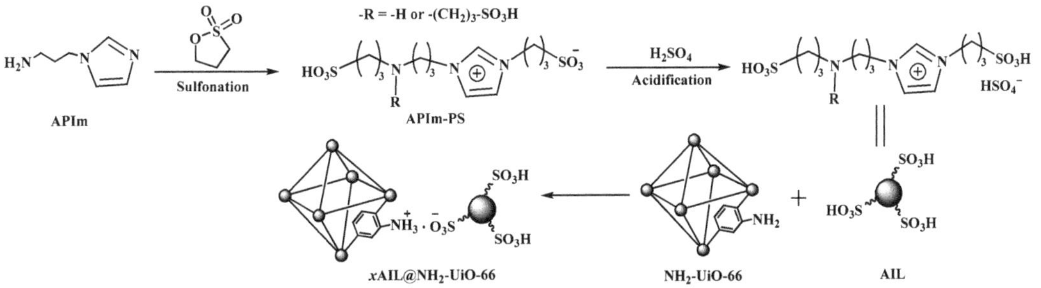

Figure 5. The suggested structures for AIL and xAIL@NH$_2$-UiO-66 (where x indicates the mass ratio of AIL to NH$_2$-UiO-66).

Apart from active ILs, heteropoly acids (HPAs) with structural diversity and tunable Brønsted/Lewis acidity have also been reported as efficient acid catalysts for biodiesel synthesis. This is despite their high activity, solubility in many polar solvents, and very low surface area, which limit their application for catalysis. Therefore, the loading of various HPAs on Zr-based MOF materials has been performed [47,48].

Xie's group [50] studied the one-pot transesterification–esterification of acidic vegetable oils to produce biodiesel by employing UiO-66-2COOH modified with HPW and sulfonated ILs as an acid catalyst (AILs/POM/UiO-66-2COOH). In their study, the surface area, pore volume, and mean pore size of the as-prepared composite catalyst were found to be 8.63 m^2/g, 0.04 cm^3/g, and 16.07 nm, respectively. Furthermore, the highest observed conversion rate was 95.8%, and the solid catalyst could maintain high activity even when 9 wt% free fatty acid and 3 wt% water were added into the feedstock.

As reported in much of the literature, Yang's group [51] synthesized HPW/UiO-66-NH$_2$ Lewis/Brønsted acid bifunctional hybrid catalyst by the electrovalent assembly of HPW and UiO-66-NH$_2$. The resulting HPW/UiO-66-NH$_2$ exhibited a highest acid density of 1.7 mmol/g, larger surface area of 301.6 m^2/g, and both Lewis and Brønsted acid sites. The biodiesel yield obtained from the (trans)esterification of *Lathyris* L. oil was more than 91.2%. Notably, the composite catalyst was reusable for four cycles with no significant decrease in activity, and hot filtration experiments showed that the composite has heterogeneous characteristics. Another study conducted by Yang's group [52] examined the synthesis of FDCA/SA-UiO-66(Zr) catalyst by a facile grinding method. Accordingly, DCA/SA-UiO-66(Zr) demonstrated superior or equivalent catalytic activity in the esterification of oleic acid due to its Lewis acidity and hydrophobicity.

Recently, our group also studied the production of biodiesel from free fatty acid with methanol over HPAs or doped HPAs incorporated into UiO-66 frameworks (e.g., ZrSiW/UiO-66, Ag$_1$(NH$_4$)$_2$PW$_{12}$O$_{40}$/UiO-66, and Ce-BDC@HSiW@UiO-66) [54–57]. All these composite catalysts exhibited good catalytic activity and reusability. Table 1 summarizes the modified Zr-based MOF catalysts used for biofuel production. As can be seen here, many researchers agree that modified Zr-based MOF catalysts can effectively catalyze esterification or transesterification processes.

2.3. Zr-Based MOF-Derived Catalysts

Recently, MOFs have also been employed as a precursor substrate and template support for derived material synthesis. The synthesis of porous carbon and metal oxide via a thermal decomposition process was first reported by Xu's group [58,59]. Since then, MOF derivatives have been attracted increasing attention as novel catalysts. In particular, the pyrolysis of defective Zr-based MOFs can provide a promising platform for various functional materials' synthesis.

Lu, et al. [60] successfully synthesized flower-like mesoporous sulfated zirconia nanosheets via the thermal decomposition of in situ sulfated Zr-MOFs as the S/Zr ratio increased to 0.5. Investigations on the sulfated zirconia nanosheets at a calcination temperature of 500 °C showed a large surface area (186.1 m^2/g) and strong interaction between the sulfate and zirconia atoms, affording excellent catalytic performance and stability for the production of biodiesel. Besides this, the mechanism of transesterification was studied, as shown in Figure 6.

Li, et al. [61] employed a UiO-66(Zr) support impregnated with calcium acetate for CaO/ZrO$_2$ catalyst synthesis via an activation process in nitrogen (UCN) and air (UCA) atmosphere. Among these catalysts, UCN650 calcined at 650 °C attained a relatively large specific surface area (24.06 m^2/g); meanwhile, the catalyst generated active sites of Ca$_x$Zr$_y$O$_{x+2y}$ and CaO inside and was shown to be effective in catalyzing palm oil transesterification, reaching a maximum conversion rate of 98.2%. Moreover, the UCN650 catalyst maintained its catalytic properties when it was recycled three times. The properties of the resulting biodiesel (density, kinetic viscosity, acid value, etc.) were also found to comply with the EN 14214 standards.

Our group also employed UiO-66 as a precursor for HSiW@ZrO$_2$ hybrid synthesis, and SEM images of HSiW@UiO-66 at different calcination temperatures (300 °C, 400 °C, 500 °C) are shown in Figure 7. Nanoporous HSiW@ZrO$_2$ was obtained by calcinating at 300 °C, exhibiting relatively high surface area (338 m^2/g), appropriate pore size (2.5 nm), strong acidity (6.2 mmol/g), and the highest catalytic activity in the esterification of oleic acid; its conversion rate was high at 94.0% and stayed above 80% after nine catalytic cycles [62].

Dimethyl ether (DME) has gained attention for its application as a second-generation fuel, and it can be synthesized through a methanol dehydration process. Goda, et al. [63] used UiO-66 as a precursor for the synthesis of ZrOSO$_4$@C catalyst. In the experiment, it was observed that ZrOSO$_4$@C has weak and intermediate acidic sites and could be effectively applied for methanol dehydration to DME, with the highest conversion (100%) and selectivity (100%).

Figure 6. Reaction mechanism of transesterification on mesoporous sulfated zirconia nanosheets. (Adapted with permission from Ref. [60]. Copyright 2020, Royal Society of Chemistry.)

 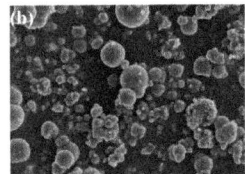 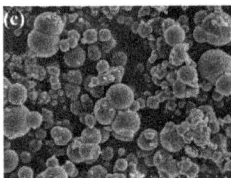 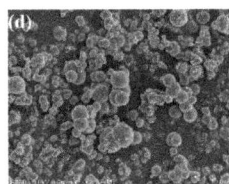

Figure 7. SEM images for (**a**) HSiW@UiO-66, (**b**) HSiW@ZrO$_2$-300, (**c**) HSiW@ZrO$_2$-400, and (**d**) HSiW@ZrO$_2$-500. (Adapted with permission from Ref. [62]. Copyright 2021, open access from Royal Society of Chemistry.)

Hong, et al. [64] prepared 3D porous Cu@ZrO$_x$ catalysts via in situ reconstruction of size-confined Cu@UiO-66 for methanol synthesis from CO$_2$ hydrogenation, and the optimized catalyst exhibited quite high methanol selectivity of 78.8% at 260 °C and 4.5 MPa, attributed to the many Cu$^+$−ZrO$_x$ interfaces present as active sites in the material framework. Zeng's group [65] also designed and prepared porous hydrous zirconia via UiO-66 pyrolysis as a support for Ni^{II} centers. In methane production from CO$_2$ hydrogenation, the resultant catalyst exhibited excellent activity and stability.

Based on the literature, Zr-based MOF-derived materials with stable porous structures and many active sites are expected to be widely used for the development of high-performance composite catalysts in the future.

3. Conclusions and Future Prospects

Herein, a comprehensive review was presented on the synthesis of Zr-based MOFs and their derived composite materials for their catalytic application in green biofuel synthesis in recent years. The current review attempted to thoroughly demonstrate the use of single Zr-based MOF catalysts, modified Zr-based MOF catalysts, and Zr-based MOF-derived catalysts in the literature. Their high surface area, adjustable pore structure, acceptable recyclability, and strong acid sites obtained by tuning structural defects make Zr-based MOFs suitable for esterification or transesterification process.

However, looking ahead, many challenges for the large-scale application of Zr-based MOFs still exist, such as the design and development of inexpensive Zr-based MOFs at an industrial scale with high yields. The self-assembly mechanism of Zr-based MOFs and their derived materials is still unclear, and further exploration via both experimental and theoretical approaches is still required. The chemical and thermal stability of Zr-based MOFs is still not adequate and needs to be further improved. Facilely tuning the defect density on nodes by exploring new modification approaches could bring beneficial changes to the catalytic performance. By combining Zr-based MOFs with appropriate active materials such as enzymes, graphene derivatives, and magnetic substances, composite materials could be synthesized to improve their catalytic performance. Controlling the structure, composition, and distribution of the active component of Zr-based MOF-derived catalysts, while aiming to maintain the original structure of the Zr-based MOFs, still needs to be further studied.

As a whole, the application of Zr-based MOFs and catalysts derived from them is important not only for green biodiesel synthesis but also for the conversion of biomass. Despite facing many challenges, hopefully, the existing issues will be resolved sooner or later, and the application prospects of biorefineries will also be very bright.

Author Contributions: Q.Z. and J.W. jointly conceived the manuscript and discussed the outline; Q.Z. wrote the manuscript; S.Z., J.M. and J.C. made preliminary revisions to the manuscript; Q.Z. and Y.Z. were in charge of project administration. All authors have read and agreed to the published version of the manuscript.

Funding: This work was supported by the National Natural Science Foundation of China (22262001), the Guizhou Science and Technology Foundation ((2020)1Y054, 20181401), the 2018 Thousand Level Innovative Talents Training Program of Guizhou Province, the Project of Anshun University supporting Doctors Research ((2021)asxybsjj01), the Creative Research Groups Support Program of Guizhou Education Department (KY (2017)049), and the Youth Growth S&T Personnel Foundation of the Guizhou Education Department (KY (2018)321).

Institutional Review Board Statement: Not applicable.

Informed Consent Statement: Not applicable.

Data Availability Statement: Not applicable.

Conflicts of Interest: The authors declare there is no conflict of interest regarding the publication of this paper.

References

1. Hoang, A.T.; Tabatabaei, M.; Aghbashlo, M.; Carlucci, A.P.; Ölçer, A.I.; Le, A.T.; Ghassemi, A. Rice bran oil-based biodiesel as a promising renewable fuel alternative to petrodiesel: A review. *Renew. Sustain. Energy Rev.* **2020**, *135*, 110204. [CrossRef]
2. Bekhradinassab, E.; Tavakoli, A.; Haghighi, M.; Shabani, M. Catalytic biofuel production over 3D macro-structured cheese-like Mn-promoted TiO_2 isotype: Mn-catalyzed microwave-combustion design. *Energy Convers. Manag.* **2021**, *251*, 114916. [CrossRef]
3. Rezania, S.; Mahdinia, S.; Oryani, B.; Cho, J.; E Kwon, E.; Bozorgian, A.; Nodeh, H.R.; Darajeh, N.; Mehranzamir, K. Biodiesel production from wild mustard (Sinapis Arvensis) seed oil using a novel heterogeneous catalyst of $LaTiO_3$ nanoparticles. *Fuel* **2022**, *307*, 121759. [CrossRef]
4. Zhang, Q.; Ling, D.; Lei, D.; Wang, J.; Liu, X.; Zhang, Y.; Ma, P. Green and Facile Synthesis of Metal-Organic Framework Cu-BTC-Supported Sn (II)-Substituted Keggin Heteropoly Composites as an Esterification Nanocatalyst for Biodiesel Production. *Front. Chem.* **2020**, *8*, 129. [CrossRef] [PubMed]
5. Cholapandian, K.; Gurunathan, B.; Rajendran, N. Investigation of CaO nanocatalyst synthesized from Acalypha indica leaves and its application in biodiesel production using waste cooking oil. *Fuel* **2022**, *312*, 122958. [CrossRef]
6. Rozina; Ahmad, M.; Elnaggar, A.Y.; Teong, L.K.; Sultana, S.; Zafar, M.; Munir, M.; Hussein, E.E.; Abidin, S.Z.U. Sustainable and eco-friendly synthesis of biodiesel from novel and non-edible seed oil of Monotheca buxifolia using green nano-catalyst of calcium oxide. *Energy Convers. Manag. X* **2022**, *13*, 100142. [CrossRef]
7. Pan, H.; Xia, Q.; Li, H.; Wang, Y.; Shen, Z.; Wang, Y.; Li, L.; Li, X.; Xu, H.; Zhou, Z.; et al. Direct production of biodiesel from crude *Euphorbia lathyris* L. Oil catalyzed by multifunctional mesoporous composite materials. *Fuel* **2021**, *309*, 122172. [CrossRef]

8. Pan, H.; Xia, Q.; Wang, Y.; Shen, Z.; Huang, H.; Ge, Z.; Li, X.; He, J.; Wang, X.; Li, L.; et al. Recent advances in biodiesel production using functional carbon materials as acid/base catalysts. *Fuel Process. Technol.* **2022**, *237*, 107421. [CrossRef]
9. Zhang, Q.; Luo, Q.; Wu, Y.; Yu, R.; Cheng, J.; Zhang, Y. Construction of a Keggin heteropolyacid/Ni-MOF catalyst for esterification of fatty acids. *RSC Adv.* **2021**, *11*, 33416–33424. [CrossRef] [PubMed]
10. Yan, W.; Zhang, D.; Sun, Y.; Zhou, Z.; Du, Y.; Du, Y.; Li, Y.; Liu, M.; Zhang, Y.; Shen, J.; et al. Structural sensitivity of heterogeneous catalysts for sustainable chemical synthesis of gluconic acid from glucose. *Chin. J. Catal.* **2020**, *41*, 1320–1336. [CrossRef]
11. Chen, B.; Yan, G.; Chen, G.; Feng, Y.; Zeng, X.; Sun, Y.; Tang, X.; Lei, T.; Lin, L. Recent progress in the development of advanced biofuel 5-ethoxymethylfurfural. *BMC Energy* **2020**, *2*, 2. [CrossRef]
12. Ji, J.; Bao, Y.; Liu, X.; Zhang, J.; Xing, M. Molybdenum-based heterogeneous catalysts for the control of environmental pollutants. *EcoMat* **2021**, *3*, e12155. [CrossRef]
13. Sahar, J.; Farooq, M.; Ramli, A.; Naeem, A.; Khattak, N.S. Biodiesel production from Mazari palm (*Nannorrhops ritchiana*) seeds oil using Tungstophosphoric acid decorated SnO_2@Mn-ZIF bifunctional heterogeneous catalyst. *Appl. Catal. A Gen.* **2022**, *643*, 118740. [CrossRef]
14. Fang, R.; Dhakshinamoorthy, A.; Li, Y.; Garcia, H. Metal organic frameworks for biomass conversion. *Chem. Soc. Rev.* **2020**, *49*, 3638–3687. [CrossRef]
15. Yadav, S.; Dixit, R.; Sharma, S.; Dutta, S.; Solanki, K.; Sharma, R.K. Magnetic metal–organic framework composites: Structurally advanced catalytic materials for organic transformations. *Mater. Adv.* **2021**, *2*, 2153–2187. [CrossRef]
16. Zhang, Q.; Zhang, Y.; Cheng, J.; Li, H.; Ma, P. An Overview of Metal-organic Frameworks-based Acid/Base Catalysts for Biofuel Synthesis. *Curr. Org. Chem.* **2020**, *24*, 1876–1891. [CrossRef]
17. Wei, Y.; Zhang, Y.; Li, B.; Guan, W.; Yan, C.; Li, X.; Yan, Y. Facile synthesis of metal-organic frameworks embedded in interconnected macroporous polymer as a dual acid-base bifunctional catalyst for efficient conversion of cellulose to 5-hydroxymethylfurfural. *Chin. J. Chem. Eng.* **2022**, *44*, 169–181. [CrossRef]
18. Li, Y.; Meng, X.; Luo, R.; Zhou, H.; Lu, S.; Yu, S.; Bai, P.; Guo, X.; Lyu, J. Aluminum/Tin-doped UiO-66 as Lewis acid catalysts for enhanced glucose isomerization to fructose. *Appl. Catal. A Gen.* **2022**, *632*, 118501. [CrossRef]
19. Ling, L.; Yang, W.; Yan, P.; Wang, M.; Jiang, H. Light-Assisted CO_2 Hydrogenation over Pd_3Cu@UiO-66 Promoted by Active Sites in Close Proximity. *Angew. Chem. Int. Ed.* **2022**, *61*, e202116396. [CrossRef] [PubMed]
20. Bai, Y.; Dou, Y.; Xie, L.-H.; Rutledge, W.; Li, J.-R.; Zhou, H.-C. Zr-based metal–organic frameworks: Design, synthesis, structure, and applications. *Chem. Soc. Rev.* **2016**, *45*, 2327–2367. [CrossRef] [PubMed]
21. Dhakshinamoorthy, A.; Santiago-Portillo, A.; Asiri, A.M.; Garcia, H. Engineering UiO-66 Metal Organic Framework for Heterogeneous Catalysis. *ChemCatChem* **2019**, *11*, 899–923. [CrossRef]
22. Cavka, J.H.; Jakobsen, S.; Olsbye, U.; Guillou, N.; Lamberti, C.; Bordiga, S.; Lillerud, K.P. A New Zirconium Inorganic Building Brick Forming Metal Organic Frameworks with Exceptional Stability. *J. Am. Chem. Soc.* **2008**, *130*, 13850–13851. [CrossRef] [PubMed]
23. Schelling, M.; Kim, M.; Otal, E.; Hinestroza, J. Decoration of Cotton Fibers with a Water-Stable Metal-Organic Framework (UiO-66) for the Decomposition and Enhanced Adsorption of Micropollutants in Water. *Bioengineering* **2018**, *5*, 14. [CrossRef]
24. Cheng, J.; Qian, L.; Guo, H.; Mao, Y.; Shao, Y.; Yang, W. A new aminobenzoate-substituted s-triazin-based Zr metal organic frameworks as efficient catalyst for biodiesel production from microalgal lipids. *Fuel Process. Technol.* **2022**, *238*, 107487. [CrossRef]
25. Zhou, F.; Lu, N.; Fan, B.; Wang, H.; Li, R. Zirconium-containing UiO-66 as an efficient and reusable catalyst for transesterification of triglyceride with methanol. *J. Energy Chem.* **2016**, *25*, 874–879. [CrossRef]
26. Caratelli, C.; Hajek, J.; Cirujano, F.G.; Waroquier, M.; i Xamena, F.X.L.; Van Speybroeck, V. Nature of active sites on UiO-66 and beneficial influence of water in the catalysis of Fischer esterification. *J. Catal.* **2017**, *352*, 401–414. [CrossRef]
27. Jrad, A.; Abu Tarboush, B.J.; Hmadeh, M.; Ahmad, M. Tuning acidity in zirconium-based metal organic frameworks catalysts for enhanced production of butyl butyrate. *Appl. Catal. A Gen.* **2019**, *570*, 31–41. [CrossRef]
28. Wei, R.; Fan, J.; Qu, X.; Gao, L.; Wu, Y.; Zhang, Z.; Hu, F.; Xiao, G. Tuning the Catalytic Activity of UiO-66 via Modulated Synthesis: Esterification of Levulinic Acid as a Test Reaction. *Eur. J. Inorg. Chem.* **2020**, *2020*, 833–840. [CrossRef]
29. Chaemchuen, S.; Heynderickx, P.M.; Verpoort, F. Kinetic modeling of oleic acid esterification with UiO-66: From intrinsic experimental data to kinetics via elementary reaction steps. *Chem. Eng. J.* **2020**, *394*, 124816. [CrossRef]
30. Desidery, L.; Chaemcheun, S.; Yusubov, M.; Verpoort, F. Di-methyl carbonate transesterification with EtOH over MOFs: Basicity and synergic effect of basic and acid active sites. *Catal. Commun.* **2018**, *104*, 82–85. [CrossRef]
31. Shaik, M.R.; Adil, S.F.; Alothman, Z.A.; Alduhaish, O.M. Fumarate Based Metal-Organic Framework: An Effective Catalyst for the Transesterification of Used Vegetable Oil. *Crystals* **2022**, *12*, 151. [CrossRef]
32. de la Flor, D.; López-Aguado, C.; Paniagua, M.; Morales, G.; Mariscal, R.; Melero, J.A. Defective UiO-66(Zr) as an efficient catalyst for the synthesis of bio jet-fuel precursors via aldol condensation of furfural and MIBK. *J. Catal.* **2021**, *401*, 27–39. [CrossRef]
33. Rapeyko, A.; Rodenas, M.; i Xamena, F.X.L. Zr-Containing UiO-66 Metal-Organic Frameworks as Highly Selective Heterogeneous Acid Catalysts for the Direct Ketalization of Levulinic Acid. *Adv. Sustain. Syst.* **2022**, *6*, 2100451. [CrossRef]
34. Cirujano, F.; Corma, A.; i Xamena, F.L. Zirconium-containing metal organic frameworks as solid acid catalysts for the esterification of free fatty acids: Synthesis of biodiesel and other compounds of interest. *Catal. Today* **2015**, *257*, 213–220. [CrossRef]
35. Cirujano, F.; Corma, A.; i Xamena, F.L. Conversion of levulinic acid into chemicals: Synthesis of biomass derived levulinate esters over Zr-containing MOFs. *Chem. Eng. Sci.* **2015**, *124*, 52–60. [CrossRef]

36. Wang, F.; Chen, Z.; Chen, H.; Goetjen, T.A.; Li, P.; Wang, X.; Alayoglu, S.; Ma, K.; Chen, Y.; Wang, T.; et al. Interplay of Lewis and Brønsted Acid Sites in Zr-Based Metal-Organic Frameworks for Efficient Esterification of Biomass-Derived Levulinic Acid. *ACS Appl. Mater. Interfaces* **2019**, *11*, 32090–32096. [CrossRef]
37. Abou-Elyazed, A.S.; Ye, G.; Sun, Y.; El-Nahas, A.M. A Series of UiO-66(Zr)-Structured Materials with Defects as Heterogeneous Catalysts for Biodiesel Production. *Ind. Eng. Chem. Res.* **2019**, *58*, 21961–21971. [CrossRef]
38. Abou-Elyazed, A.S.; Sun, Y.; El-Nahas, A.M.; Yousif, A.M. A green approach for enhancing the hydrophobicity of UiO-66(Zr) catalysts for biodiesel production at 298 K. *RSC Adv.* **2020**, *10*, 41283–41295. [CrossRef]
39. Desidery, L.; Yusubov, M.S.; Zhuiykov, S.; Verpoort, F. Fully-sulfonated hydrated UiO66 as efficient catalyst for ethyl levulinate production by esterification. *Catal. Commun.* **2018**, *117*, 33–37. [CrossRef]
40. Badoei-Dalfard, A.; Shahba, A.; Zaare, F.; Sargazi, G.; Seyedalipour, B.; Karami, Z. Lipase immobilization on a novel class of Zr-MOF/electrospun nanofibrous polymers: Biochemical characterization and efficient biodiesel production. *Int. J. Biol. Macromol.* **2021**, *192*, 1292–1303. [CrossRef]
41. Li, H.; Han, Z.; Liu, F.; Li, G.; Guo, M.; Cui, P.; Zhou, S.; Yu, M. Esterification catalyzed by an efficient solid acid synthesized from PTSA and UiO-66(Zr) for biodiesel production. *Faraday Discuss.* **2021**, *231*, 342–355. [CrossRef] [PubMed]
42. Xu, Z.; Zhao, G.; Ullah, L.; Wang, M.; Wang, A.; Zhang, Y.; Zhang, S. Acidic ionic liquid based UiO-67 type MOFs: A stable and efficient heterogeneous catalyst for esterification. *RSC Adv.* **2018**, *8*, 10009–10016. [CrossRef] [PubMed]
43. Peng, W.-L.; Mi, J.; Liu, F.; Xiao, Y.; Chen, W.; Liu, Z.; Yi, X.; Liu, W.; Zheng, A. Accelerating Biodiesel Catalytic Production by Confined Activation of Methanol over High-Concentration Ionic Liquid-Grafted UiO-66 Solid Superacids. *ACS Catal.* **2020**, *10*, 11848–11856. [CrossRef]
44. Dai, Q.; Yang, Z.; Li, J.; Cao, Y.; Tang, H.; Wei, X. Zirconium-based MOFs-loaded ionic liquid-catalyzed preparation of biodiesel from Jatropha oil. *Renew. Energy* **2021**, *163*, 1588–1594. [CrossRef]
45. Lu, P.; Li, H.; Li, M.; Chen, J.; Ye, C.; Wang, H.; Qiu, T. Ionic liquid grafted NH_2-UiO-66 as heterogeneous solid acid catalyst for biodiesel production. *Fuel* **2022**, *324*, 124537. [CrossRef]
46. Abou-Elyazed, A.S.; Sun, Y.; El-Nahas, A.; Abdel-Azeim, S.; Sharara, T.; Yousif, A. Solvent-free synthesis and characterization of Ca^{2+}-doped UiO-66(Zr) as heterogeneous catalyst for esterification of oleic acid with methanol: A joint experimental and computational study. *Mater. Today Sustain.* **2022**, *18*, 100110. [CrossRef]
47. Zhu, J.; Wang, Z.; Song, X.; Zhao, B.; Li, Y.; Wang, Y. Encapsulating Keggin-$H_3PW_{12}O_{40}$ into UiO-66(Zr) for manufacturing the biodiesel. *Micro Nano Lett.* **2021**, *16*, 90–96. [CrossRef]
48. Ma, T.; Liu, D.; Liu, Z.; Xu, J.; Dong, Y.; Chen, G.; Yun, Z. 12-Tungstophosphoric acid-encapsulated metal-organic framework UiO-66: A promising catalyst for the esterification of acetic acid with n-butanol. *J. Taiwan Inst. Chem. Eng.* **2022**, *133*, 104277. [CrossRef]
49. Xie, W.; Hu, P.; Yang, X. $Cs_{2.5}H_{0.5}PW_{12}O_{40}$ Encapsulated in Metal-Organic Framework UiO-66 as Heterogeneous Catalysts for Acidolysis of Soybean Oil. *Catal. Lett.* **2017**, *147*, 2772–2782. [CrossRef]
50. Xie, W.; Wan, F. Immobilization of polyoxometalate-based sulfonated ionic liquids on UiO-66-2COOH metal-organic frameworks for biodiesel production via one-pot transesterification-esterification of acidic vegetable oils. *Chem. Eng. J.* **2019**, *365*, 40–50. [CrossRef]
51. Tan, X.; Zhang, H.; Li, H.; Yang, S. Electrovalent bifunctional acid enables heterogeneously catalytic production of biodiesel by (trans)esterification of non-edible oils. *Fuel* **2022**, *310*, 122273. [CrossRef]
52. Li, Y.; Zhang, S.; Li, Z.; Zhang, H.; Li, H.; Yang, S. Green synthesis of heterogeneous polymeric bio-based acid decorated with hydrophobic regulator for efficient catalytic production of biodiesel at low temperatures. *Fuel* **2022**, *329*, 125467. [CrossRef]
53. Zhang, Y.; Song, X.; Li, S.; Zhao, B.; Tong, L.; Wang, Y.; Li, Y. Two-step preparation of Keggin-PW_{12}@UIO-66 composite showing high-activity and long-life conversion of soybean oil into biodiesel. *RSC Adv.* **2021**, *11*, 38016–38025. [CrossRef] [PubMed]
54. Zhang, Q.; Yang, T.; Liu, X.; Yue, C.; Ao, L.; Deng, T.; Zhang, Y. Heteropoly acid-encapsulated metal–organic framework as a stable and highly efficient nanocatalyst for esterification reaction. *RSC Adv.* **2019**, *9*, 16357–16365. [CrossRef] [PubMed]
55. Zhang, Q.; Lei, D.; Luo, Q.; Wang, J.; Deng, T.; Zhang, Y.; Ma, P. Efficient biodiesel production from oleic acid using metal–organic framework encapsulated Zr-doped polyoxometalate nano-hybrids. *RSC Adv.* **2020**, *10*, 8766–8772. [CrossRef]
56. Zhang, Q.; Yang, T.; Lei, D.; Wang, J.; Zhang, Y. Efficient Production of Biodiesel from Esterification of Lauric Acid Catalyzed by Ammonium and Silver Co-Doped Phosphotungstic Acid Embedded in a Zirconium Metal-Organic Framework Nanocomposite. *ACS Omega* **2020**, *5*, 12760–12767. [CrossRef]
57. Zhang, Q.; Yang, B.; Tian, Y.; Yang, X.; Yu, R.; Wang, J.; Deng, T.; Zhang, Y. Fabrication of silicotungstic acid immobilized on Ce-based MOF and embedded in Zr-based MOF matrix for green fatty acid esterification. *Green Process. Synth.* **2022**, *11*, 184–194. [CrossRef]
58. Liu, B.; Shioyama, H.; Akita, T.; Xu, Q. Metal-Organic Framework as a Template for Porous Carbon Synthesis. *J. Am. Chem. Soc.* **2008**, *130*, 5390–5391. [CrossRef]
59. Liu, B.; Zhang, X.; Shioyama, H.; Mukai, T.; Sakai, T.; Xu, Q. Converting cobalt oxide subunits in cobalt metal-organic framework into agglomerated Co_3O_4 nanoparticles as an electrode material for lithium ion battery. *J. Power Sources* **2010**, *195*, 857–861. [CrossRef]
60. Lu, N.; Zhang, X.; Yan, X.; Pan, D.; Fan, B.; Li, R. Synthesis of novel mesoporous sulfated zirconia nanosheets derived from Zr-based metal–organic frameworks. *CrystEngComm* **2020**, *22*, 44–51. [CrossRef]

61. Li, H.; Wang, Y.; Ma, X.; Guo, M.; Li, Y.; Li, G.; Cui, P.; Zhou, S.; Yu, M. Synthesis of CaO/ZrO_2 based catalyst by using UiO–66(Zr) and calcium acetate for biodiesel production. *Renew. Energy* **2022**, *185*, 970–977. [CrossRef]
62. Zhang, Q.; Lei, D.; Luo, Q.; Yang, X.; Wu, Y.; Wang, J.; Zhang, Y. MOF-derived zirconia-supported Keggin heteropoly acid nanoporous hybrids as a reusable catalyst for methyl oleate production. *RSC Adv.* **2021**, *11*, 8117–8123. [CrossRef] [PubMed]
63. Goda, M.N.; Abdelhamid, H.N.; Said, A.E.-A.A. Zirconium Oxide Sulfate-Carbon ($ZrOSO_4$@C) Derived from Carbonized UiO-66 for Selective Production of Dimethyl Ether. *ACS Appl. Mater. Interfaces* **2019**, *12*, 646–653. [CrossRef]
64. Liu, T.; Hong, X.; Liu, G. In Situ Generation of the Cu@3D-ZrO_x Framework Catalyst for Selective Methanol Synthesis from CO_2/H_2. *ACS Catal.* **2019**, *10*, 93–102. [CrossRef]
65. Zeng, L.; Wang, Y.; Li, Z.; Song, Y.; Zhang, J.; Wang, J.; He, X.; Wang, C.; Lin, W. Highly Dispersed Ni Catalyst on Metal-Organic Framework-Derived Porous Hydrous Zirconia for CO_2 Methanation. *ACS Appl. Mater. Interfaces* **2020**, *12*, 17436–17442. [CrossRef] [PubMed]

Review

Application of Tubular Reactor Technologies for the Acceleration of Biodiesel Production

Omojola Awogbemi * and Daramy Vandi Von Kallon

Department of Mechanical and Industrial Engineering Technology, University of Johannesburg, P.O. Box 524, Johannesburg 2006, South Africa; dkallon@uj.ac.za
* Correspondence: jolawogbemi2015@gmail.com

Abstract: The need to arrest the continued environmental contamination and degradation associated with the consumption of fossil-based fuels has continued to serve as an impetus for the increased utilization of renewable fuels. The demand for biodiesel has continued to escalate in the past few decades due to urbanization, industrialization, and stringent government policies in favor of renewable fuels for diverse applications. One of the strategies for ensuring the intensification, commercialization, and increased utilization of biodiesel is the adaptation of reactor technologies, especially tubular reactors. The current study reviewed the deployment of different types and configurations of tubular reactors for the acceleration of biodiesel production. The feedstocks, catalysts, conversion techniques, and modes of biodiesel conversion by reactor technologies are highlighted. The peculiarities, applications, merits, drawbacks, and instances of biodiesel synthesis through a packed bed, fluidized bed, trickle bed, oscillatory flow, and micro-channel tubular reactor technologies are discussed to facilitate a better comprehension of the mechanisms behind the technology. Indeed, the deployment of the transesterification technique in tubular reactor technologies will ensure the ecofriendly, low-cost, and large-scale production of biodiesel, a high product yield, and will generate high-quality biodiesel. The outcome of this study will enrich scholarship and stimulate a renewed interest in the application of tubular reactors for large-scale biodiesel production among biodiesel refiners and other stakeholders. Going forward, the use of innovative technologies such as robotics, machine learning, smart metering, artificial intelligent, and other modeling tools should be deployed to monitor reactor technologies for biodiesel production.

Keywords: tubular reactor; biodiesel; catalyst; transesterification; feedstock; reactor technologies

Citation: Awogbemi, O.; Kallon, D.V.V. Application of Tubular Reactor Technologies for the Acceleration of Biodiesel Production. *Bioengineering* 2022, 9, 347. https://doi.org/10.3390/bioengineering9080347

Academic Editor: Giorgos Markou

Received: 10 June 2022
Accepted: 23 July 2022
Published: 27 July 2022

Publisher's Note: MDPI stays neutral with regard to jurisdictional claims in published maps and institutional affiliations.

Copyright: © 2022 by the authors. Licensee MDPI, Basel, Switzerland. This article is an open access article distributed under the terms and conditions of the Creative Commons Attribution (CC BY) license (https://creativecommons.org/licenses/by/4.0/).

1. Introduction

Rapid population growth, industrialization, urbanization, economic growth, social development, and technological advancement have continued to increase energy consumption worldwide. Credible available statistics show that the global primary energy consumption was 109,583 terawatt hours (TWh) in 2000, became 140,562 TWh in 2010, and rose to 162,194 TWh in 2019 [1]. The International Association for Energy Economics has predicted that primary energy consumption will grow by over 39% over the next 20 years [2] while the International Energy Agency predicts more than a 50% increment in global primary energy consumption by 2050 [3]. The continuous energy demand is premeditated on the importance of the availability of a sustainable, reliable, affordable, and accessible energy supply to the social, economic, and industrial development of any country. However, most of the world's energy is sourced from fossil fuels. REN21, an international renewable energy policy network, reports that about 80% of the global energy mix still comes from fossil fuels [4]. Consumption of fossil-based (FB) fuels in the transportation, industrial, agricultural, commercial, household, and power generation sectors has continued to increase despite the efforts to stem the tide by various governments, organizations, and interests. The emission of carbon dioxide (CO_2) from energy-related applications is predicted to continue to increase globally. Though the CO_2 emissions from the Organization for Economic Cooperation and Development (OECD) countries have remained flat, that of the non-OECD nations are predicted to continue to rise, as shown in Figure 1 [5]. This trend is not likely to change in the foreseeable future as the share of FB fuels in the transportation sector has been predicted to be no less than 88% in 2040 [5]. This has led to the release of toxic emissions and other serious environmental degradation concerns and exacerbated global warming and climate change.

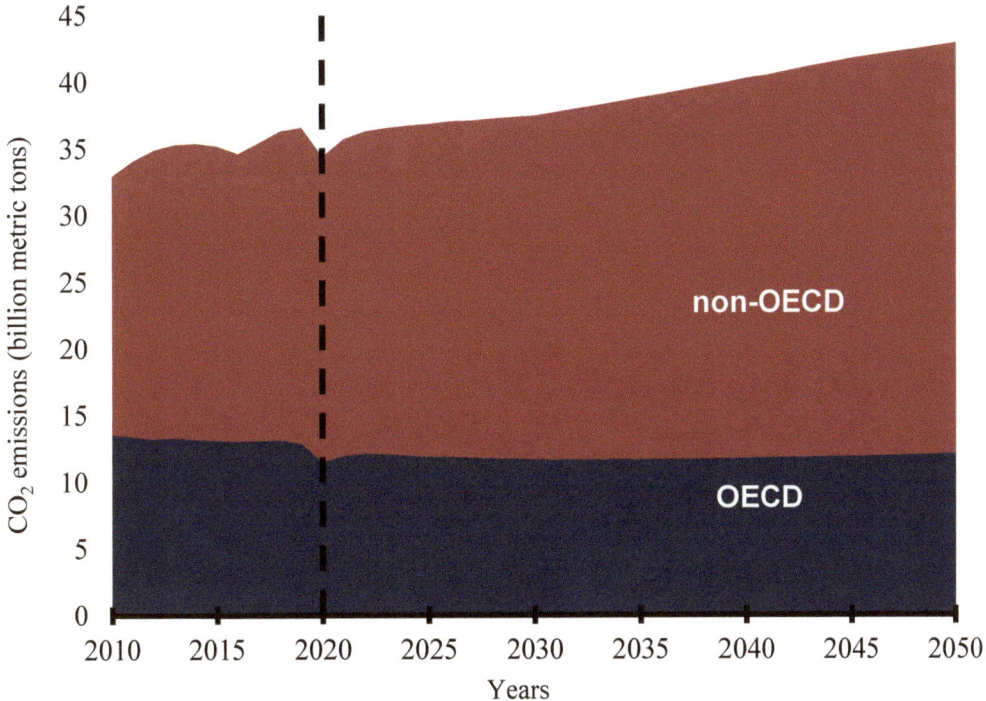

Figure 1. Energy-related CO_2 emissions (billion metric tons).

To stem this tide, researchers have shown more interest and committed more resources in terms of investments to finding sustainable and environmentally friendly alternatives to the dangerous FB fuels. Primarily, attention has been placed on finding renewable energy sources such as solar, hydroelectric, hydrogen, wind, biodiesel, green diesel, bioethanol, biofuels, and biomass [6]. Biodiesel is seen as one of the sustainable and affordable replacements for FB fuels for various applications. Biodiesel, also known as fatty acid methyl ester, is a renewable, cost-effective, and biodegradable liquid fuel synthesized from vegetable oils, recovered restaurant oil, animal fat, tallow, non-edible plant oil, waste cooking oil, microalgal plants, and other triglycerides-bearing feedstocks [7]. The use of biodiesel is to remedy the uninspiring performance of FB fuels in internal combustion engines, emission of toxic gases, and the impacts on the environment and humanity in general. It is believed that the application of biodiesel, particularly in internal combustion engines will improve engine performance, reduce tailpipe emissions, and mitigate the unpleasant effects on the environment and global health [8]. Though there are a few drawbacks related to the application of biodiesel, the fact that the benefits of biodiesel far outweigh these shortcomings makes biodiesel a sustainable alternative to FB diesel fuel (Table 1).

Table 1. Benefits and drawbacks of biodiesel [9–11].

Criteria	Benefits	Drawbacks
Renewability	Renewable and biodegradable	
Safety	Safe and non-toxic	
Environment	Ecofriendly / Environmentally sustainable	
Storage	Safer to handle, store, and transport	Can deteriorate in storage
	Compatible with FB fuel storage facilities	
Properties	High energy content	High viscosity
	Low sulfur content	High pour point
	High cetane number	
	High flash point	
Performance	Performs better than FB diesel fuel	High fuel consumption
	Contributes to power generation	Clogging of fuel filter and fuel lines
	Better thermal efficiency	
	Lower noise level	
Emission	Emits less carbon and other GHGs	Emits more NOx
	High combustion efficiency in ICEs	
	Lower smoke generation	
Combustion	Improved combustion in ICEs	Low cylinder pressure
	Better combustion speed	Reduction in heat release
Feedstocks	Readily available and low-cost feedstock	Some of the feedstocks conflict with food supply
	Synthesized from renewable feedstocks	Some feedstocks need to be cultivated
	Conversion of wastes to fuel	
Economy	Reduces fuel importation and saves foreign exchange	
	Contributes to economic growth and environmental sustainability	
	Employment generation along the value chain	
Application	Can be used without engine modifications	Unsuitable for cold temperature regions
	Contributes to power generation	Can harm rubber hoses in engines
Production	Can be produced locally by households	Unpredictable standards

There has been increased research and investment in the production of biodiesel over the past few years to be able to meet up the global demand. Indonesia, the United States, and Brazil are the three leading biodiesel producers in 2019 with 7.9 billion liters, 6.5 billion liters, and 5.9 billion liters, respectively (Figure 2) [12]. Global biodiesel production increased from 38 billion liters in 2018 to 48.3 billion liters in 2019 and 60.7 billion liters in 2022. The annual growth rate of biodiesel production increased from 4% in 2017 to 9% in 2021. The negative growth rate recorded in 2020 was due to the impact of the dreaded

COVID-19 pandemic that restricted movement and slowed down economic activities in most countries (Figure 3) [13]. Concerted efforts including research and development, infrastructure, policy framework, incentives, and investments have been put in place to ensure the increased production of biodiesel for diverse applications.

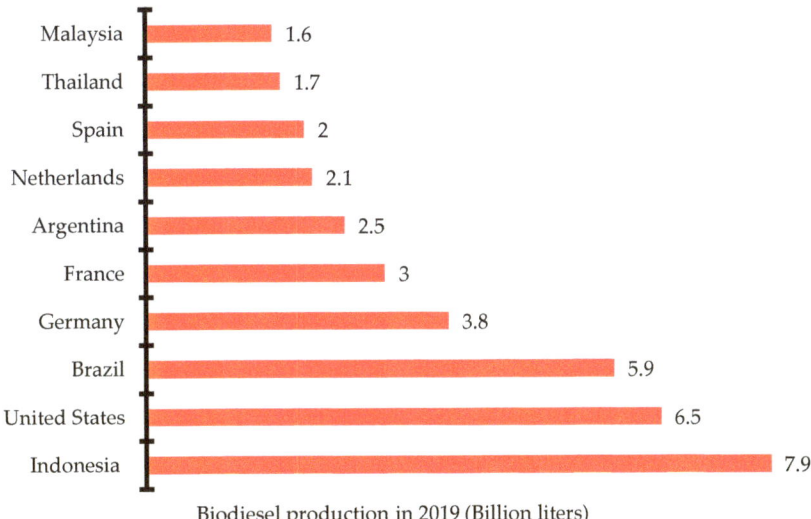

Figure 2. Ten leading biodiesel producers in 2019.

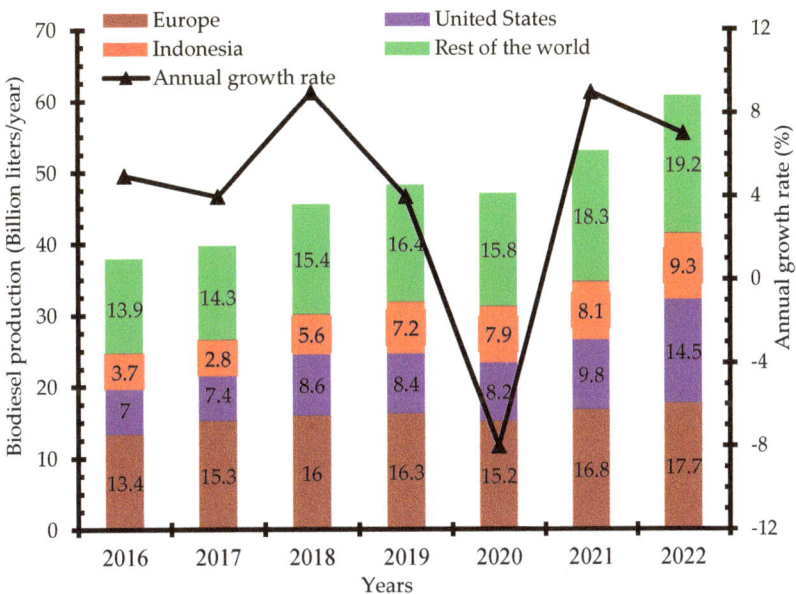

Figure 3. Global biodiesel production 2016–2022 (billion liters/year).

Motivation, Aim, and Objectives

Investigations have been conducted on the various technologies for the synthesis of biodiesel in commercial quantities. Chuah et al. [14], Kant Bhatia et al. [15], and Bashir et al. [16] studied the various advanced technologies for low-cost and high conversion yield for biodiesel production. They reported that the intensification of production techniques such as microwave, ultrasonic radiation, cavitation, plasma discharge, transesterification, pyrolysis, supercritical, and emulsification was ecofriendly and resulted in a higher product yield and lower wastes during biodiesel production. Roick et al. [17] investigated the various thermochemical, biomechanical, chemical, and other novel technologies for the commercial production of biodiesel and reported that these technologies were viable from the economic and ecological points of view. The works of Mohiddin et al. [18], Okolie et al. [19], and Lv et al. [20] investigated the impact of feedstocks, catalysts, production methods, and production infrastructure on the commercialization of biodiesel. They reported that to achieve the return on investment in biodiesel production, the choice of feedstock, catalysts, and selection of converting infrastructure plays an important role. The choice of reactor technology and other processing parameters affects the conversion rate, cost of production, and yield of biodiesel irrespective of the types of feedstock and catalyst [21].

Despite these reported cases, there are ongoing studies on how to meet the continuous demand for biodiesel for diverse applications. There is near unanimity of opinions on the need to interrogate and carry out further investigations on the mechanisms for achieving the acceleration of biodiesel production through the deployment of reactor technologies. In addition, there is an urgent need to extend the frontiers of available knowledge on the infrastructure for accelerated biodiesel generation to meet the ever-growing demand. These form the motivations for the current study. The aim of this study, therefore, is to interrogate the application of tubular reactor technologies available for the production of biodiesel with a view to further escalate further research and utilization of tubular reactors for accelerated biodiesel production. Consequently, this study presents a brief overview of biodiesel feedstocks and production techniques, modes of biodiesel production using reactor technologies, and the varieties of tubular reactor technologies for the production of biodiesel. Instances of the application of tubular reactors for biodiesel production are highlighted. The outcomes of this research will enrich scholarship by providing necessary information on the types, operations, and peculiarities of various tubular reactors for biodiesel production. Refiners, researchers, and consumers of biodiesel will be better informed on the strategies and infrastructure needed for effective, economically friendly, and environmentally sustainable biodiesel production.

2. Feedstocks and Biodiesel Conversion Techniques

2.1. Feedstock for Biodiesel Production

Over the years, various techniques have been adopted for the generation of biodiesel from diverse feedstocks. Depending on the choice of feedstocks, catalysts, and costs, refiners choose the method of converting various feedstocks to biodiesel. Biodiesel is classified into generations depending on the types of feedstocks used in synthesizing them. For example, first generation biodiesel is produced from edible oils such as palm oil, olive oil, coconut oil, etc., while second generation biodiesel is generated from rubber sed oil, castor oil, jojomba oil, karanja oil, and other non-edible oils. Waste cooking oil, animal fats, recovered fats, and chicken fats are converted into third generation biodiesel. Fourth generation biodiesel is produced from algal biomass, waste cooking oil, and genetically modified biomass.

The use of edible oils for biodiesel reduces the amount of food available for human consumption, leads to high global food prices, and exacerbates food scarcity. The deployment of non-edible oils, animal fats, and waste cooking oils eliminates food vs. fuel debates and reduces the reliance on edible food crops for the production of fuel. However, non-edible oils need a large expanse of arable land, water, and time for cultivation. These not only compete with land for growing food for human consumption but also contribute to deforestation and erosion. The use of waste cooking oil, animal fats, and recovered fats allows for the conversion of wastes to fuel, does not conflict with the food chain, and serves as a sustainable means of waste disposal [6,22]. Moreover, the use of microalgae as a feedstock for biodiesel production does not conflict with the food chain, requires no fertile land and water, and has the potential to yield about 15–300 times more than other non-edible oils [23]. Fourth generation biodiesel includes photobiological solar fuels and electro-fuels. It is a novel research area where solar energy is used to convert some feedstocks into biodiesel. Such feedstocks are renewable, broadly available, inexpensive, and ecofriendly. However, the technology is not yet fully developed and requires high financial investment [24]. Table 2 shows the feedstocks, advantages, and disadvantages of the four generations of biodiesel.

Table 2. Examples of feedstocks for the four generations of biodiesel [6,24,25].

Generations of Biodiesel	Feedstocks		Advantages	Disadvantages
	Types	Examples		
First	Edible oils	Coconut oil, Palm oil, Corn oil, Olive oil, Mustard oil, Sunflower, Rice bran, Rapeseed oil, Hazelnut oil	• Readily available • Simple conversion process • Safe handling and transportation • Easily adaptable to existing infrastructure • Easy to mix with FB diesel fuel	• Affect food security • Initiate food vs. fuel debate • Rising food costs • Cultivation of feedstocks requires land and time • Shortage of arable land for cultivation
Second	Nonedible oils	Rubber seed oil, Sapindus oil, Mukorossi oil, Thevettia peruviana oil, Jatropha curcus, Jojoba oil, Karanja oil, Neem oil, Mahua indica oil	• Do not affect the food supply • Cheap feedstock • Seed, grains, and residues are used as feedstock • Low conversion cost • Readily available • Generation of other products • Ecofriendly	• Large expanse of land and water needed to grow feedstock • Underdeveloped conversion technologies • Complicated production processes • Induce soil degradation, erosion, deforestation, and bush burning
Third	Waste oils algae	Animal tallow, Chicken fat, Poultry fat, Recovered fat, Fish oil, Waste cooking oil	• Do not require land • Do not affect food security • Cheap feedstock • Contribute to sanitation • Avenues for waste to fuel • Algae useful for water purification • Feedstocks can be engineered	• Costly production process • High energy consumption • Expensive oil extraction process • Commercial production is not sustainable • Underdeveloped technologies

Table 2. Cont.

Generations of Biodiesel	Feedstocks		Advantages	Disadvantages
	Types	Examples		
Fourth	Solar biodiesel Algae	Microalgae Synthetic cell Electronbiofuel Waste cooking oil	• Low carbon emission • Energy security • Increased carbon entrapment ability • High oil contents • Better cultivation, extraction, and production process	• High initial investment • More efforts are needed in R&D • High energy requirement • Research at infancy stage

2.2. Biodiesel Production Techniques

Biodiesel can be synthesized through two different production techniques: the physical technique and the chemical technique. The physical technique is a method of biodiesel production that does not involve any chemical reaction. This includes dilution and microemulsion. During dilution, a given volume of FB diesel fuel and other selected additives are added to natural oils to improve their viscosity and volatility. In chemical techniques, biodiesel production is achieved through the chemical modification of natural oils and fats. During the process, the physicochemical properties, and hence the behavior of the natural oils are altered. Notable examples include the pyrolysis, superfluid/ supercritical, and transesterification processes.

2.2.1. Direct Use and Dilution

During dilution, more solvent is added to the solute to reduce the concentration of the solute in the solution. In physical biodiesel production, ethanol and FB diesel fuel act as the solvents for the dilution of vegetable oil. The process generates a fuel with a lower density and viscosity than vegetable oil. For example, the addition of ethanol to FB diesel fuel produces a fuel with a better combustion efficiency, an improved brake power, and brake thermal efficiency. However, the brake torque and brake specific fuel consumption of the resulting fuel is reduced. Though dilution is an easy and non-technical process, the resulting fuel suffers from incomplete combustion and more carbon deposition in the engine cylinder. Moreover, products of dilution suffer from low volatility, poor atomization, and plugging of injector nozzles [21].

2.2.2. Micro Emulsion

Among the characteristics of vegetable oil which make it unsuitable as fuel for CI engine is its viscosity. The viscosity impediment of vegetable oil can be corrected by the microemulsion process. During the process, a co-solvent, alcohol, cetane improver, and surfactant are added to the vegetable oil to improve the viscosity and low liquidity. When butanol, hexanol, octanol, and methanol are added to vegetable oil or animal fats, the resultant fuel meets the optimum viscosity requirement for CI engine fuel. For example, when soybean oil was mixed with 2-octanol, methanol, cetane improver, and some surfactants (such as rhamnolipid), a clear, thermodynamically stable, and ecofriendly microemulsion biodiesel was produced. Biodiesel generated by microemulsion exhibits better cold flow properties, an enhanced stability and solubility, a lower activation energy, acceptable viscosity, and a shorter ignition delay. Moreover, the micro-emulsion of vegetable oils ensures a reduction in viscosity, a better cetane number, and improved spray characteristics in the CI engine. However, fuel synthesized by the micro-emulsification of vegetable oils and animal fats demonstrates incomplete combustion, a high deposit of carbon residue, and thickening of lubrication oils [21,24,26,27].

2.2.3. Pyrolysis

Pyrolysis is a chemical method of biodiesel production during which there is the thermal decomposition of materials at elevated temperatures in the absence of air and oxygen but in an inert atmosphere. Pyrolysis can also be achieved at high temperatures (usually 400 °C–1000 °C) in the presence of a catalyst which leads to the bond formation and the coming together of small molecules. The products of pyrolyzed vegetable oils, animal fats, natural fatty acids, and methyl esters of fatty acids possess physicochemical properties, fingerprints, and characteristics similar to FB diesel fuel. Products of pyrolysis from vegetable oils, animal fats, and other feedstocks demonstrate lower viscosities, flash points, pour points, and cetane numbers when compared with FB diesel fuel. Such pyrolyzate contains satisfactory sulfur, water, and sediment contents. However, their ash content, carbon residue, and pour points are unacceptable. In addition, the high cost of infrastructure for thermal cracking, high energy cost, the use of high temperature, and problems of environmental degradation are some of the drawbacks of the process [21,26,28].

2.2.4. Transesterification

Transesterification, also known as alcoholysis, is arguably the most commonly used chemical method for converting vegetable oil, natural fats, and recovered fats into biodiesel. During the process, three moles of alcohol (methanol or ethanol) stoichiometrically react with one mole of triglyceride in the presence of a catalyst to produce mono-alkyl ester and glycerol. The three steps involved, and the general equation are depicted in Figure 4. The process occurs under moderate operating conditions of about 60–80 °C, ambient pressure, and for 90 min. Other process parameters that affect the transesterification reaction include types of catalyst, a dose of catalyst, catalyst particle size, alcohol/oil molar ratio, residence time, reaction temperature, mixing/agitation speed, choice of alcohol, and composition of oil [29].

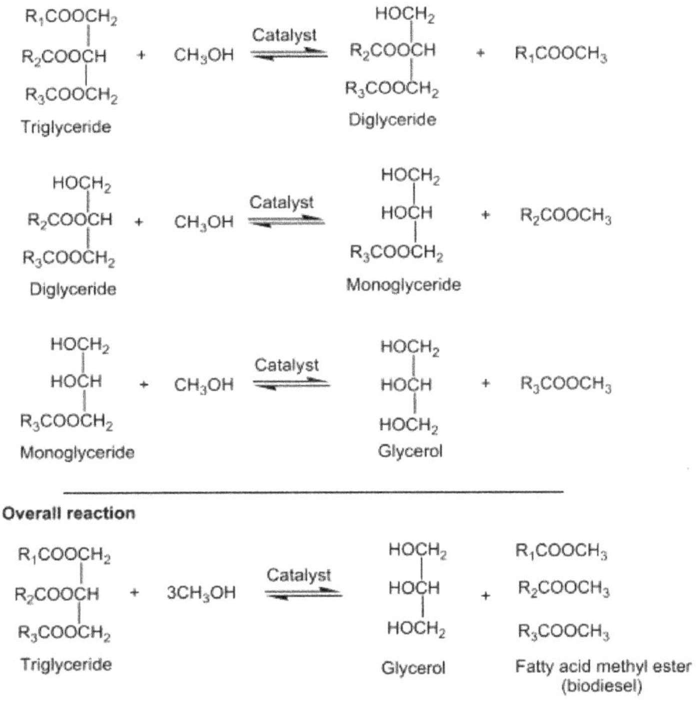

Figure 4. Transesterification reaction equation.

Methanol is the more popular alcohol for transesterification owing to its higher reactivity, cheaper cost, and lower operating temperature. If methanol is used as an alcohol, the process is called methanolysis and the product is fatty acid methyl ester (FAME). If ethanol is used as an alcohol, the process is known as ethanolysis and the product is fatty acid ethyl ester (FAEE). FAEEs are less toxic and have a better cetane number, higher oxidative stability, cloud point, pour point, lubricity properties, lower iodine value, and a higher heat capacity when compared with FAMEs. However, ethanolysis is reputed for its higher cost, energy consumption, lower transesterification reactivity, higher viscosity, formation of an azeotrope with water, and formation of more stable emulsions than methanolysis. Moreover, from the ecological point of view, FAEEs emit less exhaust gas and possess a higher biodegradability in water [30,31].

In the catalytic transesterification process, the choice of catalyst greatly affects the conversion efficiency and product yield. Generally, transesterification reactions can be catalyzed by homogeneous, heterogeneous, bio-based (enzymes) catalysts, or nanocatalysts [32]. The transesterification process can be heterogeneous when the catalyst is in a different phase from the reactants and products. In this case, solid catalysts are used. However, when liquid catalysts are used, the reactants and the products are in the same phase and the process is termed homogeneous. Biobased catalysts can be either in liquid or solid phases. Table 3 compares the examples, pros, and cons of the four major classes of catalysts for the transesterification process. Though catalytic transesterification occurs at lower temperatures and has a shorter residence time, the cost of the catalyst escalates the production cost.

Table 3. Examples, pros, and cons of classes of catalysts for the transesterification process.

Class of Catalyst	Examples	Pros	Cons	Ref.
Homogeneous	Base: NaOH, KOH, NaOCH$_3$, KOCH$_3$, NaOCH$_2$CH$_3$	• Strong catalytic activity • Fast reaction • Less energy requirements • Mild reaction conditions • Economically viable • Readily available • Not corrosive	• Not suitable for oil with a high FFA • Possible soap formation • Low biodiesel yield • Requires excessive washing • Requires water for purification • Large volume of wastewater generated • Not reusable	[33,34]
	Acid: H$_2$SO$_4$, HCl	• Strong catalytic activity • Suitable for oil with a high FFA • Not affected by oil FFA and water • Effective with low-grade oil • Esterification and transesterification occur simultaneously • No soap formation • High product yield	• Slow reaction rate • Equipment corrosion • Complex separation process • Separation and reuse of the unused catalyst	[35,36]

Table 3. Cont.

Class of Catalyst	Examples	Pros	Cons	Ref.
Heterogeneous	Base: CaO, Mg/Zr, Mg-Al hydrotalcite, ZnO/KF, ZnO/Ba, Na/BaO, MgO, $Al_2O_3/ZrO_2/WO_3$	• Reusability • Easily separatable • Fast reaction rate • Reaction occurs in moderate conditions • Low energy consumption • Long catalyst life • Non-corrosive • Comparatively cheap • Minimum effluent generation	• Prone to saponification • Generate more wastewater • Complex separation and purification process • Sensitive to the acid value of oil • Low biodiesel yield • Require a high methanol/oil molar ratio • High cost of catalyst synthesis	[37,38]
	Acid: Titanium-doped amorphous zirconia, sulfated zirconia, carbon-based solid acid catalyst	• Insensitive to the water content of the feedstock • Effective with waste oil • Easy to separate from product • High reusability • Highly recyclable • Spent catalyst can be reused	• Slow reaction rate • Expensive • Long residence time • High energy requirements • Likelihood of product contamination • Leaching of catalyst • Limited diffusion	[39,40]
Biobased	Lipase, candida Antarctica, immobilized lipase on SiO_2	• Completely bio-based • Prevent saponification • Environmentally friendly • Ecofriendly and nonpolluting • Easy product removal • Easy purification needed • Requires low temperature • Zero by-product • High reusability	• Expensive • Slow reaction rate • Sensitive to methanol • Can easily become inactive and denatured • Complexity of separation and purification	[41–43]
Nanocatalyst	Zn, Ca, Mg, Zr-based nanocatalyst	• Highly active • Strong stability • Moderate reaction conditions • High reusability • Strong resistance to saponification	• High cost of synthesis	[44,45]

2.2.5. Superfluid/Supercritical

The deployment of supercritical techniques for biodiesel production is one of the methods of biodiesel production and a possible substitute for the traditional biodiesel synthesis process. By definition, a supercritical fluid or superfluid is any substance existing above its critical pressure and temperature. It is a highly compressed fluid that combines the properties of both a gas and liquid. At a supercritical temperature and pressure conditions, there is no distinct liquid or gaseous phase of the substance. For example, the critical temperature and pressure of methanol, ethanol, acetone, methane, and ethane are 239.2 °C and 8.09 MPa, 240.9 °C and 6.14 MPa, 235.1 °C and 4.70 MPa, −82.6 °C and 4.60 MPa, and 32.3 °C and 4.88 MPa, respectively [46]. The use of any of these fluids under supercritical conditions greatly influences biodiesel production. Moreover, the choice of feedstock, reaction time, solvent/oil molar ratio, reactor type, and agitation speed influence the conversion efficiency of feedstock to biodiesel.

With the supercritical method of biodiesel production, there is no need for catalysts, and the process is guaranteed to produce high-quality biodiesel. When compared with other biodiesel generation methods, the supercritical technique allows for lower energy consumption. Available economic and energy analysis showed a reduction of about 71% in the cost of producing energy [47]. Notwithstanding the cost-effective and energy-efficient process, the supercritical method of biodiesel synthesis requires a high cost of production infrastructure and can denature the product. Table 4 compiles the different biodiesel production techniques and their advantages and disadvantages.

Table 4. Advantages and disadvantages of biodiesel production techniques [6,9,21,24,26,27].

Production Techniques	Advantages	Disadvantages
Dilution	<ul><li>Easy to produce</li><li>Does not cause pollution</li><li>Low capital and production cost</li></ul>	<ul><li>Products suffer from low volatility, poor atomization, and high viscosity</li><li>Causes the plugging of injector nozzles and fuel lines</li><li>Results in incomplete fuel combustion and increased pollution</li><li>Increased emission of smoke and CO</li><li>High engine wear and low engine durability</li><li>Gum formation</li><li>Oil deterioration</li><li>High cost of engine maintenance</li><li>Lubricating oil thickening</li><li>Inappropriate for CI engine</li><li>Products coagulate at low temperatures</li><li>High free fatty acid</li></ul>
Microemulsion	<ul><li>Lower NOx emissions</li><li>Generation of fuel with reduced viscosity and better liquidity</li><li>No generation of derivatives</li><li>Generation of quality fuel</li></ul>	<ul><li>Improper and incomplete combustion</li><li>Deposition of carbon residue in the engine</li><li>Occasional injector needle sticking</li><li>Thickening of lubricating oil</li></ul>
Pyrolysis	<ul><li>Highly versatile process</li><li>Easy process</li><li>Satisfactory physicochemical properties of products</li><li>Generation syngas and other value-added by-products</li><li>High product yield</li></ul>	<ul><li>High production cost</li><li>Complex equipment requirement</li><li>High cost of equipment</li><li>Low oxygen content of the product</li><li>Involves elevated temperatures</li><li>Product contains sulfur</li><li>No environmental benefits</li><li>High carbon residue</li><li>Lower fuel purity</li></ul>
Transesterification	<ul><li>Simple process</li><li>Allows feedstock flexibility</li><li>Moderate production conditions</li><li>Product meets international standards</li><li>Lower operation cost</li><li>Industrial-scale production</li><li>Properties of biodiesel produced similar FB diesel fuel</li><li>Flexibility in catalyst selection</li></ul>	<ul><li>Several separation processes needed</li><li>High moisture content in product</li><li>Generation of adulterated product</li><li>Expensive catalysts</li><li>Production of wastewater</li></ul>

Table 4. *Cont.*

Production Techniques	Advantages	Disadvantages
Superfluid/supercritical	• Fast reaction rate • High conversion efficiency • No need for a catalyst • Production efficiency • Low cost • Energy-efficient process	• High cost of apparatus • High reaction temperature and pressure • Denatured biodiesel generated

Moreover, since the spent catalyst must be removed from the product at the end of the reaction, time, energy, and water are expended during the catalyst removal process. To reduce the cost and environmental impact of commercial catalysts, researchers have turned attention to the use of waste-derived heterogeneous catalysts. Food wastes, crop residues, and agricultural wastes are now converted and developed into catalysts for biodiesel production. These efforts not only reduce the cost of production, and ensure the proper disposal of wastes, but also reduce the number of wastes at dumpsites and contribute to ecological sustainability [9,32]. At the end of the transesterification process, the spent solid catalysts are separated from the products by using a laboratory filter paper and reused. Products of the transesterification process must be purified to meet the established international standards, particularly the ASTM D6751 and EN 14214.

3. Modes of Biodiesel Production in Reactor Technologies

Transesterification is the most widely used method for biodiesel production. Basically, there are four steps involved in biodiesel production via transesterification. The first step is the collection of the feedstock, reagents, and other materials needed for the process. In this stage, the production reaction parameters and conditions are also determined and implemented in the reacting vessel. When the process in the reacting vessel is completed, the second step, which involves separating the slurry comprising the crude biodiesel, glycerol, catalyst, excess methanol, and other water is activated. This involves the use of the difference in densities to achieve phase separation among the resultant slurry. During this process, one of the major and predominantly low-cost gravity separation techniques including filtration, centrifugation, floatation, decantation, or sedimentation is deployed [48,49]. The heterogeneous catalyst is recovered in this stage for reuse.

In the third step, crude biodiesel is subjected to gentle heating with stirring to remove unreacted alcohol and excess moisture trapped in the biodiesel. In the fourth and final step, the biodiesel is purified to further remove any undesirable compounds such as the catalyst, soap, unconverted triglyceride, and moisture. The purification can involve the use of wet or dry washing methods, membrane filtration, and evaporation to obtain clean biodiesel. The biodiesel produced at this stage must meet the ASTM D6751 and EN 14214 standards. The wet washing purification process, though most frequently used, extends production duration, requires a large volume of clean water, and generates lots of wastewater. The treatment and disposal of wastewater and the drying of the water-washed biodiesel are energy-intensive and expensive. Drying washing is more ecofriendly, does not require water, and produces fuel of better quality when compared with wet washing. However, the cost of adsorbents and other additional apparatus makes the process uneconomical. The membrane separation technique, though still largely undeveloped and not commonly used, is environmentally benign, consumes less energy, requires no chemicals, and generates high quality products [50–52]. The biodiesel generation processes can be intensified by the use of reactor technologies. The deployment of reactor technologies contributes significantly to ensuring the mass production of biodiesel.

A reactor is a device or vessel with compartments where chemical reactions take place for the transformation of raw materials into desired products under specific and predetermined conditions. A reactor can also be an enclosed volume, an apparatus, or a specialized container where specific chemical reactions take place under a controlled atmosphere. A good reactor must contain mechanisms or facilities for the injection of the raw materials and other reagents, provide enough residence time for the chemical reaction to take place, and discharge the products. There must be facilities for heat addition and heat removal, safe operation and maintenance, and effective control to ensure operational safety, effectiveness, and an acceptable level of productivity. To achieve an optimum reactor operation, effective performance, and high product yield, the design stage must consider the configuration, construction materials, cost, reaction kinetics, heat and mass transfer, reaction parameters, and the environmental sustainability of the reactor [53,54]. A reactor can be operated either as a batch or a continuous process. In recent years, some researchers have reported an amalgamation of the batch and continuous process, which they dubbed the semi-batch/semi-continuous process to overcome some technical and operational associated with both batch and continuous production of biodiesel by transesterification.

3.1. Batch-Mode Reactors

The batch-mode reactors are the oldest, most convenient, and most popular method of biodiesel synthesis. The batch-mode reactor of biodiesel production was developed from a laboratory-scale production process by the optimization of the production parameters [16,55]. It involved the upgrading of the laboratory-scale production into commercial and industrial production scale to meet the increasing demand for biodiesel. The main feature of a batch production method is the intermittency of the process. There is no continuous flow of materials into and out of the reactor during the production period. Rather, a known quantity of raw materials is injected into the reaction and allowed to be converted into the desired product in a specified period. At the end of the process, the resultant slurry is allowed to exit the reactor and transmitted for separation, purification, and further processing [56]. When operating under the batch production mode, there is control of the inflow, adequate mixing of the reactants, and monitoring of the outflow of the materials. Despite the simplicity in the design and operation of batch reactors, the major drawbacks of the process include a longer residence time, a high operation cost, higher energy consumption, and large space requirements [57]. Figure 5 shows 20 L [58] and a 70 L [59] batch reactors for biodiesel production.

Figure 5. (a) Schematic representation of a batch-mode reactor. Fabricated batch-mode (b) 20 L; (c) 70 L reactor for biodiesel production. Adapted from [58,59].

3.2. Semi Batch-Mode Reactors

In the semi-batch/semi continuous mode reactor, there is the intermittent addition or removal of one more reagent or product during the process. There can also be a variation of the reaction parameters as the reaction proceeds. For example, more feedstock or methanol can be introduced into the reactor during the process to improve the reaction rate or product yield. In this way, the reaction equilibrium is altered in support of biodiesel formation by the gradual removal of the product during the process. Similar to the batch process, the semi-batch mode is characterized by a high operation cost, low production rate, and high energy consumption. There is a high rate of human intervention during the process leading to a highly strenuous and labor-intensive process [60–62]. Figure 6 shows the schematic diagram of a semi-continuous flow reactor for biodiesel production as reported by Malpartida et al. [63].

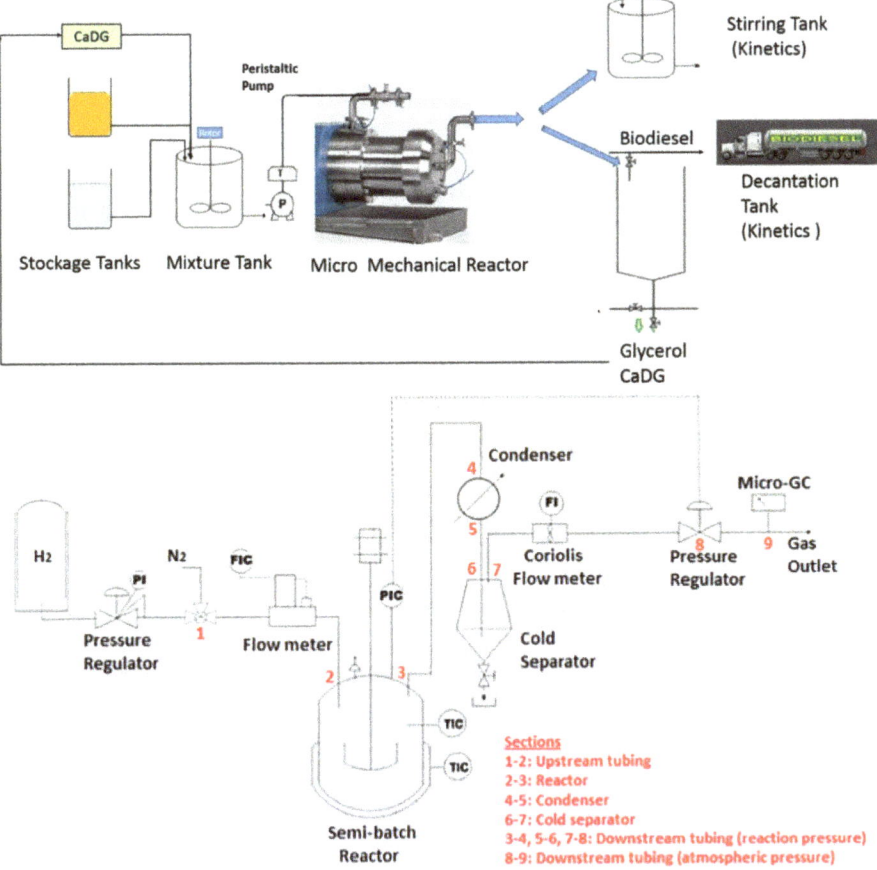

Figure 6. Schematic representations of a semi-continuous flow reactor for biodiesel production. Adapted from [62,63].

3.3. Continuous-Mode Reactors

The continuous mode reactor allows for the continuous inflow of reactants into the reactor and the simultaneous outflow of the products from the system throughout the operation period. After the initial loading of the reactor with feedstock, catalyst, and methanol, the process is initiated and agitated at the required speed to ensure a homogeneous mixing of the reactants, adequate mass, and heat transfer. At the expiration of the set residence time, the reactants are converted into products and allowed to flow out of the reactor. The process continues almost seamlessly with little or no human intervention [54,64]. The process is inbuilt with mechanisms set up to control the inflow of feedstock, catalyst, and methanol, monitoring agitation speed, residence time, and discharge of the resultant slurry.

It must be noted that the biodiesel production industry is moving towards a continuous mode of production and the use of automation and other innovative technologies to ensure large scale and industrial biodiesel production processes. When compared with the batch production process, the continuous production of biodiesel is achieved at lower operating costs, a reduced energy consumption, and with a less labor-intensive process [65]. The deployment of a continuous flow reactor for biodiesel synthesis increases the mass interfacial transport between methanol and oil leading to the synthesis of quality products at the lowest cost per unit volume of fuel [66]. Figure 7 shows the schematic representation of a continuous flow reactor for biodiesel production as presented by Buasri et al. [67] while Table 5 compares the three modes of the reactor operation for biodiesel production.

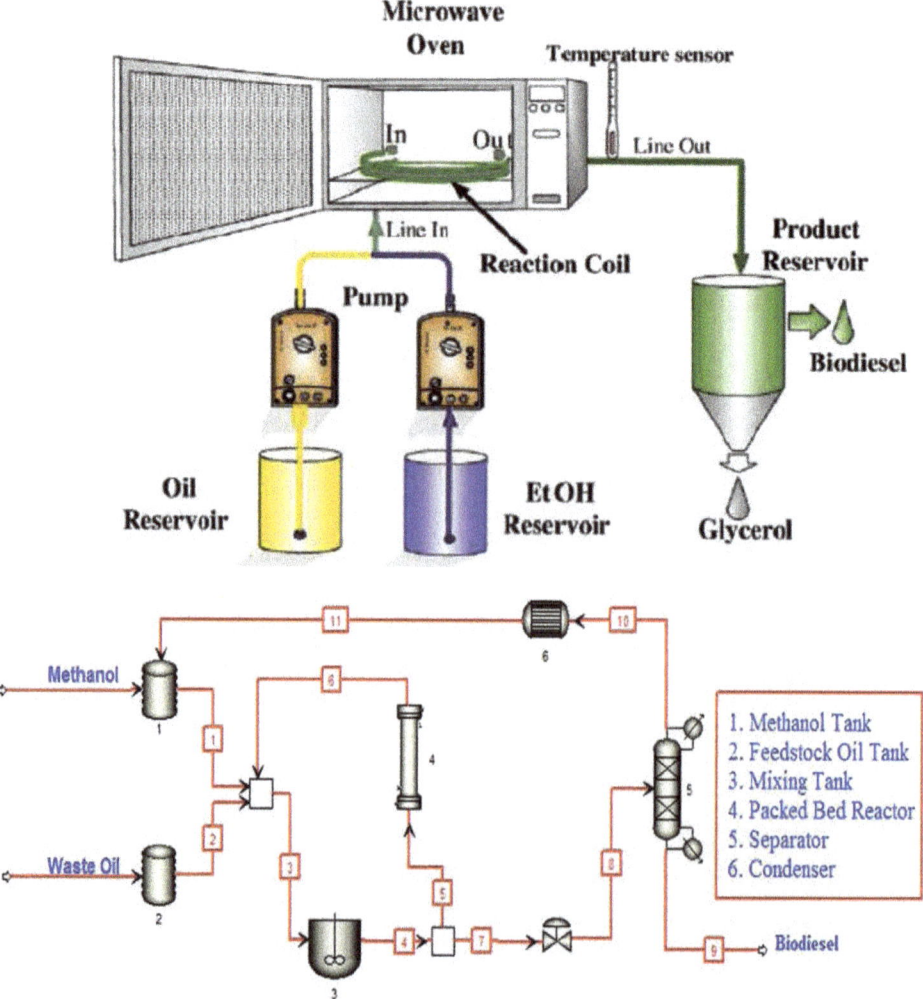

Figure 7. Schematic representations of a continuous flow reactor for biodiesel production. Adapted from [66,67].

Table 5. Merits and demerits of batch, semi-batch, and continuous reactors.

Reactor Modes	Process Description	Merits	Demerits	Ref.
Batch	• A specified quantity of reactants is allowed into the reactor • No materials added during the process • The entire slurry is emptied at the end of the process	• Simple to operate • Monitoring of inflow and outflow of materials • Can be upscaled • Adequate mixing of the reactants • Good flexibility • Enough residence time for product formation	• High operation cost • High energy consumption • Large space requirements • Slow process • Highly laborious • Product quality depends on each batch • Long residence time	[16,55–57]
Semi-batch	• Intermittent addition or removal of reactants or products during the process • Variation of production parameters during the reaction	• High production rate • Easy monitoring • Better control • Reduce material wastage • Highly flexible • Improved production rate • Moderate space requirements • Moderate operation cost • Better heat transfer • Faster production reaction • High selectivity	• Expensive operation cost • High energy consumption • Lower versatility • Highly strenuous • Labor-intensive process	[60–63]
Continuous	• Continuous inflow of raw materials • Continuous outflow of finished products • Presence of mechanisms for the control of reactants addition and residence time	• Low cost of operation • Low space requirement • Production of quality product • Less energy consumption • Better heat and mass transfer • Fast rate of reaction	• Opportunity to scale up • High selectivity • Low versatility • High initial cost of automation technologies	[64,65]

4. Tubular Reactor Technologies for Biodiesel Production

Reactor technologies for the conversion of feedstocks into biodiesel by transesterification are classified by various factors. Some of these factors include the mode of operation, operating conditions, phase numbers, mixing systems, nature of reactants and products, operating temperature and pressure, production size, residence time, mass transfer, heat exchange, and control system.

The tubular or plug-flow reactors are the simplest form of reactor technology for biodiesel production. In this type of reactor technology, reactants and reagents are fed into the reactor through the inlet and are allowed to spend some time in the reactor before being allowed to flow out from the outlet at a constant velocity. The mixing of the reactants and reagents takes place in the tubes or pipe fittings. At a constant velocity, the longer the length of the pipes, the longer the mixing time and the longer the residence time. However, the length of the mixing device and the residence time can be adjusted by altering the system pressure. Moreover, an increase in the viscosity of the mixture of the reactants and reagents will lead to laminar flow in the tubular reactors. The improvement in the reaction, length of the mixing device, and reactor size can be achieved by deploying various mixers such as in-line mechanical mixers, static mixers, and other injection devices. Moreover, the application of static mixers ensures effective radial mixing of multiple immiscible flowing liquids. Figure 8 shows the different configurations of static mixers. The suitability of a typical static mixer is determined by the type of reaction, reaction temperature, reactor configuration, Reynolds number, and viscosity of the fluids. These configurations facilitate the efficient transesterification of the different oils used as feedstock.

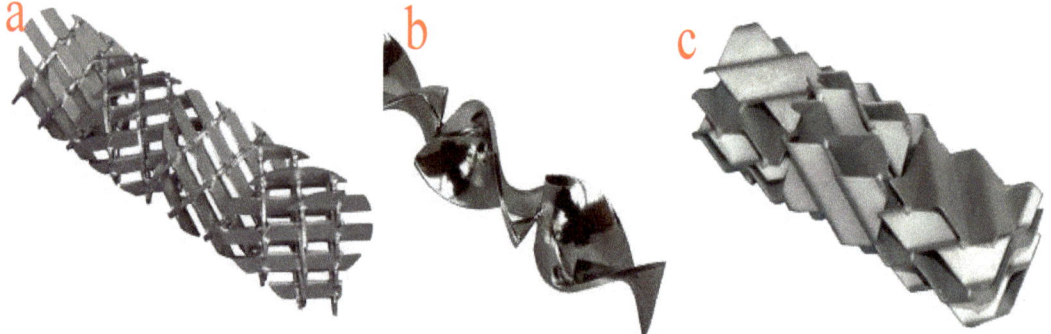

Figure 8. Different configurations of static mixers. (**a**) X-grid static mixer; (**b**) helical twist static mixer; (**c**) corrugated plate static mixer.

When compared with other reactor technologies, tubular reactors are more efficient, require minimum maintenance, and ensure the fast and homogeneous mixing of the fluids. They are not capital intensive and require less space for the construction. Reactors operating on the tubular technology can be used at high pressure and under steady-state conditions. The reactor technology allows continuous operation over a long period and easy product separation. This ensures adequate product separation and recovery of excess methanol and unreacted oils for recycling. Moreover, there is a short residence time when using tubular reactors due to the reduced length of the reactor [68,69]. However, there is a noticeable temperature and pressure drop during the reaction and between the inlet and the outlet points. Moreover, these reactors experience significant temperature changes at different points between the inlet and outlet. Moreover, the reactor requires a large length-to-diameter ratio and a limitation for the Reynolds number. In most cases, tubular reactors require a slow mixing process which often leads to large hold-ups and clogging [21,70]. Notable examples of tubular/plug flow reactors include packed bed reactors, fluidized bed reactors, trickle bed reactors, oscillatory flow reactors, and micro-channel reactors.

4.1. Packed Bed Reactors

Packed bed reactors, also known as fixed bed reactors, are one of the most used tubular reactors in chemical industries, especially for biodiesel production using heterogeneous catalysts. They can also work in the supercritical mode for improved biodiesel production. During the transesterification process for biodiesel production, the packed bed reactor provides a substrate for enzyme immobilization to improve production. Much more than the size or volume of the reactor, the amount of the solid catalyst in the tube influences the conversion of the feedstock into biodiesel [21]. The reactors are in tubular forms and the tubes are filled with packing materials including heterogeneous catalysts and activated carbon. The performance of a packed bed reactor is greatly affected by the catalyst particle size, bed structure, and the spaces between catalyst particles. The arrangement of the packing materials is governed by factors such as (i) physical attributes of the tube, (ii) the shape, size, and the surface structure of the catalyst, and (iii) the intensity, method, and speed of deposition [71]. Figure 9a shows the schematic representation of a packed bed reactor while (b) shows a packed bed reactor for biodiesel production [72].

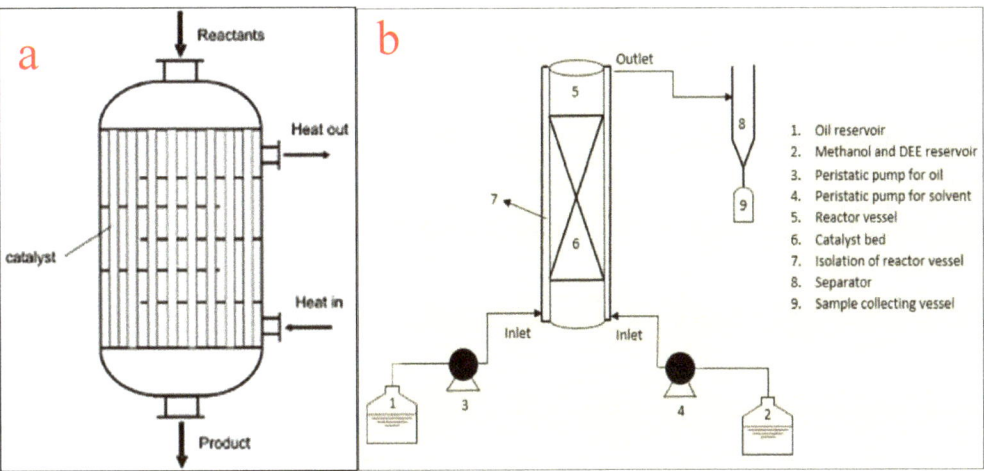

Figure 9. Packed bed reactor. (**a**) schematic representation; (**b**) a packed bed reactor for biodiesel production. Adapted from [72].

With packed bed reactors, the higher conversion efficiency of oils per unit of solid catalysts is feasible and shortens the residence time. Another major benefit of using packed bed reactors is the downstream elimination of catalysts from the product since the catalysts are packed in the tube. According to Sakdasri et al. [73], the greatest advantage of the deployment of packed bed reactors is their high conversion efficiency and ability to use heterogeneous catalysts. Despite these advantages, the reactor suffers from acute and sudden pressure drops, increased cost of operation, and high energy consumption. The pressure drops can be attributed to fluid friction, fluid viscosity, and reactor tube length. Because of these advantages, several researchers have utilized packed bed reactors for biodiesel production.

4.2. Fluidized Bed Reactors

Fluidized bed reactors, also known as expanded bed reactors, are the most popular configurations employed for the conversion of oils into biodiesel on a laboratory or commercial scale. The basic principle of the operation of a fluidized bed reactor involves a fluidization medium (gas or liquid) made to flow through the bed of solid reactants at a velocity high enough to suspend the solid and make it behave as a fluid. The reactor consists of a reservoir and a column. The reservoir is for the housing and preparation of the liquid feedstock while the column consists of a calming section, distributor, fluidized bed, and freeboard. The calming section helps to equalize the liquid feedstock flow while the distributor creates enough pressure difference across the fluidized bed. At a low fluid velocity, the particle in the vessel is stagnant, similar to the packed bed reactors. However, as the fluid velocity increases, the drag force will overcome the weight of the fluid and propel the particles into an upward movement which signifies the start of the fluidization process. At a higher fluid velocity, the particles expand and swirl around and upward in the fluidized bed. The freeboard disallows the catalyst from flowing out of the column [21,74]. Figure 10a shows the schematic representation of a packed bed reactor while (b) shows a packed bed reactor for biodiesel production as used by the authors of [75].

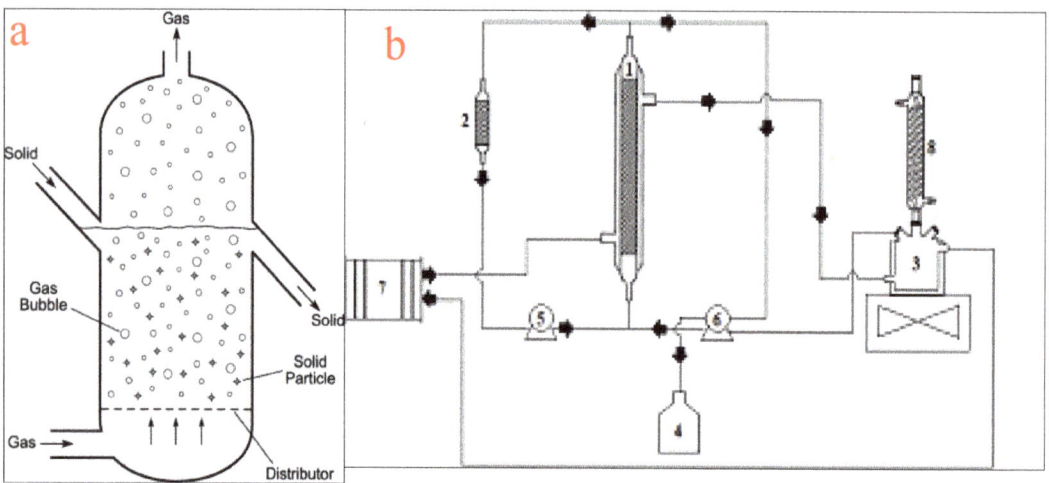

Figure 10. (a) schematic representation of a fluidized bed reactor; (b) a typical fluidized bed reactor for biodiesel production. (1 = reactor; 2 = reactor column; 3 = substrate reservoir; 4 = product vessel; 5 and 6 = peristaltic pumps; 7 = thermostatic bath; 8 = reflux condenser). Adapted from [75].

Fluidized bed reactors have become popular for the transesterification of oil into biodiesel due to their ability to ensure uniform particle mixing, uniform temperature gradients, and the ability to be operated effectively on a continuous scale [76]. However, the sudden pressure loss in the column creates a pressure loss scenario and the possible erosion of internal components. Moreover, due to the likely expansion of the bed materials in the reactor, there is a need for an increment in the reactor size and consequently, the cost of the reactor construction. Other disadvantages of fluidized bed reactors include a high operating cost, reactor wall erosion, the likelihood of particle entrainment, and high catalyst attrition [77]. The practical application of a fluidized bed reactor by Kutálek et al. [78], Fidalgo et al. [75], and Wang et al. [79] yielded a biodiesel conversion efficiency of 77%, 98.1%, and 91.5% respectively.

4.3. Trickle Bed Reactors

Trickle bed reactors are some of the most used industrial reactors in chemical and related industries including the electrochemical, petroleum, petrochemical, coal, pharmaceutical, oil and gas, waste treatment, and biochemical processes. A notable application of trickle bed reactors includes the conversion of vegetable oil into biodiesel, hydrogenation of biooils, polymerization of monomers, purification of feedstocks, and manufacturing of pharmaceuticals [80]. It is a continuous system where liquids are made to flow through a packed bed containing a packing medium. There is a platform for the solid, liquid, and gas based on gravity or pressure forces.

The trickle bed reactors for biodiesel production consist of a tubular tank and structure for solid catalysts at the base of the reactor [81]. The feedstock is introduced from the top of the column while the alcohol can be fed either top or bottom. The heating jacket mounted at the reactor wall helps to maintain the reaction temperature. The continuous heating ensures that the alcohol is vaporized while the unreacted alcohol can be separated from the product. The outlet at the top of the reactor allows for alcohol gas recycling while the outlet at the bottom of the reactor is for the products and unreacted oil to flow out [82]. Figure 11a shows the schematic representation of a trickle bed reactor while Figure 11b shows a typical trickle bed reactor for biodiesel production as used by Jindapon et al. [83].

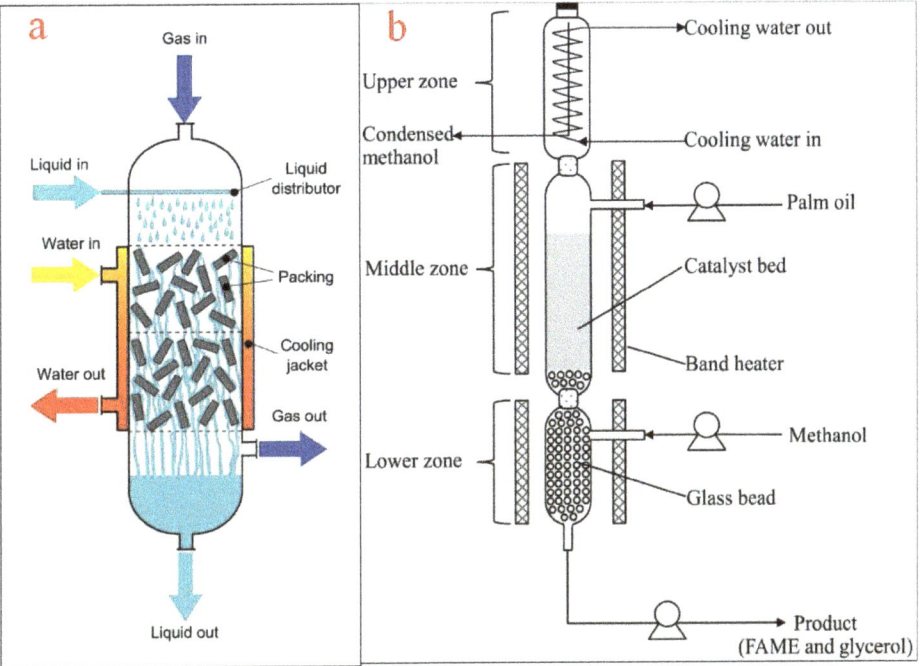

Figure 11. (**a**) Schematic representation of a trickle bed reactor; (**b**) a typical trickle bed reactor for biodiesel production. Adapted from [83].

Major advantages of using trickle bed reactors include simplicity in operation even at a high temperature and pressure, high catalyst loading per unit volume, and low capital and operating costs. Moreover, trickle bed reactors can be used for diverse applications and can accommodate a large volume of production. When used for biodiesel production, trickle bed reactors ensure a higher feedstock conversion rate and improve product productivity [81]. Despite these benefits, trickle bed reactors suffer from poor heat transfer rate, limited diffusion among particles, and unequal fluid distribution. Trickle bed reactors are difficult to scale up and controlling vessel parameters might pose a huge challenge [84]. In research, Muharam et al. [80] reported a 78.22% conversion efficiency while Jindapon et al. [83] reported a biodiesel yield of 92.3% and a product purity of 93.6%.

4.4. Oscillatory Bed Reactors

This type of tubular reactor contains equally spaced tubes with orifice plate constrictions arranged to generate oscillatory flow with intermittent changes in the flow direction using a piston drive. This unique oscillation motion produces vortex mixing that results in the filling of the entire cross-section of the baffles cavity due to fluid obstruction. The configuration of the baffles, rather than the Reynolds number of the fluid, plays an important role in the effectiveness of the reactor. Typically, baffles can be of helical, axial, integral, or wire wool configurations with a tube diameter of less than 15 mm to ensure vigorous mixing and to minimize frictional loss. For the purpose of biodiesel production from vegetable oil using heterogeneous catalysts, a tube diameter of about 5 mm is recommended to minimize the overbearing construction and feedstock costs. In the same vein, an oscillatory Reynolds number of 10 is adequate to ensure turbulent flow in the tube [85].

The use of oscillatory flow reactors ameliorates the challenges associated with the deployment of conventional flow reactors by ensuring vigorous mixing, superb heat transfer, and an excellent plug flow experience. The flow generated by the oscillatory motion is not affected by the net flow rate, the residence time, and the hydrodynamic properties of the slurry. Similarly, the moderation of the net flow rate ensures a smaller reactor volume, a compact setup, minimizes space requirement, and guarantees quality mixing [86]. To achieve efficient and economically viable biodiesel production, oscillatory flow reactors should have a short length/diameter ratio [87]. Figure 12a depicts the schematic representation of an oscillatory flow reactor while Figure 12b shows a typical oscillatory flow for biodiesel production as reported by Masngut et al. [88].

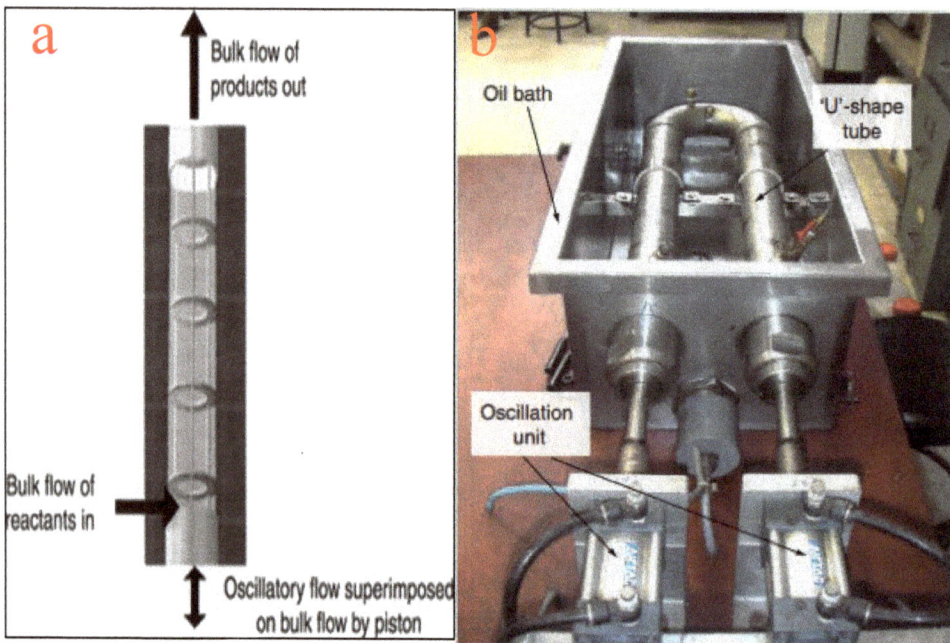

Figure 12. (**a**) Schematic representation of an oscillatory flow reactor; (**b**) a typical oscillatory flow reactor for biodiesel production. Adapted from [88].

When compared with conventional reactors, oscillatory flow reactors consume less energy and generate less waste. They also offer improved mixing efficiency and better mass and heat transfer. The reactor can be operated on both baths and continuous modes offer process flexibility and can easily be scaled up to accommodate increased production [89]. However, the oscillatory flow reactor suffers acute pressure drops as a result of persistent frictional loss. The generation of gas bubbles during operation dampens the oscillations and upsets the plug flow [90].

4.5. Micro-Channel Reactors

Micro-channel reactors are another type of tubular reactor using micro-channel technology for processing chemicals and other diverse applications. They are made up of narrow channels or tubes, in the millimeter range, which allow a high surface/volume ratio, minimize diffusion length, and improve mass and heat transfer. The flow of fluid in this type of reactor is orderly, predictable, and measurable, which also requires a lengthy pipe to ensure thorough mixing [91]. The requirement of a lengthy mixing path is a challenge that is addressed by the application of passive micromixers. The deployment of micromixers of diverse configurations and arrangements achieves an improved contact surface through the mixing of two or more liquids. For example, serial lamination micromixers split the inlet flow and merge them first horizontally and then vertically. For injection micromixers, the oil is allowed to split into substreams before the injection of methanol through a collection of nozzles. Droplet micromixers employ an internal flow field to ensure the mixing of the liquids and transport them by capillary effects, pressure gradient, and flow instability of two or more fluids [92,93].

When compared with conventional reactors, micro-channel reactors demonstrate a high surface/volume ratio, better heat and mass transfer, and improved homogeneous fluid mixing. This type of reactor also allows a shorter reaction duration, less degradation, better scalability, and easier optimization and monitoring. With micro-channel reactors, there is an opportunity for more precision reaction control, selectivities, better conversion efficiency, faster reaction speed, and improved product yield. Moreover, temperature control is easier and more precise, safer, and allows for prompt phase separation [94,95]. However, the micro-channel reactors can handle a limited volume of feedstock at a time due to the small volume of the tubes. They are also prone to intermittent clogging, fouling, tube blockage, and corrosion. Other drawbacks of this type of reactor include a high rate of leakages between channels and the prohibitive cost of building the reactor. Because the micro-channel reactors are small, they are not usually applied at an industrial scale [64,96]. Figure 13a depicts the schematic representation of a micro channel reactor while Figure 13b shows a typical micro channel reactor for biodiesel production as reported by Baydir and Aras [97].

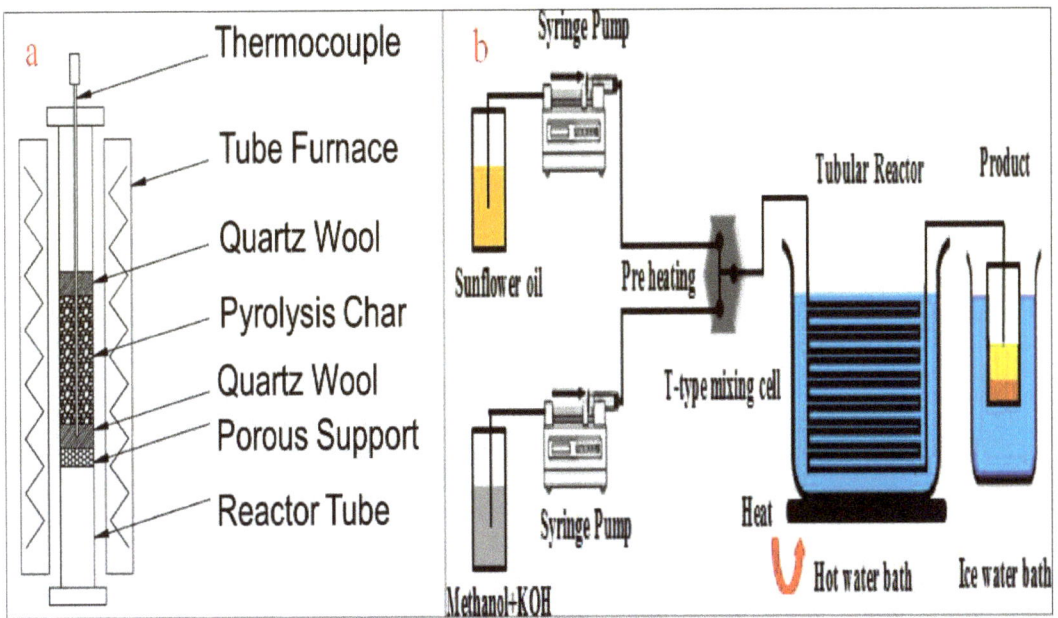

Figure 13. (a) Schematic representation of a micro channel reactor; (b) a micro channel for biodiesel production. Adapted from [97].

Generally, tubular reactors are some of the simplest and easy to construct and operate chemical reactors for biodiesel production. They are cost effective, environmentally friendly, and safe to operate. They can be operated both on batch and continuous modes, are easy to clean, and ensure a high product yield. The quality of the product generated by tubular reactors meets international standards. Though some of them suffer from sudden pressure drops, and high catalyst attrition, they nonetheless find practical industrial applications. Table 6 shows the major benefits and drawbacks of tubular reactors.

Table 6. Benefits and drawbacks of tubular reactors.

Tubular Reactor Type	Benefits	Drawbacks	Ref.
Packed bed	• Compatibility with an elevated temperature and pressure • High conversion efficiency • Better product yield • Easy and simple to operate • Cost effective • Safety	• Prone to clogging and wall erosion • Difficult to monitor and control the temperature	[76,98]
Fluidized bed	• Effective mixing • Compatible with batch and continuous modes • Low chance of tube clogging • Uniform temperature • Improved heat and mass transfer • Easy feeding of catalysts	• Expensive to build and operate • Large sudden pressure drops • Catalyst attrition • Reactor wall erosion and corrosion	[99,100]
Trickle bed	• Effective product separation • Low catalyst attrition • Simple to operate • Ease of catalyst separation	• Ineffective control of reaction parameters • Difficult scalability • Prone to clogging and wall erosion	[81,101]
Oscillatory flow	• Low construction and running costs • High product yield • Compatible with batch and continuous modes • Effective mixing • Improved heat and mass transfer	• Complex design • Not mature for industrial applications	[86,90]
Micro-channel	• High product yield • Maximum mixing achievable • Low maintenance • Easy to clean and operate • Improved product quality	• High cost of construction • Longer lengths of tubes	[102,103]

5. Recent Applications of Tubular Reactors for Biodiesel Production

Biodiesel production has been intensified through the application of reactors to meet the ever-growing demand for quality biodiesel. The use of glassware in the laboratories has not only proved inadequate and time-wasting but also not replicable in most cases. The quality of the product cannot be guaranteed after each production cycle due to many factors. Therefore, to ensure uniformity in standards, increased production efficiency, and cost-efficient production, there is critical need to adopt some proven intensification processes [54]. The use of reactors is one of the biodiesel intensification processes. For the industrial and mass production of biodiesel, the use of tubular or plug flow reactors has been massively exploited. The application of tubular reactor technologies for biodiesel production ensures cost effectiveness, improves the production rate, and allows the use of innovative technologies. Though the initial financial investment in reactor construction might be daunting, in the long run, the cost per liter is significantly lower than laboratory-scale production. There are opportunities for scaling-up, process optimization, reproducibility, quality assurance, and minimum human intervention with tubular reactor technologies [57,94]. Table 7 shows the compilation of the deployment of tubular reactor technologies for the production of biodiesel.

Table 7. Biodiesel production using tubular reactor technology.

Reactor Type	Feedstock	Catalyst Type (Dosage)	Alcohol (Dosage) [a]	Rt (h) [b]	RT (°C)	Yield (%)	Highlights	Ref.
Packed bed	WCO	CaO (0.5 wt.%)	Methanol (6:1)	4	65	98.40	• High product yield • Moderate reaction conditions	[104]
	Linseed oil	CaO (160 g)	Methanol (9.48:1)	3	30	98.08	• High biodiesel yield • Product of a high quality	[72]
	Coconut waste oil	Solid coconut waste (2.29 wt.%)	Methanol (12:1)	3	61	95	• Product meets international standards • High biodiesel yield	[105]
	WCO	Cockle shells (20 g)	Methanol (9:1)	0.75	65	72.5	• Short reaction duration • Moderate reaction conditions	[106]
	Palm oil	Ethyl acetate (6 wt.%)	Ethanol (16.7:1)	72	113	99	• High product yield • Low catalyst dosage	[107]
	Palm oil	waste seashells (10 wt.%)	Methanol (30:1)	3	65	95	• High product yield • Short reaction time	[108]
	WCO	Magnetic whole-cell biocatalysts (12 wt.%)	Methanol (3.74:1)	48	35	91.8	• High product yield • Use of biocatalyst	[109]
	Soybean oil	Magnetic chitosan microspheres (25 g)	Methanol (4:1)	72	35	82	• Effective use of biocatalyst • Low reaction temperature	[110]
Fluidized bed	Babassu oil	Novozym biocatalyst (12 wt.%)	Ethanol (12:1)	8	50	98.1	• High biodiesel yield • Simultaneous glycerol separation	[75]
	Waste frying oil	Magnetic whole-cell biocatalysts (16 wt.%)	Methanol (4:1)	48	35	89	• High conversion efficiency • Product ASTM D6751 and EN 14214 standards	[111]

Table 7. Cont.

Reactor Type	Feedstock	Catalyst Type (Dosage)	Alcohol (Dosage) [a]	Rt (h) [b]	RT (°C)	Yield (%)	Highlights	Ref.
Trickle bed	Rapeseed oil	Ca/Al oxide composite (73.8 g)	Methanol (3:1)	NS	65	94.65	• Simultaneous biodiesel and glycerol separation • High biodiesel yield	[112]
	Sunflower oil	CaO (18.5 g)	Methanol (2.9:1)	5	140	98	• High product yield • Easy separation of methanol and glycerol from biodiesel	[113]
	Palm oil	Dolomitic rock (130 g)	Methanol (12.9:1)	6	100	92.3	• Improved biodiesel yield • High glycerol purity (93.6 wt%) • Recovery of excess methanol • Easy removal of glycerol	[83]
	WCO	NaOH (1 wt.%)	Methanol (6:1)	1	60	72.5	• Product of a high standard • Product performed well in diesel engine • Easy separation and removal of glycerol	[114]
Oscillatory flow	Palm fatty acid distillate (PFAD)	Modified sulfonated glucose (2.5 wt.%)	Methanol (9:1)	0.83	60	94.21	• High product yield • High conversion efficiency (97.1%) • Biodiesel complied with ASTM D6751 standards	[115]
	WCO	KOH (1 wt.%)	Methanol (6:1)	5 min	60	81.9	• Energy efficient process • Moderate reaction conditions	[116]

Table 7. Cont.

Reactor Type	Feedstock	Catalyst Type (Dosage)	Alcohol (Dosage) [a]	Rt (h) [b]	RT (°C)	Yield (%)	Highlights	Ref.
	Rapeseed oil	KOH (1.5 wt.%)	Methanol (6:1)	10 min	60	97	• Product meets international standard • Moderate production conditions • Low energy consumption • Good mixing of the slurries	[117]
	WCO	KOH (3 wt.%)	Methanol (11:1)	1 min	65	99.7	• High product yield • Short residence time • Biodiesel complied with ASTM D6751 standards	[118]
	Soybean oil	CaO (5 wt.%)	Methanol (12:1)	5	65	52	• Less energy consumption • Better mixing	[119]
	Soybean oil	NaOH (1.2 wt.%)	Methanol (9:1)	28 s	56	99.5	• Improved mass and heat transfer	[120]
Micro-channel	Palm oil	KOH (3.5 wt.%)	Methanol (21:1)	3 min	60	100	• 100% product yield • Short residence time • Efficient mass and heat transfer	[121]
	Soybean oil	KOH (1.17 wt.%)	Methanol (8.5:1)	14.9 s	59	99.5	• Short residence time • High product yield	[103]
	Palm oil	KOH (1 wt.%)	Methanol (6:1)	5 s	60	97.14	• High biodiesel yield • Short reaction duration	[122]

[a] Alcohol: oil ratio; [b] Rt = Residence time in h, unless otherwise stated; RT = Reaction temperature; WCO = Waste cooking oil; CaO = Calcium oxide; NS = Not stated.

In recent research, Zik et al. [104] deployed a packed bed reactor to produce biodiesel from waste cooking oil (WCO) using CaO derived from chicken bones as a catalyst. Working at a reaction temperature of 65 °C, methanol:oil ratio of 6:1, and a mixing speed of 600 rpm, a biodiesel yield of 98.4% was recorded. In another study, Hashemzadeh Gargari and Sadrameli [72] used a packed bed reactor to transesterify linseed oil into biodiesel in the presence of methanol and CaO. The reaction yielded 98% biodiesel of acceptable quality and compatible with ASTM D 6751 and EN 14214 standards. Similarly, Talha and Sulaiman [105], Ani et al. [106], Akkarawatkhoosith et al. [107], and Jindapon et al. [108] recorded 95%, 72.5%, 99%, and 95% when a packed bed reactor was used to convert various feedstocks into quality biodiesel. The authors were unanimous in affirming the ease of operation, moderate operating conditions, high product yield, and operational advantages derived from the use of a packed bed reactor for effective biodiesel production.

The use of a fluidized bed reactor for biodiesel production has been investigated and found to be operationally feasible by various researchers in recent times. Using a magnetic whole-cell biocatalyst, Chen et al. [109] converted pretreated WCO into quality biodiesel in a novel magnetically fluidized bed reactor with a column internal diameter of 100 mm and 950 mm length. With a reaction temperature of 35 °C and a catalyst concentration of 12 wt%, a biodiesel yield of 91.8% was achieved. In another study, Zhou et al. [110] utilized a fluidized bed reactor to synthesize soybean oil into biodiesel using a magnetic chitosan microspheres-immobilized lipase as a catalyst. The reactor consisted of a glass column of 30 mm inner diameter and 400 mm height maintained at 35 °C for 72 h and a 25 mL/min fluid flow rate. At the end of the reaction, a biodiesel yield of 82% was recorded. A product yield of 98.1% and simultaneous glycerol separation and removal were recorded when a 42.4 cm^3 glass tube fluidized bed reactor was deployed for the transesterification of babassu oil into biodiesel [75]. The generated biodiesel was of good quality and in line with international standards. Similar results were recorded by Liu et al. [111] when methanol and 16 wt% magnetic whole-cell biocatalysts were used to transesterify waste frying oil into biodiesel in a fluidized bed reactor. The results of these investigations confirm the feasibility of a fluidized bed reactor to improve biodiesel yield, glycerol removal, and catalyst stability in tolerable reaction conditions. The generated biodiesel was of acceptable quality and complied with the ASTM D6751 and EN 14214 standards.

A 91 mL thermally insulated trickle bed reactor was applied for the production of biodiesel from rapeseed oil, methanol, and a heterogeneous Ca/Al oxide composite catalyst. The transesterification process progressed significantly well in the reactor leading to simultaneous biodiesel and glycerol separation and collection with over 94% biodiesel yield [112]. Moreover, Son and Kusakabe [113] used a trickle bed reactor (internal diameter = 16 mm, height = 200 mm) for the conversion of low-grade used sunflower oil into biodiesel using CaO as a catalyst. The authors reported a 98% biodiesel yield and the continuous separation of excess methanol and glycerol from the product. Jindapon et al. [83] achieved a 92.3% product yield when methanol and calcined dolomitic rock were used as raw materials to transesterify palm oil to quality biodiesel in a trickle bed reactor (glass column with an internal diameter of 3 cm and 40 cm long). The authors listed high biodiesel yield, better glycerol quality, recovery of excess methanol, and removal of glycerol as the novel benefits of the process.

Using a 15 L cylindrical oscillatory flow reactor, NaOH catalyst, and methanol, García-Martín et al. [114] converted WCO into biodiesel and tested the fuel in a 140 hp EURO4 test bed engine. The easy to operate and energy-efficient reactor achieved 72.5% product yield, easy separation of glycerol from the product, and the generated fuel met ASTM D6751 and EN 14214 standards. In a similar vein, Kefas et al. [115] deployed an oscillatory flow reactor to convert palm fatty acid distillate and modified sulfonated glucose catalyst into biodiesel using methanol as alcohol. At the end of 50 min residence time and a reaction temperature of 60 °C, a biodiesel yield and conversion efficiency of 94.21% and 97.1%, respectively, were achieved. The same pattern of results was obtained when Soufi et al. [116], Phan et al. [117], and Santikunaporn et al. [118] achieved 81.9%, 97%, and 99.7%, respectively, when they engaged an oscillatory flow reactor to generate biodiesel from vegetable oils. The authors listed an enhanced product yield, uniform mixing, low and uniform shear, improved heat and mass transfer, compact reactor design, and linear scalability as some of the benefits of using oscillatory flow reactors for biodiesel production [90].

The use of a microchannel reactor for biodiesel generation has been investigated by various researchers. Mohammadi et al. [119] used soybean oil, 5 wt.% calcined CaO, and methanol (methanol:oil ration of 12:1), at 65 °C for 5 min. The reactor had a 52% biodiesel yield after 5 min, which was a significant improvement from the conventional reactor. Wen et al. [120] used a 316 L stainless steel micro-channel reactor to generate methyl ester from soybean oil using NaOH and methanol. The process was maintained at 56 °C and a biodiesel yield of 99.5% was achieved after 28 s. This is a significant improvement over conventional reactors, in terms of short residence time and product yield. The works of Azam et al. [121], Dai et al. [103], and Kaewchada et al. [122] in achieving biodiesel yields of 100%, 99.5%, and 97.14%, respectively, attest to the viability of the micro-channel reaction for biodiesel production.

6. Chemical Kinetics of Biodiesel Production by a Tubular Reactor

The tubular reactor is the simplest reactor to achieve the transesterification of triglycerides to biodiesel [123]. Transesterification is a slowly reversible reaction. The forward reaction is achieved with a lower activation energy than the backward reaction, and therefore favors the conversion of feedstock to biodiesel [124]. As shown by the reactions in Figure 4, the two intermediate compounds formed during transesterification are diglycerides (DG) and monoglycerides (MG). The final products, i.e., biodiesel and glycerol (GL) are generated during the third stage due to the consecutive reversible reaction involved in the transesterification of triglyceride (TG) to biodiesel. However, many factors can contribute to reducing the activation energy, influencing the forward reaction, and consequently increasing product formation.

The transesterification reaction is greatly influenced by the quantity of alcohol present in the reactor. Raheem et al. [125] reported that excess methanol in the reacting chamber enhances the speedy conversion of DG and MG to FAME particularly when the ratio of methanol to oil exceeds the stoichiometric value. The chemical kinetic study of transesterification is a process of three orders. The compound of the second order chemical kinetic of transesterification is developed from the first order. Equations (1)–(3) show the kinetics of the chemical process [125]:

$$I_1 = k_1 \, C_{TG} \cdot C_{MOH} - k_{-1} \, C_{DG} \cdot C_{FAME} \qquad (1)$$

$$I_2 = k_2 \, C_{DG} \cdot C_{MOH} - k_{-2} \, C_{MG} \cdot C_{FAME} \qquad (2)$$

$$I_3 = k_3 \, C_{MG} \cdot C_{MOH} - k_{-3} \, C_{GL} \cdot C_{FAME} \qquad (3)$$

where I = Specific reaction rate (mol g^{-1} min^{-1}),
C = Concentration (mol^{-1}),
k_i = Reaction rate constant forward reaction (I$_2$ mol^{-1} min^{-1} g^{-1}),
k_{-i} = Reaction rate constant for reverse reaction (I$_2$ mol^{-1} min^{-1} g^{-1})
The conversion of Equations (1)–(3) into yields Equations (4)–(6), respectively.

$$-\frac{d[TG]}{dt} = k_1[TG][CH_3OH] - k_{-1}[DG][FAME] \qquad (4)$$

$$-\frac{d[DG]}{dt} = k_2[DG][CH_3OH] - k_{-2}[MG][FAME] - k_1[TG][CH_3OH] - k_{-1}[DG][FAME] \qquad (5)$$

$$-\frac{d[MG]}{dt} = k_3[MG][CH_3OH] - k_{-3}[GL][FAME] - k_2[DG][CH_3OH] - k_{-2}[MG][FAME] \qquad (6)$$

The combination of Equations (4)–(6) yields Equation (7):

$$K_1 = \frac{k_1}{k_{-1}} \qquad (7)$$

$$K_2 = \frac{k_2}{k_{-2}} \qquad (8)$$

$$K_3 = \frac{k_3}{k_{-3}} \qquad (9)$$

The first order chemical kinetic design model measures the effect of time and reaction temperature on the conversion process. With the negligible impacts of the catalyst concentration, the first step of the process is assumed to be the forward reaction only while the backward reaction is neglected. The conversion rate for the irreversible first order kinetics is shown in Equation (10). When t = 0, $\ln[TG]_0 = 0$ but as the $[TG]_0$ approaches 1.0 mol/dm^3, the $[TG]_t$ is the percentage of biodiesel yield at time, t. The conversion of TG to FAME is defined as X_{FAME} (Equation (11)) [125]:

$$-r_a = \frac{-d[TG]}{dt} = k[TG] \times [CH_3OH]^3 \qquad (10)$$

$$X_{FAME} = 1 - \frac{[TG]}{[TG]_0} \qquad (11)$$

The irreversible second order chemical kinetic model gives the conversion rate as shown in Equations (12) and (13), with k as the rate constant.

$$-r_A = \frac{-d[TG]}{dt} = k \times [TG]^2 \qquad (12)$$

$$\frac{dX_{FAME}}{dt} = k[TG]_0(1 - X_{FAME})^2 \qquad (13)$$

The reversible second order chemical kinetic model ensures there is adequate collision and reactions between the reactants in the overall reaction process for the continuous production of biodiesel. These collisions must be at the required process conditions. The combination of Equations (14) and (15) helps to achieve Equation (16) [126,127].

$$[TG] = ([TG]_0 - X_{FAME}) \text{ and } [CH_3OH] = [CH_3OH] - X_{FAME} \qquad (14)$$

$$\frac{d[X_{FAME}]}{dt} = \frac{k_1 k_3}{k_2}([TG]_0)g([CH_3OH] - X_{FAME}) \qquad (15)$$

$$y(t) = \frac{1}{[CH_3OH]_0 - [TG]_0} \ln \frac{[TG]_0[CH_3OH]_0 - X_{FAME}}{[CH_3OH]_0[TG]_0 - X_{FAME}} = \frac{k_1 k_3}{k_2} t \qquad (16)$$

Generally, a higher temperature increases the appropriate energy required to ensure a more effective collision among the reacting molecules and ensure the digestibility of the reactants. In a tubular reactor for biodiesel production, the tubes must be properly lagged and none of the reactants must be exhausted. Designing and understanding the chemical kinetics that govern the transesterification reaction will accelerate biodiesel production, especially when combined with the use of tubular reactor technologies.

7. Implications and Future Perspectives

While it is incontrovertible that the deployment of a tubular reactor will escalate the production of biodiesel, the impact of such an increment should not be lost. Undoubtedly, the use of tubular reactor technologies will ensure the democratization of biodiesel production and utilization, popularize the use of biodiesel for diverse applications, allow countries to move away from the use of FB fuels, create employment and other socioeconomic benefits, and ultimately slow down environmental degradation [128]. However, there are direct and indirect impacts of the deployment of tubular reactor technologies for the acceleration of biodiesel production on various aspects of our lives and environment.

Acceleration of biodiesel production will increase the use of biodiesel as internal combustion engine fuel thereby improving engine performance and mitigating the emission of CHGs and the environmental pollution associated with FB fuels. The use of waste cooking oil, waste animal fats, recovered kitchen fats, and other forms of wastes as feedstock for biodiesel production will improve sanitation, contribute to waste management, and reduce the contamination of aquatic and terrestrial habitats. Converting these wastes to biodiesel will also improve air quality, eliminate unpleasant odors, prevent the breeding of flies and other disease-causing pathogens, and improve human health [129,130]. Acceleration of production and utilization of biodiesel will, however, increase the emission of NOx from internal combustion engine tailpipes. There will also be increased water utilization for biodiesel purification.

Though the cost per liter might reduce, the cost of capital investment will skyrocket as a result of the deployment of tubular reactor technologies for the acceleration of biodiesel generation. The total cost of biodiesel production includes the cost of raw materials, cost of the plant, labor costs, cost of energy, etc. These are dependent of the production scale, production technique, type of raw materials chosen, and location of the plant, among others [131]. The cost of raw materials (feedstock, catalyst, chemicals, etc.) accounts for about 70–80% of the total production cost of biodiesel. However, with the utilization of used vegetable oil and waste animal fats as feedstock and the use of waste-derived catalysts which are reusable for many production cycles, the cost of raw materials has been significantly reduced. For example, the plant cost, consisting of reactor purchasing, land acquisition, piping, instrumentation, installation, and electrical and auxiliary facilities requires huge financial commitments and is second only to the cost of raw materials. Bokhari et al. [132] and Karmee et al. [133] agreed that adapting biodiesel intensification approaches will escalate the cost of production.

Other implications of tubular reactor technologies for biodiesel production are the anticipated increment in net GHGs from land-use change, deforestation, food security, water scarcity, and resource utilization. Though the tubular reactor ensures the quick and effective mixing of fluids, a high product yield, and better heat and mass transfer, there are legitimate concerns about process control and sustainability criteria. The increased need to use a reactor for biodiesel production is to reduce the production cost, energy and water requirements, reaction times, human intervention, and labor costs. The current ecological burden, unpredictable product quality, lack of ability to meet ASTM and EN standards, low conversion efficiency, and safety concerns associated with laboratory-scale biodiesel production will be addressed with industrial-scale production.

To meet the expected market share of biodiesel, feedstock price, availability, volatility, and accessibility are major factors. Feedstock that does not affect food change, nor requires land and water should be developed, tested, and nurtured to maturity to guarantee massive biodiesel production. Such feedstock should be subjected to effective pretreatment techniques to aid its digestibility and improve the conversion efficiency [134]. Compact biodiesel production plants with less energy consumption, and a minimum carbon footprint but using innovative technologies are needed. Such plants must require minimum human intervention and rely on robotics technology, artificial intelligence, and smart metering. Novel manufacturing techniques and practices must be developed to construct state-of-the-art reactors for biodiesel production.

Going forward, more investigations are needed to discover more advanced transesterification reactors with the capability to convert used vegetable oil, animal fats, and natural oils into biodiesel. The use of locally available construction materials and methods for reactors should be encouraged. Innovative, economic, and eco-friendly technologies must be embraced to replace conventional methods with a view to improving the sustainability of the process. The selection of appropriate, locally available, low-cost, and high-yielding feedstock is one of the most crucial criteria for sustainable and large-scale biodiesel production. There must be deliberate efforts for the targeted sensitization of the populace to improve the acceptability of biodiesel. The identified factors responsible for negative attitudes and other social repellents of biodiesel among people from diverse socioeconomic statuses must be addressed.

8. Conclusions

The use of tubular reactor technology for the intensification of biodiesel production is key to ensuring improved production, commercialization, and the better-quality production of biodiesel. The increased production of biodiesel will reduce the challenges associated with the adaptation and utilization of biodiesel for diverse applications. The commercialization, increased production, and utilization of biodiesel will benefit the environment, assist in the diversification of the fuel base, provide viable alternatives to FB fuels, limit the emission of GHGs, and slow down environmental pollution. The deployment of the tubular reactor for biodiesel production will create employment, ensure environmental cleanliness, prevent contamination of surface and underground water bodies, and increase the utilization of the quality of biodiesel for diverse applications.

The production of biodiesel must be incentivized through provision subsidies, and tax exemptions to encourage biodiesel refiners. Land must be made available for investors to build industrial-scale tubular reactors to ensure the commercial generation of biodiesel. More wastes must be brought into the feedstock basket to further bring down the cost of raw materials for biodiesel production. The large-scale generation of biodiesel under a circular economy must be escalated to enjoy the technical, sanitary, socioeconomic, and environmental benefits associated with the application of reactor technologies for biodiesel generation. The use of innovative technologies such as robots, smart metering, artificial intelligence, machine learning, genetic algorithms, cloud computing, smart cameras, and other modeling tools must be introduced into biodiesel research.

Author Contributions: Conceptualization, O.A.; methodology, O.A. and D.V.V.K.; software, O.A. and D.V.V.K.; investigation, O.A. and D.V.V.K.; resources, O.A. and D.V.V.K.; data curation, O.A.; writing—original draft preparation, O.A.; writing—review and editing, O.A. and D.V.V.K.; supervision, D.V.V.K.; project administration, O.A. and D.V.V.K.; funding acquisition, D.V.V.K. All authors have read and agreed to the published version of the manuscript.

Funding: This research received no external funding.

Acknowledgments: The authors are grateful to the Library department and Faculty of Engineering and Built Environment, University of Johannesburg, South Africa for their invaluable support.

Conflicts of Interest: The authors declare no conflict of interest.

Abbreviations

ASTM	American Society for Testing and Materials
CI	Compression ignition
DG	Diglyceride
FAEE	Fatty Acid Ethyl Ester
FAME	Fatty Acid Methyl Ester
FB	Fossil-based
FFA	Free fatty acid
GHG	Greenhouse gas
GL	Glycerol
ICE	Internal combustion engine
MG	Monoglyceride
OECD	Organization for Economic Cooperation and Development
R & D	Research and Development
RT	Reaction temperature
Rt	Residence time

TG	Triglyceride
TWh	Terawatt hour
WCO	Waste cooking oil

References

1. Primary Energy Consumption, 2019. Available online: https://ourworldindata.org/grapher/primary-energy-cons?tab=chart&time=2000..latest&country=~{}OWID_WRL (accessed on 1 April 2022).
2. BP Energy Outlook 2030. Available online: https://www.iaee.org (accessed on 1 April 2022).
3. Today in Energy. Available online: https://www.eia.gov/todayinenergy/detail.php?id=49876 (accessed on 1 April 2022).
4. Renewable Energy Global Status Report 2021. Available online: https://www.ren21.net/five-takeaways-from-ren21s-renewables-2021-global-status-report/ (accessed on 1 April 2022).
5. Transportation Sector Energy Consumption. Available online: https://www.eia.gov/outlooks/ieo/pdf/transportation.pdf (accessed on 1 April 2022).
6. Awogbemi, O.; Kallon, D.V.V.; Onuh, E.I.; Aigbodion, V.S. An overview of the classification, production and utilization of biofuels for internal combustion engine applications. *Energies* **2021**, *14*, 5687. [CrossRef]
7. Long, F.; Liu, W.; Jiang, X.; Zhai, Q.; Cao, X.; Jiang, J.; Xu, J. State-of-the-art technologies for biofuel production from triglycerides: A review. *Renew. Sust. Energ. Rev.* **2021**, *148*, 111269. [CrossRef]
8. Anwar, M. Biodiesel feedstocks selection strategies based on economic, technical, and sustainable aspects. *Fuel* **2021**, *283*, 119204. [CrossRef]
9. Awogbemi, O.; Kallon, D.V.V.; Aigbodion, V.S. Trends in the development and utilization of agricultural wastes as heterogeneous catalyst for biodiesel production. *J. Energy Inst.* **2021**, *98*, 244–258. [CrossRef]
10. Setiyo, M.; Yuvenda, D.; Samue, O.D. The Concise latest report on the advantages and disadvantages of pure biodiesel (B100) on engine performance: Literature review and bibliometric analysis. *Indones. J. Sci. Technol.* **2021**, *6*, 469–490. [CrossRef]
11. Firoz, S. A review: Advantages and disadvantages of biodiesel. *Int. J. Eng. Res. Technol.* **2017**, *4*, 530–533.
12. Global Biodiesel Production by Country 2019. Available online: https://www.statista.com/statistics/271472/biodiesel-production-in-selected-countries (accessed on 1 April 2022).
13. Biofuel Production by Country/Region and Fuel Type, 2016–2022. Available online: https://www.iea.org/data-and-statistics/charts/biofuel-production-by-country-region-and-fuel-type-2016-2022 (accessed on 1 April 2022).
14. Chuah, L.F.; Klemeš, J.J.; Yusup, S.; Bokhari, A.; Akbar, M.M. A review of cleaner intensification technologies in biodiesel production. *J. Clean. Prod.* **2017**, *146*, 181–193. [CrossRef]
15. Kant Bhatia, S.; Kant Bhatia, R.; Jeon, J.M.; Pugazhendhi, A.; Kumar Awasthi, M.; Kumar, D.; Kumar, G.; Yoon, J.J.; Yang, Y.H. An overview on advancements in biobased transesterification methods for biodiesel production: Oil resources, extraction, biocatalysts, and process intensification technologies. *Fuel* **2021**, *285*, 119117. [CrossRef]
16. Bashir, M.A.; Wu, S.; Zhu, J.; Krosuri, A.; Khan, M.U.; Aka, R.J.N. Recent development of advanced processing technologies for biodiesel production: A critical review. *Fuel Process. Technol.* **2022**, *227*, 107120. [CrossRef]
17. Roick, C.; Okonye, L.; Diankanua, N.; Joshua, G. Commercial Technologies for Biodiesel Production. In *Biodiesel Technology and Applications*; Inamuddin, M.I., Ahamed, R.B., Mashallah, R., Eds.; Scrivener Publishing: Beverly, MA, USA, 2021; Volume 3, pp. 195–213. [CrossRef]
18. Mohiddin, M.N.B.; Tan, Y.H.; Seow, Y.X.; Kansedo, J.; Mubarak, N.M.; Abdullah, M.O.; Chan, Y.S.; Khalid, M. Evaluation on feedstock, technologies, catalyst and reactor for sustainable biodiesel production: A review. *J. Ind. Eng. Chem.* **2021**, *98*, 60–81. [CrossRef]
19. Okolie, J.A.; Escobar, J.I.; Umenweke, G.; Khanday, W.; Okoye, P.U. Continuous biodiesel production: A review of advances in catalysis, microfluidic and cavitation reactors. *Fuel* **2022**, *307*, 121821. [CrossRef]
20. Lv, L.; Dai, L.; Du, W.; Liu, D. Progress in enzymatic biodiesel production and commercialization. *Processes* **2021**, *9*, 355. [CrossRef]
21. Tabatabaei, M.; Aghbashlo, M.; Dehhaghi, M.; Panahi, H.K.S.; Mollahosseini, A.; Hosseini, M.; Soufiyan, M.M. Reactor technologies for biodiesel production and processing: A review. *Prog. Energy Combust. Sci.* **2019**, *74*, 239–303. [CrossRef]
22. Mofijur, M.; Rasul, M.G.; Hassan, N.; Nabi, M. Recent development in the production of third generation biodiesel from microalgae. *Energy Procedia* **2019**, *156*, 53–58. [CrossRef]
23. Hossain, N.; Hasan, M.; Mahlia, T.; Shamsuddin, A.; Silitonga, A. Feasibility of microalgae as feedstock for alternative fuel in Malaysia: A review. *Energy Strategy Rev.* **2020**, *32*, 100536. [CrossRef]
24. Singh, D.; Sharma, D.; Soni, S.L.; Sharma, S.; Kumar Sharma, P.; Jhalani, A. A review on feedstocks, production processes, and yield for different generations of biodiesel. *Fuel* **2020**, *262*, 116553. [CrossRef]
25. Awogbemi, O.; Kallon, D.V.V.; Aigbodion, V.S.; Panda, S. Advances in biotechnological applications of waste cooking oil. *Case Stud. Chem. Environ. Eng.* **2021**, *4*, 100158. [CrossRef]
26. Gebremariam, S.N.; Marchetti, J.M. Biodiesel production technologies. *AIMS Energy* **2017**, *5*, 425–457. [CrossRef]
27. Ayoub, M.; Yusoff, M.H.M.; Nazir, M.H.; Zahid, I.; Ameen, M.; Sher, F.; Floresyona, D.; Budi Nursanto, E. A comprehensive review on oil extraction and biodiesel production technologies. *Sustainability* **2021**, *13*, 788. [CrossRef]

28. Shaah, M.A.H.; Hossain, M.S.; Allafi, F.A.S.; Alsaedi, A.; Ismail, N.; Ab Kadir, M.O.; Ahmad, M.I. A review on non-edible oil as a potential feedstock for biodiesel: Physicochemical properties and production technologies. *RSC Adv.* **2021**, *11*, 25018–25037. [CrossRef]
29. Awogbemi, O.; Inambao, F.; Onuh, E.I. Modelling and Optimization of Transesterification of Waste Sunflower Oil to Fatty Acid Methyl Ester: A case of Response Surface Methodology vs Taguchi Orthogonal Approach. *Int. J. Eng. Res. Technol.* **2019**, *12*, 2346–2361.
30. Troter, D.Z.; Todorović, Z.B.; Đokić-Stojanović, D.R.; Veselinović, L.M.; Zdujić, M.V.; Veljković, V.B. Choline chloride-based deep eutectic solvents in CaO-catalyzed ethanolysis of expired sunflower oil. *J. Mol. Liq.* **2018**, *266*, 557–567. [CrossRef]
31. Đokić Stojanović, D.R.; Todorović, Z.B.; Troter, D.; Stamenković, O.S.; Veselinović, L.; Zdujić, M.; Manojlović, D.D.; Veljković, V.B. Influence of various cosolvents on the calcium oxide-catalyzed ethanolysis of sunflower oil. *J. Serbian Chem. Soc.* **2019**, *84*, 253–265. [CrossRef]
32. Catarino, M.; Ramos, M.; Dias, A.; Santos, M.T.; Puna, J.; Gomes, J. Calcium rich food wastes based catalysts for biodiesel production. *Waste Biomass Valorization* **2017**, *8*, 1699–1707. [CrossRef]
33. de Lima, A.L.; Ronconi, C.M.; Mota, C.J. Heterogeneous basic catalysts for biodiesel production. *Catal. Sci. Technol.* **2016**, *6*, 2877–2891. [CrossRef]
34. Alcañiz-Monge, J.; El Bakkali, B.; Trautwein, G.; Reinoso, S. Zirconia-supported tungstophosphoric heteropolyacid as heterogeneous acid catalyst for biodiesel production. *Appl. Catal. B Environ.* **2018**, *224*, 194–203. [CrossRef]
35. Silitonga, A.S.; Shamsuddin, A.H.; Mahlia, T.M.I.; Milano, J.; Kusumo, F.; Siswantoro, J.; Dharma, S.; Sebayang, A.H.; Masjuki, H.H.; Ong, H.C. Biodiesel synthesis from Ceiba pentandra oil by microwave irradiation-assisted transesterification: ELM modeling and optimization. *Renew. Energy* **2020**, *146*, 1278–1291. [CrossRef]
36. Esmi, F.; Borugadda, V.B.; Dalai, A.K. Heteropoly acids as supported solid acid catalysts for sustainable biodiesel production using vegetable oils: A Review. *Catal. Today* **2022**, in press. [CrossRef]
37. Kaur, M.; Malhotra, R.; Ali, A. Tungsten supported Ti/SiO_2 nanoflowers as reusable heterogeneous catalyst for biodiesel production. *Renew. Energy* **2018**, *116*, 109–119. [CrossRef]
38. Kesserwan, F.; Ahmad, M.N.; Khalil, M.; El-Rassy, H. Hybrid CaO/Al_2O_3 aerogel as heterogeneous catalyst for biodiesel production. *Chem. Eng. J.* **2020**, *385*, 123834. [CrossRef]
39. Mukhtar, A.; Saqib, S.; Lin, H.; Shah, M.U.H.; Ullah, S.; Younas, M.; Rezakazemi, M.; Ibrahim, M.; Mahmood, A.; Asif, S. Current status and challenges in the heterogeneous catalysis for biodiesel production. *Renew. Sustain. Energy Rev.* **2022**, *157*, 112012. [CrossRef]
40. Guo, L.; Xie, W.; Gao, C. Heterogeneous H6PV3MoW8O40/AC-Ag catalyst for biodiesel production: Preparation, characterization and catalytic performance. *Fuel* **2022**, *316*, 123352. [CrossRef]
41. Amini, Z.; Ilham, Z.; Ong, C.; Mazaheri, H.; Chen, W.H. State of the art and prospective of lipase-catalyzed transesterification reaction for biodiesel production. *Energy Convers. Manag.* **2017**, *141*, 339–353. [CrossRef]
42. Rizwanul Fattah, I.; Ong, H.C.; Mahlia, T.M.I.; Mofijur, M.; Silitonga, A.S.; Rahman, S.M.; Ahmad, A. State of the art of catalysts for biodiesel production. *Front. Energy Res.* **2020**, *8*, 101. [CrossRef]
43. Yaashikaa, P.; Kumar, P.S.; Karishma, S. Bio-derived catalysts for production of biodiesel: A review on feedstock, oil extraction methodologies, reactors and lifecycle assessment of biodiesel. *Fuel* **2022**, *316*, 123379. [CrossRef]
44. Rezania, S.; Korrani, Z.S.; Gabris, M.A.; Cho, J.; Yadav, K.K.; Cabral-Pinto, M.M.S.; Alam, J.; Ahamed, M.; Nodeh, H.R. Lanthanum phosphate foam as novel heterogeneous nanocatalyst for biodiesel production from waste cooking oil. *Renew. Energy* **2021**, *176*, 228–236. [CrossRef]
45. Vahid, B.R.; Haghighi, M.; Alaei, S.; Toghiani, J. Reusability enhancement of combustion synthesized MgO/MgAl2O4 nanocatalyst in biodiesel production by glow discharge plasma treatment. *Energy Convers. Manag.* **2017**, *143*, 23–32. [CrossRef]
46. Adewale, P.; Dumont, M.J.; Ngadi, M. Recent trends of biodiesel production from animal fat wastes and associated production techniques. *Renew. Sust. Energ. Rev.* **2015**, *45*, 574–588. [CrossRef]
47. Qadeer, M.U.; Ayoub, M.; Komiyama, M.; Daulatzai, M.U.K.; Mukhtar, A.; Saqib, S.; Ullah, S.; Qyyum, M.A.; Asif, S.; Bokhari, A. Review of biodiesel synthesis technologies, current trends, yield influencing factors and economical analysis of supercritical process. *J. Clean. Prod.* **2021**, *309*, 127388. [CrossRef]
48. De Paola, M.G.; Mazza, I.; Paletta, R.; Lopresto, C.G.; Calabrò, V. Small-scale biodiesel production plants—An overview. *Energies* **2021**, *14*, 1901. [CrossRef]
49. Ghazvini, M.; Kavosi, M.; Sharma, R.; Kim, M. A review on mechanical-based microalgae harvesting methods for biofuel production. *Biomass Bioenergy* **2020**, *158*, 106348. [CrossRef]
50. Arenas, E.; Villafán-Cáceres, S.M.; Rodríguez-Mejía, Y.; García-Loyola, J.A.; Masera, O.; Sandoval, G. Biodiesel dry purification using unconventional bioadsorbents. *Processes* **2021**, *9*, 194. [CrossRef]
51. Jariah, N.F.; Hassan, M.A.; Taufiq-Yap, Y.H.; Roslan, A.M. Technological advancement for efficiency enhancement of biodiesel and residual glycerol refining: A mini review. *Processes* **2021**, *9*, 1198. [CrossRef]
52. Hayyan, A.; Ng, Y.S.; Hadj-Kali, M.K.; Junaidi, M.U.M.; Ali, E.; Aldeehani, A.K.; Alkandari, K.H.; Alajmi, F.D.H.; Yeow, A.T.H.; Zulkifli, M.Y. Natural and low-cost deep eutectic solvent for soap removal from crude biodiesel using low stirring extraction system. *Biomass Convers. Biorefin.* **2022**, *12*, 113–121. [CrossRef]

53. Athar, M.; Zaidi, S. A review of the feedstocks, catalysts, and intensification techniques for sustainable biodiesel production. *J. Environ. Chem. Eng.* **2020**, *8*, 104523. [CrossRef]
54. Akubude, V.; Jaiyeoba, K.; Oyewusi, T.; Abbah, E.; Oyedokun, J.; Okafor, V. Overview on Different Reactors for Biodiesel Production. In *Biodiesel Technology and Applications*; Inamuddin, M.I., Ahamed, R.B., Mashallah, R., Eds.; Scrivener Publishing: Beverly, MA, USA, 2021; pp. 341–359. [CrossRef]
55. Topare, N.S.; Patil, K.D.; Khedkar, S.V.; Inamdar, N. Lab Scale Batch Reactor Design, Fabrication and Its Application for Biodiesel Production. In *Techno-Societal 2020*; Pawar, P.M., Balasubramaniam, R., Ronge, B.P., Salunkhe, S.B., Vibhute, A.S., Melinamath, B., Eds.; Springer: Berlin/Heidelberg, Germany, 2021; pp. 819–828. [CrossRef]
56. Mehboob, A.; Nisar, S.; Rashid, U.; Choong, T.S.Y.; Khalid, T.; Qadeer, H.A. Reactor designs for the production of biodiesel. *Int. J. Chem. Biochem. Sci.* **2016**, *10*, 87–94.
57. Zahan, K.A.; Kano, M. Technological Progress in biodiesel production: An overview on different types of reactors. *Energy Procedia* **2019**, *156*, 452–457. [CrossRef]
58. Bello, E.I.; Daniyan, I.A.; Akinola, A.O.; Ogedengbe, I.T. Development of a biodiesel processor. *Res. J. Eng. Appl. Sci.* **2013**, *2*, 182–186.
59. Abbaszaadeh, A.; Najafi, G.; Ghohadian, A. Design, fabrication and evaluation of a novel biodiesel processor system. *Int. J. Renew. Energy Technol. Res.* **2013**, *2*, 249–255.
60. Costa, W.A.; Bezerra, F.W.F.; Oliveira, M.S.; Silva, M.P.; Cunha, V.M.B.; Andrade, E.H.A.; Carvalho, R.N. Appliance of a high pressure semi-batch reactor: Supercritical transesterification of soybean oil using methanol. *Food Sci. Technol.* **2019**, *39*, 754–773. [CrossRef]
61. Silva, M.G.; Oliveira, G.S.; Carvalho, J.C.R.; Nobre, L.R.P.; Deus, M.S.; Jesus, A.A.; Oliveira, J.A.; Souza, D.F.S. Esterification of oleic acid in a semi-batch bubble reactor for biodiesel production. *Braz. J. Chem. Eng.* **2019**, *36*, 299–308. [CrossRef]
62. Silva, M.G.; Nobre, L.R.P.; Santiago, L.F.P.; Deus, M.S.; Jesus, A.A.; Oliveira, J.A.; Souza, D.F.S. Mathematical modeling and simulation of biodiesel production in a semibatch bubble reactor. *Energy Fuels* **2018**, *32*, 9614–9623. [CrossRef]
63. Malpartida, I.; Maireles-Torres, P.; Vereda, C.; Rodríguez-Maroto, J.M.; Halloumi, S.; Lair, V.; Thiel, J.; Lacoste, F. Semi-continuous mechanochemical process for biodiesel production under heterogeneous catalysis using calcium diglyceroxide. *Renew. Energy* **2020**, *159*, 117–126. [CrossRef]
64. Tran-Nguyen, P.L.; Ong, L.K.; Go, A.W.; Ju, Y.H.; Angkawijaya, A.E. Non-catalytic and heterogeneous acid/base-catalyzed biodiesel production: Recent and future developments. *Asia-Pac. J. Chem. Eng.* **2020**, *15*, e2490. [CrossRef]
65. Qiao, B.Q.; Zhou, D.; Li, G.; Yin, J.Z.; Xue, S.; Liu, J. Process enhancement of supercritical methanol biodiesel production by packing beds. *Bioresour. Technol.* **2017**, *228*, 298–304. [CrossRef]
66. Tran, N.N.; Gelonch, M.E.; Liang, S.; Xiao, Z.; Sarafraz, M.M.; Tišma, M.; Federsel, H.J.; Ley, S.V.; Hessel, V. Enzymatic pretreatment of recycled grease trap waste in batch and continuous-flow reactors for biodiesel production. *Chem. Eng. J.* **2021**, *426*, 131703. [CrossRef]
67. Buasri, A.; Chaiyut, N.; Loryuenyong, V.; Rodklum, C.; Chaikwan, T.; Kumphan, N. Continuous process for biodiesel production in packed bed reactor from waste frying oil using potassium hydroxide supported on Jatropha curcas fruit shell as solid catalyst. *Appl. Sci.* **2012**, *2*, 641–653. [CrossRef]
68. Bogatykh, I.; Osterland, T. Characterization of residence time distribution in a plug flow reactor. *Chem. Ing. Tech.* **2019**, *91*, 668–672. [CrossRef]
69. Sungwornpatansakul, P.; Hiroi, J.; Nigahara, Y.; Jayasinghe, T.K.; Yoshikawa, K. Enhancement of biodiesel production reaction employing the static mixing. *Fuel Process. Technol.* **2013**, *116*, 1–8. [CrossRef]
70. Farobie, O.; Sasanami, K.; Matsumura, Y. A novel spiral reactor for biodiesel production in supercritical ethanol. *Appl. Energy* **2015**, *147*, 20–29. [CrossRef]
71. Jakobsen, H.A. Packed bed reactors. In *Chemical Reactor Modeling*; Jakobsen, H.A., Ed.; Springer: Berlin/Heidelberg, Germany, 2009; pp. 953–984. [CrossRef]
72. Hashemzadeh Gargari, M.; Sadrameli, S.M. Investigating continuous biodiesel production from linseed oil in the presence of a Co-solvent and a heterogeneous based catalyst in a packed bed reactor. *Energy* **2018**, *148*, 888–895. [CrossRef]
73. Sakdasri, W.; Ngamprasertsith, S.; Daengsanun, S.; Sawangkeaw, R. Lipid-based biofuel synthesized from palm-olein oil by supercritical ethyl acetate in fixed-bed reactor. *Energy Convers. Manag.* **2019**, *182*, 215–223. [CrossRef]
74. Antunes, F.; Machado, F.; Rocha, T.; Melo, Y.; Santos, J.; da Silva, S. Column reactors in fluidized bed configuration as intensification system for xylitol and ethanol production from napier grass (Pennisetum Purpureum). *Chem. Eng. Process. Process Intensif.* **2021**, *164*, 108399. [CrossRef]
75. Fidalgo, W.R.R.; Ceron, A.; Freitas, L.; Santos, J.C.; de Castro, H.F. A fluidized bed reactor as an approach to enzymatic biodiesel production in a process with simultaneous glycerol removal. *J. Ind. Eng. Chem.* **2016**, *38*, 217–223. [CrossRef]
76. Borges, M.; Díaz, L. Catalytic packed-bed reactor configuration for biodiesel production using waste oil as feedstock. *Bioenergy Res.* **2013**, *6*, 222–228. [CrossRef]
77. Wang, K.; Brown, R.C.; Homsy, S.; Martinez, L.; Sidhu, S.S. Fast pyrolysis of microalgae remnants in a fluidized bed reactor for bio-oil and biochar production. *Bioresour. Technol.* **2013**, *127*, 494–499. [CrossRef]
78. Kutálek, K.; Čapek, L.; Smoláková, L.; Kubička, D. Aspects of Mg–Al mixed oxide activity in transesterification of rapeseed oil in a fixed-bed reactor. *Fuel Process. Technol.* **2014**, *122*, 176–181. [CrossRef]

79. Wang, Y.; Yang, G.; He, J.; Sun, G.; Sun, Z.; Sun, Y. Preparation of biochar catalyst from black liquor by spray drying and fluidized bed carbonation for biodiesel synthesis. *Process Saf. Environ. Prot.* **2020**, *141*, 333–343. [CrossRef]
80. Muharam, Y.; Aufa, T.; Santoso, T.B. Modeling of partial hydrogenation of polyunsaturated fatty acid methyl esters in a trickle bed reactor. *Eng. J.* **2020**, *24*, 195–204. [CrossRef]
81. Restrepo, J.B.; Bustillo, J.A.; Bula, A.J.; Paternina, C.D. Selection, Sizing, and Modeling of a Trickle Bed Reactor to Produce 1, 2 Propanediol from Biodiesel Glycerol Residue. *Processes* **2021**, *9*, 479. [CrossRef]
82. Azarpour, A.; Rezaei, N.; Zendehboudi, S. Performance analysis and modeling of catalytic trickle-bed reactors: A comprehensive review. *J. Ind. Eng. Chem.* **2021**, *103*, 1–41. [CrossRef]
83. Jindapon, W.; Ruengyoo, S.; Kuchonthara, P.; Ngamcharussrivichai, C.; Vitidsant, T. Continuous production of fatty acid methyl esters and high-purity glycerol over a dolomite-derived extrudate catalyst in a countercurrent-flow trickle-bed reactor. *Renew. Energy* **2020**, *157*, 626–636. [CrossRef]
84. Degirmenci, V.; Rebrov, E.V. Design of catalytic micro trickle bed reactors. *Phys. Sci. Rev.* **2016**, *1*, 1–29. [CrossRef]
85. McDonough, J.R.; Phan, A.N.; Harvey, A.P. Rapid process development using oscillatory baffled mesoreactors—A state-of-the-art review. *Chem. Eng. J.* **2015**, *265*, 110–121. [CrossRef]
86. Bianchi, P.; Williams, J.D.; Kappe, C.O. Oscillatory flow reactors for synthetic chemistry applications. *J. Flow Chem.* **2020**, *10*, 475–490. [CrossRef]
87. Thangarasu, G.R.V.; Vinayakaselvi, M.; Ramanathan, A. A critical review of recent advancements in continuous flow reactors and prominent integrated microreactors for biodiesel production. *Renew. Sust. Energ. Rev.* **2022**, *154*, 111869. [CrossRef]
88. Masngut, N.; Harvey, A.P.; Ikwebe, J. Potential uses of oscillatory baffled reactors for biofuel production. *Biofuels* **2010**, *1*, 605–619. [CrossRef]
89. Oliva, J.A.; Pal, K.; Barton, A.; Firth, P.; Nagy, Z.K. Experimental investigation of the effect of scale-up on mixing efficiency in oscillatory flow baffled reactors (OFBR) using principal component based image analysis as a novel noninvasive residence time distribution measurement approach. *Chem. Eng. J.* **2018**, *351*, 498–505. [CrossRef]
90. Abbott, M.; Harvey, A.; Perez, G.V.; Theodorou, M. Biological processing in oscillatory baffled reactors: Operation, advantages and potential. *Interface Focus* **2013**, *3*, 20120036. [CrossRef]
91. Natarajan, Y.; Nabera, A.; Salike, S.; Tamilkkuricil, V.D.; Pandian, S.; Karuppan, M.; Appusamy, A. An overview on the process intensification of microchannel reactors for biodiesel production. *Chem. Eng. Process. Process Intensif.* **2019**, *136*, 163–176. [CrossRef]
92. Mohd Laziz, A.; KuShaari, K.; Azeem, B.; Yusup, S.; Chin, J.; Denecke, J. Rapid production of biodiesel in a microchannel reactor at room temperature by enhancement of mixing behaviour in methanol phase using volume of fluid model. *Chem. Eng. Sci.* **2020**, *219*, 115532. [CrossRef]
93. Kumar, Y.; Das, L.; Biswas, K.G. Biodiesel: Features, Potential Hurdles, and Future Direction. In *Status and Future Challenges for Non-Conventional Energy Sources*; Joshi, J., Sen, R., Sharma, A., Salam, P.A., Eds.; Springer: Singapore, 2022; Volume 2, pp. 99–122. [CrossRef]
94. Tiwari, A.; Rajesh, V.M.; Yadav, S. Biodiesel production in micro-reactors: A review. *Energy Sustain. Dev.* **2018**, *43*, 143–161. [CrossRef]
95. Kiani, M.R.; Meshksar, M.; Makarem, M.A.; Rahimpour, E. Catalytic membrane micro-reactors for fuel and biofuel processing: A mini review. *Top. Catal.* **2021**. [CrossRef]
96. Gupta, J.; Agarwal, M.; Dalai, A.K. An overview on the recent advancements of sustainable heterogeneous catalysts and prominent continuous reactor for biodiesel production. *J. Ind. Eng. Chem.* **2020**, *88*, 58–77. [CrossRef]
97. Baydir, E.; Aras, O. Increasing biodiesel production yield in narrow channel tubular reactors. *Chem. Eng. Process. Process Intensif.* **2022**, *170*, 108719. [CrossRef]
98. Budžaki, S.; Miljić, G.; Sundaram, S.; Tišma, M.; Hessel, V. Cost analysis of enzymatic biodiesel production in small-scaled packed-bed reactors. *Appl. Energy* **2018**, *210*, 268–278. [CrossRef]
99. He, P.Y.; Zhang, Y.J.; Chen, H.; Han, Z.C.; Liu, L.C. Low-energy synthesis of kaliophilite catalyst from circulating fluidized bed fly ash for biodiesel production. *Fuel* **2019**, *257*, 116041. [CrossRef]
100. e Silva, W.C.; Teixeira, L.F.; Carvalho, A.K.F.; Mendes, A.A.; de Castro, H.F. Influence of feedstock source on the biocatalyst stability and reactor performance in continuous biodiesel production. *J. Ind. Eng. Chem.* **2014**, *20*, 881–886. [CrossRef]
101. Asimakopoulos, K.; Kaufmann-Elfang, M.; Lundholm-Høffner, C.; Rasmussen, N.B.K.; Grimalt-Alemany, A.; Gavala, H.N.; Skiadas, I.V. Scale up study of a thermophilic trickle bed reactor performing syngas biomethanation. *Appl. Energy* **2021**, *290*, 116771. [CrossRef]
102. Oh, P.P.; Lau, H.L.N.; Chen, J.; Chong, M.F.; Choo, Y.M. A review on conventional technologies and emerging process intensification (PI) methods for biodiesel production. *Renew. Sust. Energy Rev.* **2012**, *16*, 5131–5145. [CrossRef]
103. Dai, J.Y.; Li, D.Y.; Zhao, Y.C.; Xiu, Z.L. Statistical optimization for biodiesel production from soybean oil in a microchannel reactor. *Ind. Eng. Chem. Res.* **2014**, *53*, 9325–9330. [CrossRef]
104. Zik, N.A.F.A.; Sulaiman, S.; Jamal, P. Biodiesel production from waste cooking oil using calcium oxide/nanocrystal cellulose/polyvinyl alcohol catalyst in a packed bed reactor. *Renew. Energy* **2020**, *155*, 267–277. [CrossRef]
105. Talha, N.S.; Sulaiman, S. In situ transesterification of solid coconut waste in a packed bed reactor with CaO/PVA catalyst. *Waste Manag.* **2018**, *78*, 929–937. [CrossRef] [PubMed]

106. Ani, F.N.; Nur, H.; Said, M.F. Optimization of Biodiesel Production using a Stirred Packed-bed Reactor. *Int. J. Technol.* **2018**, *9*, 219–228.
107. Akkarawatkhoosith, N.; Kaewchada, A.; Ngamcharussrivichai, C.; Jaree, A. Biodiesel production via interesterification of palm oil and ethyl acetate using ion-exchange resin in a packed-bed reactor. *Bioenergy Res.* **2020**, *13*, 542–551. [CrossRef]
108. Jindapon, W.; Ashokkumar, V.; Rashid, U.; Rojviriya, C.; Pakawanit, P.; Ngamcharussrivichai, C. Production of biodiesel over waste seashell-derived active and stable extrudate catalysts in a fixed-bed reactor. *Environ. Technol. Innov.* **2020**, *20*, 101051. [CrossRef]
109. Chen, G.; Liu, J.; Yao, J.; Qi, Y.; Yan, B. Biodiesel production from waste cooking oil in a magnetically fluidized bed reactor using whole-cell biocatalysts. *Energy Convers. Manag.* **2017**, *138*, 556–564. [CrossRef]
110. Zhou, G.; Chen, G.; Yan, B. Biodiesel production in a magnetically-stabilized, fluidized bed reactor with an immobilized lipase in magnetic chitosan microspheres. *Biotechnol. Lett.* **2014**, *36*, 63–68. [CrossRef]
111. Liu, J.; Chen, G.; Yan, B.; Yi, W.; Yao, J. Biodiesel production in a magnetically fluidized bed reactor using whole-cell biocatalysts immobilized within ferroferric oxide-polyvinyl alcohol composite beads. *Bioresour. Technol.* **2022**, *355*, 127253. [CrossRef]
112. Meng, Y.L.; Tian, S.J.; Li, S.F.; Wang, B.Y.; Zhang, M.H. Transesterification of rapeseed oil for biodiesel production in trickle-bed reactors packed with heterogeneous Ca/Al composite oxide-based alkaline catalyst. *Bioresour. Technol.* **2013**, *136*, 730–734. [CrossRef]
113. Son, S.M.; Kusakabe, K. Transesterification of sunflower oil in a countercurrent trickle-bed reactor packed with a CaO catalyst. *Chem. Eng. Process. Process Intensif.* **2011**, *50*, 650–654. [CrossRef]
114. García-Martín, J.F.; Barrios, C.C.; Alés-Álvarez, F.J.; Dominguez-Sáez, A.; Alvarez-Mateos, P. Biodiesel production from waste cooking oil in an oscillatory flow reactor. Performance as a fuel on a TDI diesel engine. *Renew. Energy* **2018**, *125*, 546–556. [CrossRef]
115. Kefas, H.M.; Yunus, R.; Rashid, U.; Taufiq-Yap, Y.H. Enhanced biodiesel synthesis from palm fatty acid distillate and modified sulfonated glucose catalyst via an oscillation flow reactor system. *J. Environ. Chem. Eng.* **2019**, *7*, 102993. [CrossRef]
116. Soufi, M.D.; Ghobadian, B.; Najafi, G.; Mohammad Mousavi, S.; Aubin, J. Optimization of methyl ester production from waste cooking oil in a batch tri-orifice oscillatory baffled reactor. *Fuel Process. Technol.* **2017**, *167*, 641–647. [CrossRef]
117. Phan, A.N.; Harvey, A.P.; Eze, V. Rapid production of biodiesel in mesoscale oscillatory baffled reactors. *Chem. Eng. Technol.* **2012**, *35*, 1214–1220. [CrossRef]
118. Santikunaporn, M.; Techopittayakul, T.; Echaroj, S.; Chavadej, S.; Chen, Y.H.; Yuan, M.H.; Asavatesanupap, C. Optimization of biodiesel production from waste cooking oil in a continuous mesoscale oscillatory baffled reactor. *Eng. J.* **2020**, *24*, 19–28. [CrossRef]
119. Mohammadi, F.; Parvareh, A.; Rahimi, M. Transesterificatin of Soybean Oil to Biodiesel Using CaO as a Heterogeneous Catalyst in a Micro-channel Reactor. In Proceedings of the 8th International Chemical Engineering Congress & Exhibition (IChEC 2014), Kish, Iran, 24–27 February 2014; pp. 1–5.
120. Wen, Z.; Yu, X.; Tu, S.T.; Yan, J.; Dahlquist, E. Intensification of biodiesel synthesis using zigzag micro-channel reactors. *Bioresour. Technol.* **2009**, *100*, 3054–3060. [CrossRef]
121. Azam, N.A.M.; Uemura, Y.; Kusakabe, K.; Bustam, M.A. Biodiesel production from palm oil using micro tube reactors: Effects of catalyst concentration and residence time. *Procedia Eng.* **2016**, *148*, 354–360. [CrossRef]
122. Kaewchada, A.; Pungchaicharn, S.; Jaree, A. Transesterification of palm oil in a microtube reactor. *Can. J. Chem. Eng.* **2016**, *94*, 859–864. [CrossRef]
123. Talebian-Kiakalaieh, A.; Amin, N.A.S.; Mazaheri, H. A review on novel processes of biodiesel production from waste cooking oil. *Appl. Energy* **2013**, *104*, 683–710. [CrossRef]
124. Janajreh, I.; Almazrouei, M. Chemical Kinetic and High Fidelity Modeling of Transesterification. In *Biofuels-Challenges and Opportunities*; Qubeissi, M.A., Ed.; IntechOpen: London, UK, 2018; pp. 47–64. [CrossRef]
125. Raheem, I.; Mohiddin, M.N.B.; Tan, Y.H.; Kansedo, J.; Mubarak, N.M.; Abdullah, M.O.; Ibrahim, M.O. A review on influence of reactor technologies and kinetic studies for biodiesel application. *J. Ind. Eng. Chem.* **2020**, *91*, 54–68. [CrossRef]
126. Feyzi, M.; Hosseini, N.; Yaghobi, N.; Ezzati, R. Preparation, characterization, kinetic and thermodynamic studies of MgO-La$_2$O$_3$ nanocatalysts for biodiesel production from sunflower oil. *Chem. Phys. Lett.* **2017**, *677*, 19–29. [CrossRef]
127. Chua, S.Y.; Periasamy, L.A.P.; Goh, C.M.H.; Tan, Y.H.; Mubarak, N.M.; Kansedo, J.; Khalid, M.; Walvekar, R.; Abdullah, E.C. Biodiesel synthesis using natural solid catalyst derived from biomass waste—A review. *J. Ind. Eng. Chem.* **2020**, *81*, 41–60. [CrossRef]
128. Awogbemi, O.; Kallon, D.V.V.; Owoputi, A.O. Biofuel Generation from Potato Peel Waste: Current State and Prospects. *Recycling* **2022**, *7*, 23. [CrossRef]
129. Awogbemi, O.; Kallon, D.V.V.; Bello, K.A. Resource Recycling with the Aim of Achieving Zero-Waste Manufacturing. *Sustainability* **2022**, *14*, 4503. [CrossRef]
130. Awogbemi, O.; Kallon, D.V.V.; Aigbodion, V.S.; Mzozoyana, V. Property Determination, FA Composition and NMR Characterization of Palm Oil, Used Palm Oil and Their Methyl Esters. *Processes* **2021**, *10*, 11. [CrossRef]
131. Gebremariam, S.N.; Marchetti, J.M. Economics of biodiesel production: Review. *Energy Convers. Manag.* **2018**, *168*, 74–84. [CrossRef]

132. Bokhari, A.; Chuah, L.F.; Yusup, S.; Klemeš, J.J.; Akbar, M.M.; Kamil, R.N.M. Cleaner production of rubber seed oil methyl ester using a hydrodynamic cavitation: Optimisation and parametric study. *J. Clean. Prod.* **2016**, *136*, 31–41. [CrossRef]
133. Karmee, S.L.; Patria, R.D.; Lin, C.S.K. Techno-economic evaluation of biodiesel production from waste cooking oil—A case study of Hong Kong. *Int. J. Mol. Sci.* **2015**, *16*, 4362–4371. [CrossRef] [PubMed]
134. Awogbemi, O.; Kallon, D.V.V. Pretreatment techniques for agricultural waste. *Case Stud. Chem. Environ. Eng.* **2022**, *6*, 100229. [CrossRef]

MDPI
St. Alban-Anlage 66
4052 Basel
Switzerland
www.mdpi.com

Bioengineering Editorial Office
E-mail: bioengineering@mdpi.com
www.mdpi.com/journal/bioengineering

Disclaimer/Publisher's Note: The statements, opinions and data contained in all publications are solely those of the individual author(s) and contributor(s) and not of MDPI and/or the editor(s). MDPI and/or the editor(s) disclaim responsibility for any injury to people or property resulting from any ideas, methods, instructions or products referred to in the content.

www.ingramcontent.com/pod-product-compliance
Lightning Source LLC
LaVergne TN
LVHW070419100526
838202LV00014B/1489